PHOTODYNAMIC ACTION
and
DISEASES CAUSED BY LIGHT

By

HAROLD FRANCIS BLUM, PH.D.

The Washington Biophysical Institute
(Formerly Associate Professor of Physiology,
University of California Medical School)

American Chemical Society
Monograph Series

REINHOLD PUBLISHING CORPORATION
330 WEST FORTY-SECOND STREET, NEW YORK, U. S. A.

1941

Printed in the United States of America by
INTERNATIONAL TEXTBOOK PRESS, SCRANTON, PA.

TO

SAMUEL STEEN MAXWELL

GENERAL INTRODUCTION

American Chemical Society Series of Scientific and Technologic Monographs

By arrangement with the Interallied Conference of Pure and Applied Chemistry, which met in London and Brussels in July, 1919, the American Chemical Society was to undertake the production and publication of Scientific and Technologic monographs on chemical subjects. At the same time it was agreed that the National Research Council, in coöperation with the American Chemical Society and the American Physical Society, should undertake the production and publication of Critical Tables of Chemical and Physical Constants. The American Chemical Society and the National Research Council mutually agreed to care for these two fields of chemical development. The American Chemical Society named as Trustees, to make the necessary arrangements for the publication of the monographs, Charles L. Parsons, secretary of the society, Washington, D. C.; the late John E. Teeple, then treasurer of the society, New York; and Professor Gellert Alleman of Swarthmore College. The Trustees arranged for the publication of the A. C. S. series of (a) Scientific and (b) Technologic Monographs by the Chemical Catalog Company, Inc. (Reinhold Publishing Corporation, successors) of New York.

The Council, acting through the Committee on National Policy of the American Chemical Society, appointed editors (the present list of whom appears at the close of this introduction) to have charge of securing authors, and of considering critically the manuscripts submitted. The editors endeavor to select topics of current interest, and authors recognized as authorities in their respective fields.

The development of knowledge in all branches of science, especially in chemistry, has been so rapid during the last fifty years, and the fields covered by this development so varied that it is difficult for any individual to keep in touch with progress in branches of science outside his own specialty. In spite of the facilities for the examination of the literature given by Chemical Abstracts and by such compendia as Beilstein's Handbuch der Organischen Chemie, Richter's Lexikon, Ostwald's Lehrbuch der Allgemeinen Chemie, Abegg's and Gmelin-Kraut's Handbuch der Anorganischen Chemie, Moissan's Traité de Chimie Minérale Générale, Friend's and Mellor's Textbooks of Inorganic Chemistry and Heilbron's Dictionary of Organic Compounds, it often takes a great deal of time to coördinate the knowledge on a given topic. Consequently when men who have spent years in the study of important subjects are willing to coördinate their knowledge and present it in concise, readable form, they perform a service of the highest value. It was with a clear recognition of the usefulness of such work that the American Chemical Society undertook to sponsor the publication of the two series of monographs.

Two distinct purposes are served by these monographs: the first, whose fulfillment probably renders to chemists in general the most important service, is to present the knowledge available upon the chosen topic in a form intelligible to those whose activities may be along a wholly different line. Many chemists fail to realize how closely their investigations may be connected with other work which on the surface appears far afield from their own. These monographs enable such men to form closer contact with work in other lines of research. The second purpose is to promote research in the branch of science covered by the monograph, by furnishing a well-digested survey of the progress already made, and by pointing out directions in which investigation needs to be extended. To facilitate the attainment of this purpose, extended references to the literature enable anyone interested to follow up the subject in more detail. If the literature is so voluminous that a complete bibliography is impracticable, a critical selection is made of those papers which are most important.

Preface

Several years ago, when I wrote a short review of photodynamic action for *Physiological Reviews,* I was amazed at the voluminous literature which had accumulated in this supposedly restricted field. At the same time I was introduced to an even more extensive mass of papers concerning diseases caused by light. It was obvious that the two fields were closely associated in the minds of most investigators, but that the relationship between them might be more apparent than real in many instances. It was impossible to do more than touch upon the latter subject in the original review, but a book covering both fields was considered at that time.

It was recognized that such a book must be more than a review of the evidence and points of view presented by the different workers in the field. It should attempt a synthesis of various phases, which seemed possible if one took as a starting point certain elementary principles of physics and photochemistry. At some points the material was so uncertain and fragmentary that no definite conclusions could be drawn. In others the experimental testing of hypotheses evolved in the course of writing was so tempting that the completion of the book was delayed to that end, but this could not go on indefinitely. Hence, many lacunae have been left, hoping that what has been written will stimulate experiment which will eventually eliminate them.

I have tried to maintain a consistent point of view throughout the book, which involves the risk, when dealing with such fragmentary material as presented in some parts, that the data may be twisted to fit the point of view. On the other hand, such treatment has the great advantage that, in the event that the conclusions reached are subsequently found to be erroneous, they can be directly attacked and demolished. If any of the conclusions or tentatives which are presented in this book are shown to be false, no one will be happier than myself, as this can only mean a step forward in the clarification of the subject.

In examining the literature, I have discovered numerous instances of misquotation and misinterpretation of the original sources, which, once started, have been carried through subsequent publications. I can only hope that I have not committed more such errors than I have corrected.

Thanks to a fellowship from the John Simon Guggenheim Memorial Foundation, I was able to spend some months in Europe in 1937, which gave me opportunity to examine literature not available at home, and to discuss various problems with other workers in the field. Among these were some of its pioneers who are no longer active: Walther Hausmann of Vienna, whose career ended with his suicide in 1938 at a time of political stress; Gunni Busck of Copenhagen, whose researches ended over thirty years ago while he was still a young man, leaving the field the poorer for his absence; and Walther Straub of Munich, who saw clearly the relationship between photosensitized oxidation and photodynamic action at a very early date, but who left the field for other endeavors. It is a pleasure to acknowledge the courtesy and helpfulness of these men. Every worker in the field must acknowledge his debt to them, together with that owed to Otto Raab, Herman von Tappeiner, and Adolph Jodlbauer, whose early efforts provided the basis for all later studies.

I must also thank Drs. H. Haxthausen and Friedrich Ellinger of Copenhagen, the latter now of New York, and Drs. Hubert Jausion and A.-C. Guillaume of Paris for their kindness and assistance during the same period.

The part played by certain secondary references is not adequately shown in the book itself because, with few exceptions, the original references have been examined and cited. Hausmann's "Grundzüge der Lichtbiologie und Lichtpathologie" (1923); Guillaume's "Radiations Lumineuse en Physiologie" (1927); Hausmann and Haxthausen's "Lichterkrankungen der Haut" (1929); Laurens' "Physiological Effects of Radiant Energy" (1933); Jausion and Pagès' "Les Maladies de Lumière" (1933); and Ellinger's "Die biologischen Grundlagen der Strahlenbehandlung" (1935) have been frequently consulted. Though differences in viewpoint with various parts of these works may appear at times, the writer wishes to express his appreciation of these valuable studies, and of their usefulness to him in the preparation of the following pages.

But for the stimulus and opportunity for study provided by a Fellowship from the John Simon Guggenheim Memorial Foundation, the book might not have developed beyond the preliminary stages. Without assistance from the Washington Biophysical Institute at a critical time, it might never have been completed.

Space and facilities have been provided at various times by The National Institute of Health, Division of Industrial Hygiene, U. S. Public Health Service, Washington; The Biological Laboratory, Naples; The Laboratory of the College de France, Concarneau; and The Biological Laboratory, Cold Spring Harbor. I am happy to express my gratitude to these institutions and their directors.

I am indebted for valuable criticism to Dr. Harold A. Abramson, Dr. Joseph C. Aub, Dr. F. S. Brackett, Dr. Dean Burk, Dr. P. A. Cole, Prof. G. S. Forbes, Dr. Alexander Hollaender, Prof. F. A. Jenkins, Prof. W. A. Noyes, Jr., Prof. A. C. Redfield, Dr. Louis Strait, and Dr. George Wald who read parts or the whole of the manuscript. Thanks are due to Dr. J. F. Danielli, Dr. Hugh Davson, Dr. E. Luce-Clausen, Dr. Gordon Mackinney, Dr. A. E. Mirsky, Mr. Nello Pace, Dr. Eric Ponder, Dr. H. P. Rusch, Mr. Frederick Shelden, and Dr. W. J. Turner for permission to use unpublished material.

The author wishes to express his thanks to the following for permission to reproduce figures: The editors of the *British Journal of Dermatology and Syphilis, American Journal of Physiology, The Journal of Clinical Investigation, Plant Physiology,* The University of California Experiment Station, the *Bureau of Standards Journal of Research,* and the Smithsonian Institution; to Dr. T. Caspersson; and the publishing firms H. K. Lewis & Company, Ltd., London; Julius Springer, Berlin; Walter de Gruyter, Berlin; The Central News Agency, Ltd., Johannesburg; W. B. Saunders Company, Philadelphia; and Williams and Wilkins Company, Baltimore.

HAROLD F. BLUM.

Washington, D. C.
February, 1940.

Contents

Part I
Introduction

Chapter 1

Introduction

During the winter of 1897-98, Oscar Raab, a student in the laboratory of Professor Tappeiner at Munich, was set to study the toxicity of a dye, acridine, for paramecia. To this end he measured the time required to kill these organisms when they were placed in solutions of different concentrations of dye, and found, to his surprise, that the time varied widely for different experiments in which the same concentration was used. He quickly recognized that the time required to kill was related to the intensity of light in the laboratory, and, following this lead, was able to show that paramecia swimming in solutions containing dye were rapidly killed when exposed to the sun's rays, whereas when no dye was present they survived for long periods in sunlight.* Obviously, the dye rendered the organisms sensitive to light in a manner which appeared analogous to sensitization of the photographic plate.

Raab's findings led to a long series of researches, principally by Tappeiner and Jodlbauer in the Munich laboratories. In a relatively short time it was discovered that many dyes and pigments, such as, for example, the dye eosin and the pigment chlorophyll, could sensitize a wide variety of living organisms to light; and many of the fundamental characteristics of the phenomenon were described within a few years following the original discovery. All this came at a time when interest in the biological effects of light was at a high point following the pioneer work of Finsen, whose therapeutic use of light, and other photobiological studies, had aroused widespread interest among biologists and physicians; and it is only natural that attempts were made to explain all manner of photobiological phenomena in terms of photosensitization by dyes and pigments. It is probable that when Tappeiner gave the name "photodynamic action" (*photodynamische Erscheinung*) to the phenomenon, he thought it to be the basis of photobiological processes in general, as that name might imply. Although this early hope was not fulfilled, and the phenomenon has been found to have a more limited significance in photobiology, the name photodynamic action has persisted.

It seems best to continue to use this name exclusively to describe phenomena of the type discovered by Raab. These make up a clear-cut entity since the fundamental process is in all instances the same, namely, the sensitization of a biological system to light by a substance which

* While Marcacci reported what may have been an example of the same phenomenon in 1888, he failed to recognize its essential nature. Raab's experiments were first published in 1900, following a preliminary note by Tappeiner in the same year.

serves as a light absorber for photochemical reactions in which molecular oxygen takes part. These reactions have nothing in common with the normal oxygen metabolism of living systems. In fact, they probably represent oxidations of structural components of the cell which do not take place thermally, at ordinary temperatures, since their occurrence would be incompatible with the stability of the cell, but which do take place photochemically, granted the presence of an appropriate photosensitizer. Evidence for this mechanism will be presented at length in the second part of this book. Clearly this is an instance of photosensitization in the photochemical sense; but since it is possible that other types of photosensitization occur in biological systems, it is convenient to give it a special name. The choice of "photodynamic action" is not altogether a happy one, but has the advantage of priority and usage; it is certainly more appropriate than "optical sensitization," which has sometimes been employed because of superficial resemblance to sensitization of the photographic plate.

The early wave of interest in photobiology, which had given such prominence to photodynamic action, waned, and when study of the effects of radiation was revived some years later it was directed along other lines. Much of the fundamental work on photodynamic action had been forgotten, but many of the false concepts remained. The subsequent literature contains many vague or incorrect statements, and experiments which are often unwitting repetitions of earlier and more thorough studies. Throughout, there is a tendency to ascribe a more general significance to the phenomenon than is justified.

As a model system, photodynamic action is of interest for the elucidation of other photobiological processes, but while the examples reported from the laboratory are legion, authentic instances in nature are relatively few. However, these few are not without considerable importance. There can be little doubt that photodynamic action is a casual factor in certain diseases of domestic animals, which have distinct economic importance. For example, the disease known locally as geeldikkop, which has been shown within the last few years to be an instance of photodynamic action, is the most destructive disease among sheep in the Karoo veldt of South Africa, where sheep raising is the major industry. It seems probable that this particular disease is not restricted to this area, but is to be found in other parts of the world where its nature is not as yet recognized. Sensitization to light following feeding on buckwheat or on St. John's Wort has been recognized as of importance to stock raisers for over a century, and the latter is a serious problem in Australia, New Zealand, and our own western states. It is probable that a number of similar diseases occur which are not commonly recognized. Although most veterinarians are aware of the existence of diseases which are precipitated by exposure to sunlight, effective diagnosis and control can come only with a better acquaintance with the underlying principles which govern the occurrence of such diseases. Part III of this book will be devoted to a statement of present knowledge of diseases produced by light in domestic animals.

Sensitivity to light among human beings is usually regarded as a very rare oddity, and has received little attention from the medical profession in general. Considering this fact, the number of papers devoted to this subject is surprisingly large, and it seems probable that cases are not extremely uncommon. Recent studies indicate that light may play an important role in the occurrence of skin cancer, which fact gives special interest to this group of diseases. Although several types of sensitivity to light may be distinguished, it is, unfortunately, common practice to assume that all cases fall into a single group, and, quite erroneously, that photodynamic action is the etiological basis for all. It is hoped that the discussion of these diseases in Part IV will help to clarify this subject. It should at least point out that progress can be expected only when more careful and more appropriate studies are made than have been the rule in the past.

Sensitivity to light has occurred as a result of the administration of photodynamic substances as therapeutic or diagnostic agents, as well as from contact with such substances in manufacturing processes where, in certain instances, such contact may be regarded as an important industrial risk. Sometimes it has been deliberately produced as a therapeutic measure, unfortunately without a proper understanding of the underlying principles in some cases, and certainly with questionable results in most. Later chapters will be devoted to the discussion of these subjects.

Obviously, a study of any phase of photobiology demands an elementary knowledge of light, and of photochemistry. Within recent years the introduction of the quantum concept into these subjects has had revolutionary effects, and has provided a better basis for the study of photobiology than was available heretofore. Since many readers may not be acquainted with these concepts, the following chapter will be devoted to a brief summary of those elementary principles necessary for an understanding of the subsequent material. This treatment cannot be exhaustive, and will cover only the points essential to the present study, for which a thorough knowledge of photochemistry is not required.

For those interested in only one phase of the subject matter, it is intended that each part of the book may be read independently of the others, although some reference to earlier pages will be required for a full understanding of the later parts.

Chapter 2

The Nature of Radiation and its Effects*

According to modern physical theory, radiation of all kinds, including light, has properties of both waves and particles. It is useful for some purposes to consider radiation as waves and for other purposes as discrete corpuscles, or quanta. To study the refraction of light through the lenses of a microscope the wave principle must be applied, for the quantum aspects do not enter into such a problem; but in the study of photochemistry the quantum aspects are essential. Photobiological processes are essentially photochemical, and here the quantum point of view must be adopted. The reader interested in the relationship of the two concepts is referred to the lucid introductory account by Darrow (in Duggar, 1936).

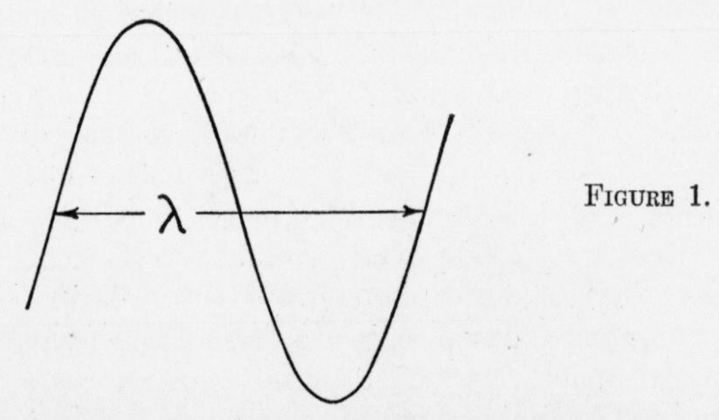

FIGURE 1.

The Wave Aspects. Electromagnetic radiation may be considered to proceed as transverse waves. In the simplest case, it may be represented as a sine wave as in Figure 1. It is common usage to characterize any radiation by its wave-length, λ, *i. e.*, the distance between nearest corresponding points on the wave. Light† is ordinarily made up of many wave-lengths; when composed of virtually only one wave-length it is called monochromatic.

The wave-length units in common use are:

The micron, symbol $\mu = 10^{-4}$ cm. $= 10^{-3}$ mm.
The milli-micron, symbol mμ‡ $= 10^{-7}$ cm. $= .001\mu$.
The Ångstrom unit, symbol Å or A.U. $= 10^{-8}$ cm. $= 0.1$ mμ.
The X unit, symbol X.U. $= 10^{-11}$ cm. $= .001$Å.

* This chapter deals only with principles and material which will be used in subsequent discussion. It is not intended as an exhaustive treatment.

† The term *light* is limited by some to that radiation which is capable of stimulating the human eye; others include a somewhat wider wave-length range.

‡ In the past, the symbol $\mu\mu$ has sometimes been used incorrectly for mμ.

The particular unit used is selected for convenience according to the magnitude of the wave-lengths dealt with. Thus, for the visible region which lies between approximately 3900 and 7400Å, the wave-length is generally expressed in mμ or Å, for the x-ray region in X.U., and for the infrared in μ. The Ångstrom unit will be used almost exclusively in this book.

Radiation may also be characterized by its frequency, ν, *i.e.*, the number of waves of a given wave-length which will be produced in a given time. Thus:

$$\nu = \frac{c}{\lambda} \tag{1}$$

where c is the velocity of light (3×10^{10} cm. per sec.*). The frequency remains constant regardless of the medium through which the radiation passes, but since the velocity of light varies according to the medium, the wave-length must also vary. Thus frequency is a more fundamental characteristic than wave-length; but frequencies must be calculated from wave-lengths which may be measured directly by the method of interference, and for this reason wave-lengths (measured in air) are most often used in characterizing radiation. A more or less arbitrary subdivision of radiation into wave-length regions is in common use, and may be conveniently employed. Those wave-lengths perceived by the human eye are called *visible,* the range of wave-lengths immediately short of this *ultraviolet,* and the region just longer *infrared.* The x-rays and gamma rays occupy a region of very short wave-lengths, and the very long wave-lengths are known as Hertzian or radio waves. There are no sharp distinctions between these regions, their boundaries being purely arbitrary and inexact, but they are useful for purposes of description. For convenience they are mapped in Figure 2.

The Quantum Aspects. In dealing with any problem of the interaction of radiation with matter, it is necessary to assume that radiation consists of discrete packets of energy, or quanta. The existence of a relationship between the wave and quantum aspects is indicated by the fact that the magnitude of the energy quantum, q, is proportional to the frequency of the radiation, so that the quantity of energy making up the quantum might be used to characterize a particular radiation quite as well as wave-length or frequency. This quantity of energy may be calculated from the equation:

$$q = h\nu \tag{2}$$

where h is Planck's constant, which has a value of 6.6236×10^{-27} erg seconds. The origin of Planck's constant cannot be discussed here, but the reader is referred to such works as Richtmyer (1928), or Andrade (1924).

* This value is very nearly exact for either air or vacuum, the most recent value for vacuum being 2.99776×10^{10} cm. per sec. This value and those for Planck's constant and Avogadro's number are recent estimates by Prof. R. T. Birge.

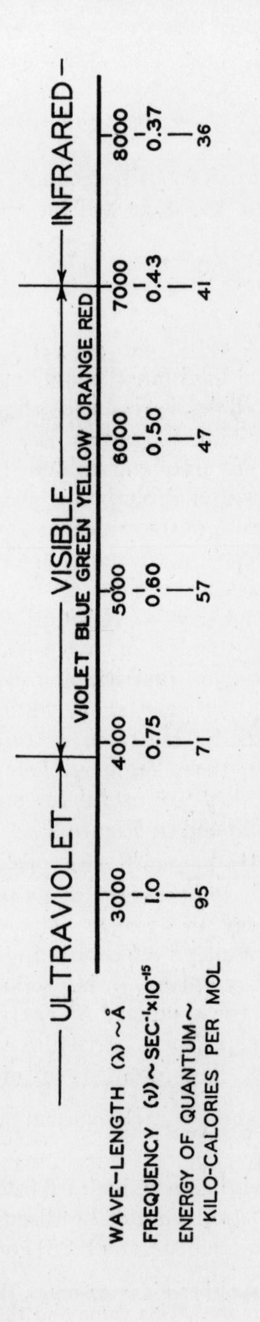

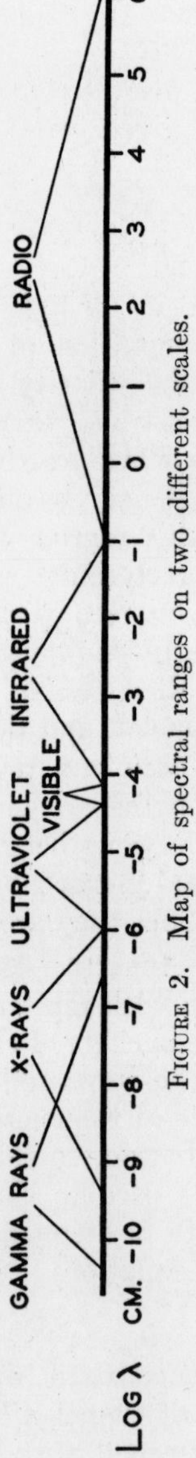

FIGURE 2. Map of spectral ranges on two different scales.

If it is not at once apparent that the quantity $h\nu$ is an energy quantity, this becomes clear when the following calculation is made for the quantum corresponding to $\lambda = 5000\text{Å}$:

$$\nu = \frac{c}{\lambda} = \frac{3 \times 10^{10}\ \text{cm. sec.}^{-1}}{5.0 \times 10^{-5}\ \text{cm.}} = \frac{3 \times 10^{10}\ \text{sec.}^{-1}}{5.0 \times 10^{-5}}$$

$$h\nu = 6.6236 \times 10^{-27}\ \text{erg sec.} \times \frac{3 \times 10^{10}\ \text{sec.}^{-1}}{5.0 \times 10^{-5}}$$

$$= \frac{6.6236 \times 10^{-27} \times 3 \times 10^{10}}{5.0 \times 10^{-5}}\ \text{ergs} = 3.97 \times 10^{-12}\ \text{ergs}$$

The magnitude of the quantum is here expressed in units of energy, ergs, which can be translated into calories or any other convenient energy unit.

The Emission and Absorption of Radiant Energy

Emission and Absorption by Atoms. The isolated atom may exist in a limited number of "quantized" states, each representing a definite internal energy of the atom. These different quantized states, or "energy levels," may be represented by a diagram such as Figure 3, which describes the mercury atom. The energy levels are shown as horizontal lines; the most probable, or "normal" level is at the bottom of the diagram, the highest energy level at the top. To change from one energy level to another, the atom must emit or absorb a quantum of energy equal to the difference in energy between the two levels. This fixes the size of the quantum, and hence the frequency and wave-length of the emitted or absorbed radiation.

When one looks at a mercury arc, one sees light which is emitted in this way from mercury atoms inside the arc. The atom can enjoy only one energy level at a given instant, so that it is possible for an atom to emit only one quantum of radiant energy in a single act. In the mercury arc, the atom is first raised to an unstable energy level, or excited state, by receiving energy which is provided by electron impact in the arc. After emitting a quantum of radiant energy by returning to a lower energy level, the atom may be again raised to a higher energy level by receiving energy from electron impact, and may again emit a quantum in returning to a lower energy level. In this way the arc emits radiation, indefinitely, so long as the electrical field is maintained.

By means of a spectroscope the wave-lengths of light which are given off by the mercury arc may be spread out and separated. One sees, then, only a few lines which are separate images of the source as it comes through the entrance slit of the spectroscope. These lines represent the restricted wave-lengths which the mercury atoms emit, which, of course, correspond to definite quantum sizes. The mercury arc emits a large number of quantum sizes in the ultraviolet which the eye cannot see, but which may be detected by photographic methods, using a proper quartz spectrograph. A photograph showing lines given off by the mercury arc appears as Figure 4. This is called an emission spectrum, and the kind shown by mercury or any other atom, a line spectrum, because the light is emitted only in definite, discrete wave-lengths and thus appears as lines on the spectrogram. Figure 7 (p. 26) shows the intensities of these

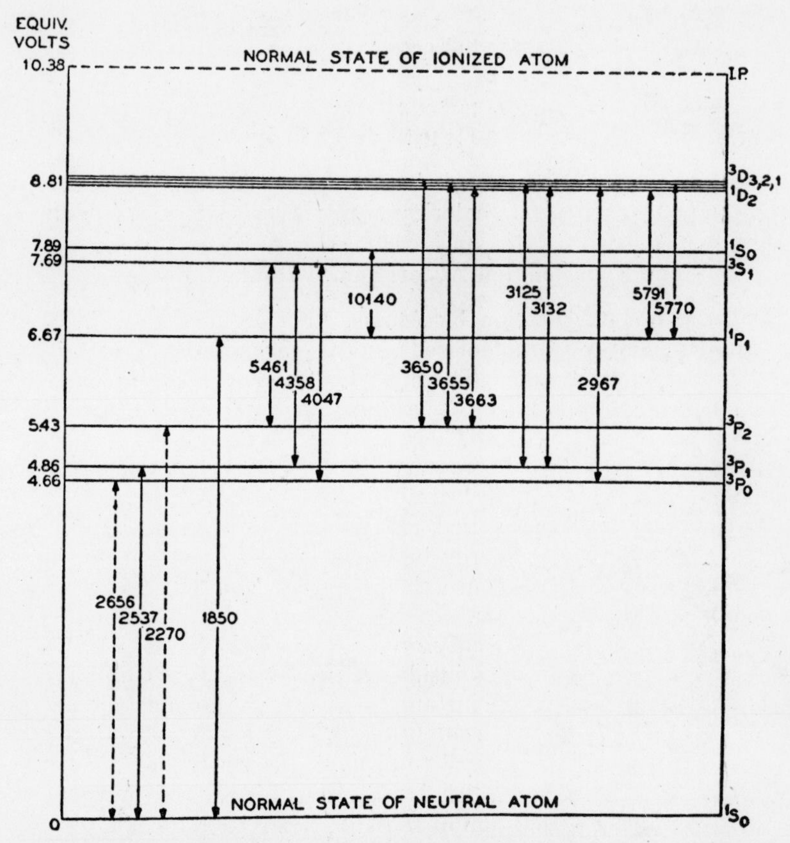

FIGURE 3. Energy levels of mercury. The horizontal lines represent energy levels of the atom in equivalent volts. The vertical arrows indicate possible jumps from one level to another, and are marked according to the wave-length of the radiation emitted or absorbed when such a jump occurs. The notations at the right are spectroscopic descriptions of the corresponding states of the atom. (From K. K. Darrow's "Electrical Phenomena in Gases," Williams and Wilkins Co., 1930.)

various lines for a given mercury arc. The relative intensities of the different lines may be modified in a number of ways, and will differ somewhat from one arc to another; but the wave-lengths of the lines remain the same.* The vertical arrows in Figure 3 indicate "jumps"

* This is strictly true only for low-pressure arcs; slight shifts may appear in the spectra of high-pressure arcs.

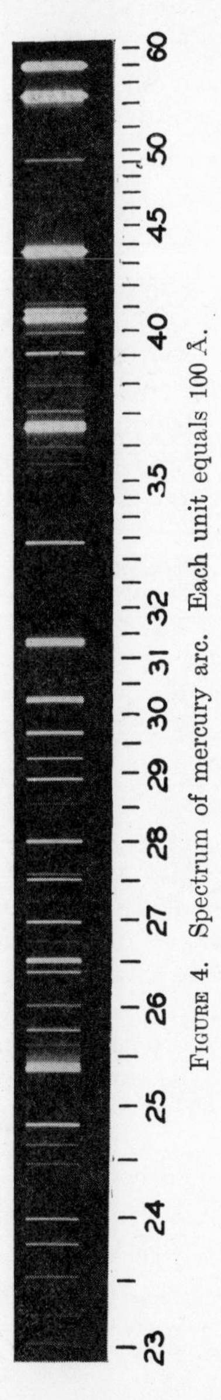

FIGURE 4. Spectrum of mercury arc. Each unit equals 100 Å.

between energy levels, and each is labelled with the wave-length of the radiation emitted when the indicated jump is made. Comparison will show that these jumps correspond to the spectral lines represented in Figures 4 and 7.

Thus far the spectrum of the mercury atom has been used as an example. The number and position of the energy levels is different for every kind of atom, and atoms of a given element may emit only certain sized quanta. Thus, every element has a characteristic pattern of emission lines, which may be used to identify that element. The mercury spectrum is relatively simple, for atomic spectra may be highly complex, as, for example, in the case of the iron arc spectrum, which consists of thousands of lines.

The energy levels of the atom were at one time supposed to represent orbit positions of electrons rotating about a central positively charged nucleus, each orbit representing a particular energy level of the atom, *e.g.* in the original Bohr atom model for hydrogen. While this is a simple and convenient model, it seems best to use the less explicit picture provided by the energy level diagram. The energy levels of the atom are undoubtedly related to the electronic pattern of the atom, however, and may be referred to as electronic levels, to distinguish them from the rotation and vibration levels of molecules, which will be discussed below.

An atom can accept only quanta which will supply the exact energy required to raise it from one energy level to another; thus the atom may absorb just the same sized quanta which it may emit. If the spectrum of a source which emits all the visible wave-lengths, such as a tungsten filament lamp, is dispersed by means of a spectroscope, it forms a continuous band of light running through the spectral colors from violet at one end to red at the other. If a chamber containing sodium vapor is interposed somewhere between the source and the eye, two dark lines are seen in the yellow region of the spectrum; these are called absorption lines, and constitute the absorption spectrum of sodium in this region. Upon comparison, the wave-lengths of these absorption lines are found to correspond to the emission lines of sodium in the same spectral region. This is due to the fact that the sodium atoms accept only those quanta which will serve to bridge the gaps between the same energy levels which were traversed in the act of emission of quanta of radiation. The same principle holds in all cases; certain wave-lengths corresponding to certain quantum sizes are possible for absorption and emission. However, at ordinary temperatures, absorption at many wave-lengths is negligible, so that a much smaller number of absorption lines than of emission lines is ordinarily observed. For example, at ordinary temperatures mercury vapor does not show absorption lines in the visible corresponding to the emission lines of the mercury arc in this region. When observable, the absorption spectrum of a given element is, like its emission spectrum, typical of the element.

Absorption by Molecules. The initial act in any photobiological process is the capture of a quantum of radiant energy by a molecule. Absorption by molecules is hence more interesting with regard to the present studies than absorption by atoms. The two processes are essentially the same, however, the combination of atoms into molecules merely introducing greater complexity. This results because the electronic levels of the atoms are modified by their mutual effect upon one another in the molecule, thus increasing the complexity of the energy levels beyond those of the individual atoms. This results in a greater number of quantum sizes, and hence wave-lengths, which can be absorbed by the molecule. The atoms of the molecule are able to rotate about one another, and also to vibrate toward and away from one another; these motions of rotation and vibration increase the number of quantized energy states. The quanta represented by transitions among these energy states are smaller than those associated with changes in electronic levels. Pure molecular rotation spectra are found in the far infrared, rotation vibration spectra are found in the near infrared, and both are superimposed on electronic spectra in the visible and ultraviolet.

As a result of the presence of superimposed vibration and rotation, molecular spectra show *bands* of absorption rather than *lines*. In the case of gases, these bands are made up of lines of so nearly the same wave-length that they may be very difficult to separate into the component lines by means of an ordinary spectroscope. The presence of neighboring molecules may effect the energy levels within the molecule, and hence the absorption spectrum. Such effects are, of course, much greater in the case of solutions than in the case of gases, and should be still more important in living systems, where surface effects may bring molecules into very close relationship. For this reason, principally, the absorption spectra of liquids and solutions are made up of continuous bands rather than bands separable into discrete lines. There is often a characteristic spectral pattern for a given molecular species, however, in which definite regions may be characteristic of given atomic arrangements within the molecule, so that closely related molecular species may show absorption spectra which agree closely in certain spectral regions. The absorption spectra of prophyrins, shown in Figure 45 (p. 216), offer a nice illustration of this; all show very similar absorption in the ultraviolet and violet with maxima near 4000Å, although there is considerable variation in their absorption in the visible.

Needless to say, if the solvent and solute molecules combine chemically, the absorption spectrum may be greatly modified. In the case of compounds capable of ionization, the absorption may be quite different when the compound is dissolved in a polar solvent, which allows it to separate into its ions, than when dissolved in a non-polar solvent which does not permit dissociation.

Results of Absorption of Light by Atoms and Molecules

In the following pages the absorption of quanta will be of greater interest than their emission. The act of absorption may be represented as follows:

$$A + h\nu \longrightarrow A'$$

where A is the normal, and A' the excited atom or molecule. The lifetime of the excited atom or molecule is very short, about 10^{-9} to 10^{-6} second in most cases in which it has been measured (see p. 68). The events of greatest interest to us are those which follow absorption, that is, the way in which the atom or molecule gets rid of the energy which it gains in the act of absorption. While one cannot predict exactly what will happen to a given atom or molecule which absorbs a quantum of radiation, a number of possibilities may be recognized, and in some cases a prediction as to the probability of the occurrence of a particular event may be made. The events which may follow the absorption of a quantum of radiant energy may be summarized as follows:

Fluorescence. An excited atom or molecule, that is, one which has just acquired a quantum, may get rid of the energy by emitting another quantum of radiation, *i.e.*, by fluorescence. The simplest example of fluorescence is resonance, in which the absorption of a quantum takes place with a resulting change in energy level (for example, the 2537Å line, Figure 3), and a quantum of the same size is subsequently emitted concomitant with transfer back to the original energy level. More commonly, however, the emitted quantum is smaller than the exciting quantum, so that the radiation emitted as fluorescence has longer wave-length than the absorbed (Stokes' rule); the remaining energy must be got rid of by radiation of a second quantum or in some other way, *e.g.*, degradation to heat. More rarely, the emitted radiation may have shorter wave-length than the exciting radiation ("anti-Stokes'" fluorescence); in this case the extra energy is provided by the transition to a lower energy state than that enjoyed by the atom before excitation.

Chemical Reaction. An excited atom or molecule may take part in chemical reaction. Some of the principles governing such reactions will be discussed below.

Collisions of the Second Kind. An excited atom or molecule which meets another particle * may transfer energy by increasing the kinetic energy or by producing excitation (*i.e.*, increasing the potential energy) of the second particle. This is called a collision of the second kind, in contrast to a collision of the first kind, in which an unexcited particle adds energy to a second particle by collision. In the event that energy transferred by collision of the second kind causes excitation of the receiving particle, that particle may increase the kinetic energy of other particles by collision, may emit its energy as fluorescence, or may take part in a chemical reaction. The last case would be a photosensitized reaction, a type which will be of particular interest in later chapters.

Ionization. If the absorbed quantum is sufficiently large, it may produce ionization by providing enough energy to separate an electron from the particle, resulting in a charged ion. For each kind of atom or molecule, the quantum must have a certain minimum energy, the ionization potential, in order that such ionization may occur. Hence, it is to be expected that ionization will be more often produced by short wave-

* The term particle is used here to mean either an atom or a molecule.

lengths than by long ones. Ionization is a common occurrence in the gamma and x-ray regions and in the far ultraviolet, but is certainly not common in the visible or even in the near ultraviolet.

In the results of quantum absorption mentioned above, the nucleus of the atom is not affected. X-rays and gamma rays affect principally the inner electrons of the atom but not the nucleus, whereas bombardment of the atom with high-speed particles of subatomic size, such as neutrons, etc., results in disturbance of the nucleus itself, and is a different type of event. The study of such effects in living systems is receiving considerable attention at the present moment, but like the effects of gamma and x-rays, is quite outside the province of this book and should not be confused with the type of phenomenon which will be dealt with herein.

Photochemical Principles

The Equivalence Law. Photochemistry received a sudden impetus by the introduction of the equivalence law, generally associated with the name of Einstein. The law is an outcome of the quantum theory, and states that the first act in any photochemical reaction is the absorption of a quantum of energy by an atom or molecule in the reacting system. This does not mean that every absorbed quantum causes chemical reaction.

The Grotthus-Draper Law. This principle is actually contained in the equivalence law, since it merely states that only those wave-lengths of light which are absorbed by a system may produce photochemical reaction in that system. It was established early in the last century, long before the advent of the quantum theory, by the two independent investigators whose names it bears. It is of great fundamental importance, and has been called the first law of photochemistry. It is rather disappointing to find that biologists and medical investigators sometimes neglect the implications of this law even to the present day; it should, after all, be obvious that energy cannot alter a system by merely passing through it, but must be absorbed. The converse principle is very useful in the study of photobiological processes, for it is necessary to assume that those wave-lengths of light which serve to bring about a given photobiological response must be absorbed by some particular component of the living system in question (see p. 31).

The Reciprocity Law. This is often known as the Bunsen-Roscoe law. It states that the product of the duration, t, and the intensity, I, of light flux which will produce a given quantity of a given chemical reaction is a constant:

$$I \times t = k \tag{3}$$

It follows from the equivalence law, since the number of quanta absorbed in unit time by a system irradiated with a light flux of given intensity must be constant unless the absorption changes during the reaction.* In other words, as the intensity of the light flux is changed, the time required to absorb the same number of quanta must vary inversely as the intensity. The deviations from this simple relationship which occur experimentally must be due to various complicating factors in the absorbing system inherent in the reactions following the absorption of the quantum. The reciprocity law is useful in the study of photobiological processes, but is often masked by some biological response (*e.g.*, see pp. 49-51).

In the following chapters it will often be convenient to use the term *dosage* to indicate the product of time and intensity, or, in other words, the quantity of energy applied.

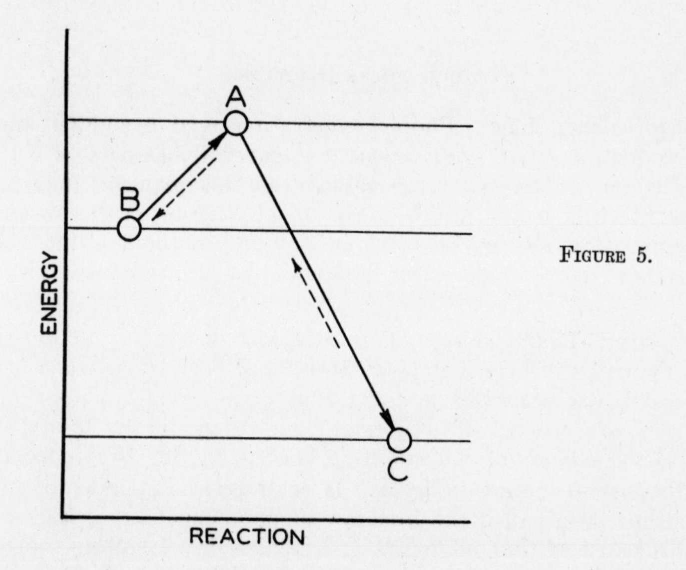

FIGURE 5.

Energy Relationships. The accompanying diagram, Figure 5, illustrates certain aspects of the energy changes in chemical reaction. The reactants are represented by position B, which has a greater potential energy than that of the products, represented by position C. In order to go from B to C, the higher energy position, A, must be traversed. This requires the addition of energy equivalent to the difference in energy between A and B, the "activation" energy, which has a specific value for a given reaction. Once this energy is supplied, the reaction may run "down hill" (*i.e.*, in the directions of the solid arrow) of its own accord. In a given system there is a random distribution of energies among the molecules, and only a certain proportion will have the required energy of activation at any instant. Only these molecules will be capable of

* Intensity is the rate of flow of radiation through a unit area normal to the light beam. Obviously $I \times t$ has dimensions of energy.

reacting. Reaction in such a system will be accelerated by increasing the temperature, thus raising the total energy, and hence the proportion of molecules having the required energy of activation. Such a reaction is termed a *thermal reaction.*

In photochemical reactions the energy of activation is supplied directly to an atom or molecule when it is raised to an excited state by the absorption of a quantum of radiant energy. After this primary act the reaction may proceed "down hill," as in a thermal reaction. In some instances enough energy may be supplied by the quantum to cause an actual increase in the potential energy, which corresponds to going "up hill" from C to B through position A, in the direction of the broken arrows. The latter type of reaction is extremely important in biology, since it includes photosynthesis by the plant, which supplies the energy needs of virtually all living systems (see, *e.g.*, Blum, 1937 a, b). In either type of photochemical reaction, excess of energy must be dissipated to heat, *i.e.*, must increase the kinetic energy of other molecules in the system by collisions.

It should be emphasized at this point that activation, either thermal or by absorption of a quantum of radiant energy, does not render any molecule capable of entering into chemical reaction. Activation is much more specific, and the chemical reaction which follows activation depends on the other atoms and molecules in the system as well as on the activated molecule or atom.*

Effect of Temperature. In discussing the effect of temperature on biological processes it is convenient to use the term *temperature coefficient* as defined by the following equation:

$$Q_{10} = \frac{K_{t+10}}{K_t} \tag{4}$$

where Q_{10} is the temperature coefficient, K_t is the rate of a chemical reaction or photobiological process at a given temperature, and K_{t+10} is the rate at a temperature 10 °C higher.

Since the activation of atoms or molecules by absorption of quanta is independent of temperature ($Q_{10} = 1$), many photochemical reactions are virtually unaffected by temperature. The rates of thermal reactions, on the other hand, are, as a rough rule, doubled or trebled by a rise in temperature of 10 °C ($Q_{10} = 2$ to 3).†

* For a discussion of activation and chemical kinetics see Daniels (1938) or Letort (1937).

† This rule applies only for a limited temperature range, such as is dealt with in biological processes, and values of Q_{10} have significance only under such limited conditions. A more rigid characterization of a thermal reaction is given by the equation:

$$\ln \frac{K_2}{K_1} = \frac{E}{R}\left(\frac{T_2 - T_1}{T_2 T_1}\right) \tag{5}$$

where K_1 and K_2 are the rates of the reaction at absolute temperatures T_1 and T_2,

If a photochemical reaction displays a high overall temperature coefficient, it means that the rate of the reaction is dominated by a thermal reaction which occurs after the primary reaction. In the case of photobiological effects, high overall temperature coefficients may be observed in which a biological response having a high temperature coefficient is set off by a photochemical reaction which itself has a low temperature coefficient. On this basis it is sometimes possible to separate experimentally two parts of the photobiological response (see pp. 179 and 198). A biological response having a high temperature coefficient is undoubtedly dominated by a thermal chemical reaction, but must often be treated differently from a secondary chemical reaction in a photochemical process.

It must be kept in mind that absorption of radiation may increase the temperature of a system, so that it is possible that thermal reactions in biological systems may sometimes be accelerated in this way, thus simulating photochemical processes (see pp. 39-40).

Photochemical Mechanisms. The photochemist is much concerned with the mechanism of photochemical reactions, which term he uses in a restricted sense to describe the series of atom and molecule combinations leading to the final product. In this study he has made great progress, having demonstrated a variety of mechanisms, which must be examined in detail to appreciate their significance. A recent book by Rollefson and Burton (1939) describes many of these mechanisms and the experimental means by which they have been unravelled. All deal, however, with far less complex systems than those found in living organisms.

In atomic reactions, which are confined to gaseous systems, the activated atom may combine with another atom, or proceed by various combinations and recombinations to the final end products. In the case of molecules there are other possible steps by which chemical reaction may proceed after the absorption of a quantum. For example, the energy of the quantum may serve indirectly to increase the amplitude of vibration of the atoms with respect to one another, so that they may actually be separated and the molecule dissociated into component parts. The result of this may be ions, atoms, or, in the case of the polyatomic molecules, free radicals. The products of this primary disruption of the molecule are capable of entering into chemical reactions which the molecule itself could not have entered into before the absorption of a quantum of radiant energy.

In most photobiological phenomena the elucidation of intimate mechanisms is more difficult, or at least the tools for its accomplishment are

ln represents the natural logarithm, and R is the gas constant. The constant E is the "activation" energy which has been discussed above. This equation is based on the distribution of energies among the molecules of a system, in which no energy is applied directly to the molecules, as contrasted to photochemical reactions where energy is added to the molecules as quanta of radiant energy. Hence it is best to use Q_{10} which has no theoretical significance, in distinguishing between photochemical and thermal processes in living systems.

less developed. Unfortunately the methods which have been most useful to the photochemist can seldom be directly applied in photobiology. Hence in most instances little is known of the steps between absorption of the light quantum and the final products or response.

It is necessary, in most cases, to confine ourselves to consideration of the initial and end results, the study of which must in any event precede more careful elucidation of the mechanism. It is extremely important, however, to keep in mind the primary act of the photochemical reaction, for its quantum aspect recalls various factors which must be taken into consideration in the study of any photobiological reaction as a whole. A photochemical reaction involving molecules might be symbolized in the following fashion:

$$M + h\nu \longrightarrow M'$$
$$M' + B \longrightarrow MB$$

But this would convey the impression that the primary reaction is known to be the activation of a molecule M, after which the activated molecule M' reacts with the molecule B. This would certainly be too much to say about any photobiological reaction known at present, and it would seem wiser to adopt some less rigid form of expression of the hypothetical events following the primary act. I suggest, therefore, the following scheme which I shall employ at times in this book:

$$M + h\nu \longrightarrow M_r$$
$$M_r + B \longrightarrow MB$$

where M_r represents the molecule M which has received a quantum and is in some way rendered reactive so that it may combine with the molecule B. The equation so written is intended to convey no opinion as to the manner in which the molecule M has been rendered reactive; M_r may be an activated molecule, a free radical, an ion, or a series of steps, but the equation carries no implication as to which. The equation then gives a clear picture of our knowledge if we know that MB is formed in a system containing M and B by the action of light, and where, from the nature of the light absorbed (*e.g.*, correspondence to the absorption spectrum), the molecule M is known to be the one which is absorbing the radiation. Certainly little more than this is known about most photobiological reactions, and the convention given above will allow the formulation of useful, tentative hypotheses without implying a more intimate knowledge of the process than is actually at command.

One type of reaction, photosensitization, is of particular interest, because it includes the basic process in photodynamic action. In such reactions the light is absorbed by atoms or molecules which are themselves found unchanged at the end of the reaction. A photosensitized reaction may be represented according to the above convention as follows:

$$M + h\nu \longrightarrow M_r$$
$$M_r + B \longrightarrow B_r + M$$
$$B_r + D \longrightarrow BD$$

where M is the light-absorbing sensitizer molecule, which is not used up in the reaction, but which renders the second molecule B capable of reacting with molecule D. In some instances of photosensitization, the light-absorbing particle merely transfers its energy of activation by collision to another particle, which then participates in chemical reaction. The equation given above is not intended to imply such a process specifically, but to include this possibility among others. The photochemical mechanism of certain reactions of this type will be treated in Chapter 6.

Quantum Yields. Following the equivalence law, a definite relationship between the number of quanta absorbed and the number of molecules transformed might be expected. For example, in the simple case of a reaction where the absorption of a quantum always resulted in the dissociation of the absorbing molecule, the ratio of molecules transformed to quanta absorbed would be unity. This ratio between the number of molecules of the end product (or sometimes of the initial reactants) and the number of absorbed quanta is called the quantum yield.* It has been measured for a considerable number of reactions and in a few cases has been found to approximate unity or some small integer, as is to be expected if the equivalence law holds.

The quantum yields of many reactions deviate widely from unity, however, being sometimes very large numbers, and sometimes less than unity. This may be accounted for in numerous ways. In some instances the quantum absorbed by one molecule may set off a long chain reaction involving many molecules and resulting in a very high quantum yield. In others, the conditions of experiment may be such that most of the molecules which receive quanta do not react but lose their energy by fluorescence or in other ways; this results in low quantum yields.

The determination of quantum yields has given much information regarding photochemical mechanisms, but in living systems, their direct determination is rarely possible. Quantum requirements may sometimes be estimated indirectly, however, as in the case of photodynamic action (see Chapter 7).

Kinetics. Another valuable tool in elucidating photochemical mechanisms is the study of reaction rates. Unfortunately, the biologist is again unable to avail himself of this tool in most instances. Rates of photobiological processes usually include a biological response which masks the rate of the basic photochemical process, and as a result it is usually

* In this calculation a unit, the Einstein, is often employed. This is the amount of energy which would be absorbed if every molecule in a mole absorbed a light quantum, *i.e.,* $h\nu \times 6.0227 \times 10^{23}$ ergs (6.0227×10^{23} is Avogadro's number, the number of molecules in one mole). Values of the Einstein are given in Figure 2 in kilo-calories.

necessary to resort to indirect methods to study the kinetics of the photochemical reaction. Illustrations of this will be found in the studies of photodynamic hemolysis described in Chapters 4 and 7.

For detailed discussion of the broad field of photochemistry the following references are suggested: Rollefson and Burton (1939), Kistiakowsky (1928), Griffith and McKeown (1929), and Leighton (1938).

Quantitative Absorption Laws

When a system is illuminated with monochromatic light (*i.e.*, quanta all of the same magnitude) the number of quanta absorbed is a function of the number of absorbing particles, and hence of the thickness of the system through which the radiation is directed. This function is logarithmic since, ideally, each layer of the system absorbs a definite proportion of the incident radiation. These facts are described by the Bougeur-Lambert law, which may be expressed as follows:

$$I = I_0 e^{-kl} \tag{6}$$

or

$$\ln \frac{I_0}{I} = kl \tag{7*}$$

where I_0 is the intensity of the incident light, and I that of the light after traversing a thickness l of the system; k is a constant for the particular system at a given wave-length λ; and e the base of natural logarithms. Obviously k is a function of the proportion of molecules in the initial energy state and of the probability that they are capable of absorbing light quanta of the magnitude corresponding to wave-length λ. This law is rigidly true for monochromatic radiation in all homogeneous systems.

The Bougeur-Lambert law is important in the consideration of penetration of radiation into a living system. For example, in the case of penetration into human skin, it is found that the ultraviolet radiation which produces sunburn (wave-lengths shorter than about 3300Å) penetrates only a very short distance. The absorption follows the Bougeur-Lambert law at least approximately [see the data of Bachem and Reed (1931), and Figure 41 (p. 202)], and k is so large that by far the greater part of the radiation is absorbed within the first small fraction of a millimeter. By using very sensitive instruments radiation might be detected at greater depths in the skin, but the quantity would certainly be negligible. False concepts have arisen from the notion that ultraviolet radiation penetrates tissues to a considerable depth.

It is also obvious that the quantity of light absorbed should be directly proportional to the concentration of molecules in the system. This is expressed by Beer's law:

$$I = I_0 e^{-klc} \tag{8}$$

* Throughout this book ln will represent the natural logarithm and log the logarithm to the base 10.

$$\text{or} \qquad \ln\frac{I_0}{I} = klc \qquad (9)$$

where c is the concentration of the absorbing molecules and k the absorption coefficient corresponding to k above. This law holds in a large number of cases over considerable ranges of concentration. It is applicable to solutions of absorbing substances dissolved in transparent solvents and is the fundamental principle applying to all colorimetry. There are a number of exceptions to this law all of which are not clearly explained. It must, therefore, be used with caution when applied to living systems. For comparative measurements the following expressions are in common use.

$$E = \log\frac{I_0}{I}, \text{ when } l = 1 \text{ cm., and } c = 1 \text{ mol. per liter} \qquad (10)$$

$$\epsilon = \log\frac{I_0}{I} \text{ when } l = 1 \text{ cm., and } c = 100 \text{ per cent} \qquad (11)$$

Both E and ϵ are called the *extinction coefficient.**

When light passes from air into a liquid, a part of it is reflected back into the air, while the remainder passes into the medium. This is true for passage from any transparent medium into another. The light which enters, the second medium may be either transmitted, absorbed, or scattered, depending upon the nature of the second medium. Reflection and scattering are generally important factors in biological systems, and are difficult to estimate.†

Sources of Radiation

To the living organism the most important source of radiation is the sun, but in photobiological studies it is sometimes necessary to use other sources which can be controlled for experimentation, or to obtain more intense radiation in the ultraviolet region of the spectrum. It is necessary to have a clear knowledge of the characteristics of the sources which are employed, or serious mistakes may be made in the interpretation of results. Unfortunately, the failure to recognize this important and elemental factor has been the cause of grave errors which appear in the literature of photobiology.

Temperature or Black-body Radiation. The unordered motion of atoms or molecules results in collisions in which some of the kinetic energy of the molecules is dissipated in the form of radiation. Thus all systems emit some radiation, a system at equilibrium emitting the same quantity of energy which it receives from its surroundings. Such radiation

* Sometimes expressed as the natural logarithm, as in (9); k is usually called the absorption coefficient in this case. Various units for l and c are used in different laboratories.

† See Mestre (1935).

consists of quanta of a great number of sizes, and appears as a wide, continuous spectrum. As the temperature of a system increases, the kinetic energy and number of collisions of the molecules is increased, and as a result the total radiation increases. The region of maximum intensity of the spectrum shifts toward shorter wave-lengths as the temperature increases, because the average molecule achieves higher energy and hence emits larger quanta. Below 500 °C. the radiation from most solids and liquids consists of small quanta belonging to the infrared portion of the spectrum, but as the temperature rises above this value, larger quanta are emitted which belong to the visible spectrum. Thus bodies below 500 °C. may be detected by the skin through the sensation of heat, while bodies above this temperature may be detected by the eye as well. Gases and vapors emit visible radiation only at very high temperatures (unless activated by electrical energy).

The emission of most systems approaches the theoretical conditions for a black body—the hypothetical perfect radiator—in which emission is a function of temperature alone. The region of maximum emission for a black body moves toward shorter wave-lengths as the temperature increases, the relationship being given by Wien's displacement law which is expressed in the following equation:

$$\lambda_{max} \times T = 0.2885 \text{ cm. degree} \tag{12}$$

where T is the absolute temperature. The distribution of the emission spectrum of a black body as a function of temperature is described by the more complicated equation of Planck.* A simpler equation for the distribution of black body radiation which holds sufficiently well in the visible and near ultraviolet region for general purposes,† and which we shall use later, is Wien's equation, which in a simple and usable form may be written:

$$\log I_\lambda = 26 - 5 \log \lambda - \text{antilog}\,(7.7939 - \log \lambda - \log T) \tag{13}$$

where I_λ is the relative intensity for a given wave-length λ, and T is the absolute temperature.

The distribution of radiation from a tungsten filament lamp is similar to that of a black body, and may be so considered for most conditions of experimentation if a slight correction is made in the temperature. Following Wien's displacement law (12) the higher the temperature of the tungsten filament, the shorter the wave-length at which the maximum emission occurs; thus lamps which burn at high temperatures emit more of their energy in the visible than do lamps burning at lower temperatures. Figure 6 shows distribution curves for a black body at 3000 °K., about the temperature of the filament in a 500-watt tungsten lamp of the usual

* See Richtmyer (1928).

† See Harrison (1929).

commercial type; the maximum emission occurs in the infrared, so that only a small part of the total emission is visible. In tungsten lamps the apparent, or "color," temperature is somewhat higher than the true temperature. The color temperature, T_c, may be determined and used in Wien's equation (13) in place of T to calculate the distribution for the visible spectrum.*

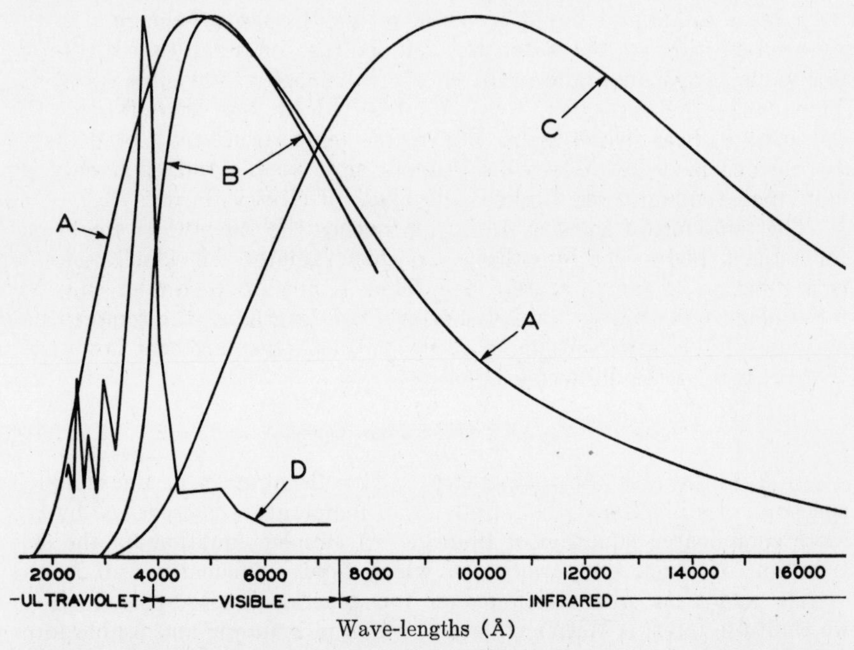

FIGURE 6. Spectral distribution of energy from different sources. Ordinates arbitrarily chosen to put all maxima at the same value. The curves must not be taken to represent relative intensities.

A, Emission from a black-body at 6000 °K. B, Sunlight at the surface of the earth; an approximate curve. C, Emission from a black-body at 3000 °K, about the temperature of a 500-watt tungsten filament lamp. D, Emission from a National Carbon Company arc using "Therapeutic C" carbons.

Figure 6 also shows the distribution of the emission from a black body at 6000 °K., a very significant curve; the maximum occurs at 4800 Å, well within the visible. This is the region at which the maximum of the sun's emission occurs, and the temperature of the outer layers of the sun may therefore be estimated as approximately 6000 °K. Estimates of the distribution of the energy in sunlight outside the earth's atmosphere show that it fits quite well the black body curve for this temperature.

The radiation from the sun must pass, however, through a number of absorbing layers before reaching the earth, so that the spectral dis-

* See Harrison (1929).

tribution of sunlight at the surface of the earth is quite different from that of a black body at 6000 °K. The first of these absorbing layers lies at the surface of the sun itself, where the atoms of the various elements exist in the gaseous state. These elements remove the lines characteristic of their absorption spectra, so that dark lines, known as the Fraunhofer lines, interrupt the continuous spectrum. There is no further change in the spectral distribution of the sun's radiation in passing through space, but a considerable modification takes place in the earth's atmosphere due to specific absorption by atmospheric constituents. Ozone in the outer layers of the atmosphere removes a considerable amount of the shorter wave-lengths, and water vapour absorbs much of the infrared. Other atmospheric constituents modify the spectrum to a less extent. In this way sunlight is reduced to the approximate limits 2900 to 18,500 Å. In Figure 6 is shown, for comparison, an approximate average curve for the spectral distribution of solar radiation at the surface of the earth. The intensity and spectral distribution of sunlight vary with the season of the year because of the difference in the distance of the sun from the earth, and the difference in the thickness of the earth's atmosphere through which the radiation must pass. The intensity and spectral distribution also vary with latitude and time of day because of variations in the thickness of earth's atmosphere traversed. The quantity of water vapor in the atmosphere modifies the spectrum of sunlight, particularly in the infrared, and for this reason curve B in Figure 6 is not carried far into this region. Variable factors, such as clouds and smoke, may further modify the composition of sunlight. Since both relative spectral distribution and total intensity vary over a wide range, it is impossible to give a true average curve for the spectral distribution of sunlight, and the curve reproduced in Figure 6 is only an approximation.

Line Emission. Pure line emission is characteristic of atoms in the gaseous state and results from jumps from one energy level to another. The spectrum of the mercury arc is of importance in experimental studies as well as being the most common source of ultraviolet for therapeutic purposes. Mercury vapor is excited in such an arc by means of the energy of an electric current; it is usually enclosed in quartz which transmits important lines in the ultraviolet, and hence is often referred to as the quartz-mercury arc, or by the misnomer "quartz lamp." Several types of such arcs are on the market; the wave-lengths of emitted lines are virtually the same for all, although the relative intensities may vary to a certain extent. The relative intensities of the various lines for a given therapeutic arc are shown in Figure 7; they were measured for a given current and temperature and may vary to a considerable extent with these factors. Smudging or aging of the lamp may affect the relative and absolute intensities.

The common mercury arc operates at a relatively high temperature so that it emits a certain amount of thermal radiation which is principally in the infrared. It is thus a "hot" lamp. By exciting the mercury vapor at low pressure an arc is obtained which emits most of its energy at 2537 Å. This has been called the "cold quartz arc." Another type of mercury-vapor lamp which has been recently developed is the so-called "sun lamp," which combines the thermal emission of the tungsten filament with mercury-vapor lines.

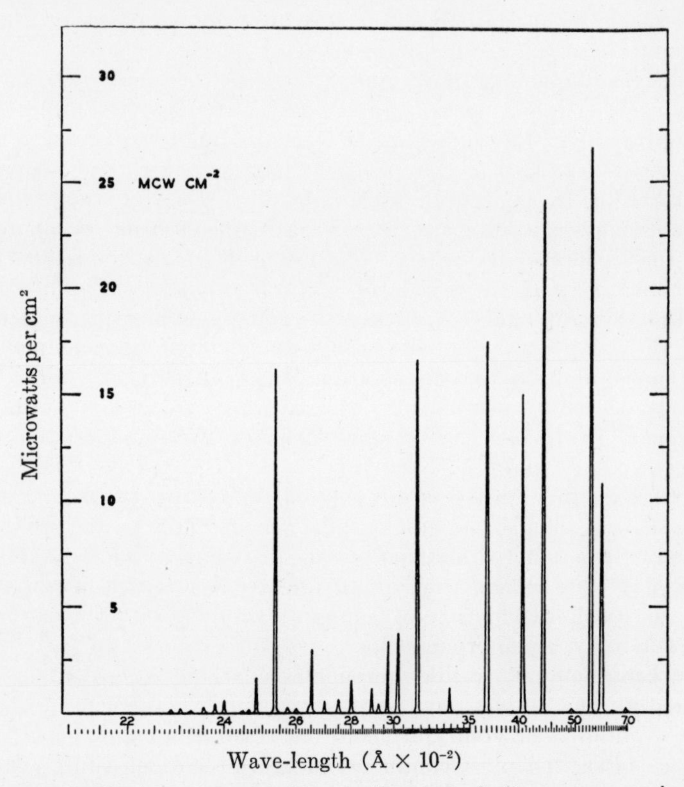

FIGURE 7. Distribution of energy from a "low-intensity" mercury arc in quartz. [From E. D. McAlister, *Smithsonian Miscellaneous Collection,* **87,** No. 17 (1933).]

Within the last few years, high-pressure mercury arcs have been developed which emit a strong continuum, and hence greater intensity at wave-lengths longer than 2600 Å. Such arcs have not yet received extensive use. In certain later chapters, particularly 18-20, frequent reference will be made to the "mercury arc," which should always be understood as the therapeutic type, since it may be assumed with certainty that the arcs used in the experiments therein described were always of this type.

The carbon arc, which is also used as a source for therapeutic purposes, operates at high temperature, and a part of its emission is of the thermal or black-body type. However, a great amount of band emission is present; characteristically the principal emission is in the ultraviolet between 3500 and 4300 Å in the region of the cyanogen band emission. By impregnating or coring the carbons with certain metals the emission may be enhanced by line emission in other regions. This is done with various "therapeutic" carbons to increase emission shorter than 3300 Å. Numerous other factors influence the emission spectrum.* An example of carbon arc emission is shown in Figure 6.

Other sources are possible and will, no doubt, become available in the future; those discussed above are most commonly used.

The Measurement of Radiation. The measurement of radiation is by no means a simple problem, and has served as the stumbling block for many attempts at quantitative study of photobiological processes. Many instruments are available, each having its particular use. The bolometer and thermocouple are uniformly sensitive throughout the spectrum; they are useful as standard instruments, and for many other purposes. Various types of photoelectric cells are available which have greater sensitivity in restricted regions of the spectrum, and which are useful for certain purposes. Different schemes for measurement or estimation of radiation, where special methods have been employed for specific purposes, will be discussed in the following chapters. For a discussion of the measurement of radiation, the review by Brackett (in Duggar, 1936) may be recommended, or the more extensive compilation of Forsythe (1937).

It is frequently desirable to study the effect of different intensities of radiation on some biological system, *e.g.*, to test the reciprocity law; and this may often be adequately accomplished without resorting to monochromatic radiation, since only relative intensities need be determined. For such purposes it is essential that the method used for varying the intensity does not alter the spectral distribution of the radiation.

For example, suppose a tungsten filament lamp is used as a source. The intensity of the radiation incident upon the biological system to be studied may be varied by altering its distance from the source. If the diameter of the filament is small enough to be regarded as a point source, the relative intensity may be estimated from the inverse square law:

$$I \propto \frac{1}{d^2} \tag{14}$$

where I is the intensity, and d the distance from the source. Although only a certain fraction of the spectrum emitted by the tungsten filament will affect the biological system, this fraction will be the same at all intensities. Thus values of I obtained for different distances may be taken as relative measures of the radiant energy affecting the biological system. Similarly, a photocell may be made to replace the biological system and used to measure the relative intensity at that point, because the photo-

* See Coblenz, Dorcas and Hughes (1926).

cell is also sensitive to a definite fraction of the total spectrum emitted by the source. It is not necessary that the photocell be sensitive to the same spectral region as the biological system, so long as only relative intensities are required.

On the other hand, if the light intensity were varied by altering the current through the lamp, this would alter the color temperature and the spectral distribution of the emitted radiation. In this case the intensity of the radiation affecting the biological system would bear no simple relationship to the total intensity, nor to the intensity measured by any instrument which did not have exactly the same spectral sensitivity as the biological system.

The Eye and its Fallacies. The most commonly used instrument for the estimation of intensity has been neglected in the preceding discussion because it merits special treatment. Properly used, as an instrument of comparison within the spectral region in which its sensitivity is relatively high, the human eye is extremely satisfactory for measurement of intensity of radiation. On the other hand, it is entirely unreliable for estimation of intensity when no comparison standard is at hand. In a spectrophotometer, where monochromatic light is used and comparison is made with a relative standard, and in a photometer where sources having the same spectral distribution are compared, the eye may be used to make very accurate measurements. However, if one attempts to compare the intensities of two sources of light of different colors, or the extent of absorption by two substances of different colors, the error may be tremendous, because the sensitivity of the eye varies throughout the spectrum (see Figure 8 I, J, p. 30). Again, if one attempts to estimate the intensity of sunlight at different times of the day, serious errors may result because of the fact that the eye adapts its sensitivity to the intensity of the prevailing light. Needless to say, the eye gives no criterion of the amount of radiation in the ultraviolet or infrared.

Chapter 3

General Biological Effects of Radiation

The present chapter deals principally with the delimitation of spectral regions of sunlight which elicit photobiological responses, and their correlation with absorption spectra. The material has been arranged with this in mind, and with particular regard to the needs of subsequent chapters, rather than to illustrate the relative importance of the individual photobiological processes. The general principles involved, are, however, of fundamental interest in all photobiological study.

At one time it was believed that only the ultraviolet, and the short wave-length region of the visible spectrum are photochemically active, or at least that this is the region of greatest photochemical activity. This idea probably arose with the observation that the unsensitized photographic plate responds only to these spectral regions, a fact which finds its explanation in other ways. As a result, the terms "chemically active rays" and "actinic rays" were applied to a vaguely defined region extending from the blue to shorter wave-lengths. Similarly, it was found that little, if any, photochemical activity was elicited by radiation of wave-lengths longer than those included in the visible spectrum, and the term "heat rays" was applied, again rather vaguely, to the infrared. The term "light" was applied only to the spectral region perceived by the human eye. With the development of quantum theory and modern photochemistry it became apparent that these delimitations had no real physical meaning, and the terms disappeared completely from photochemical literature. They lingered longer in biology, and are still to be found in current publications, although rarely. However, the concept of rather rigidly delimited spectral regions of photobiological activity persists, as shown, for example, in the tendency to speak of the "ultraviolet" as a spectral region sharply separable from the visible as regards its ability to affect living systems; and a tacit belief in distinct differences in the quality of radiations of different wave-lengths seems rather common. To be sure, certain very inexact limits may be set, but photobiological processes as a whole cover a wide spectral range (see Figure 8, p. 30), and the limits for a given photobiological process are set by specific characteristics of the system, for example, the absorption spectrum of some specific light-absorbing substance.

Normally the living organism is subjected only to the spectrum of sunlight,* and only these wave-lengths are of great importance in photo-

* For convenience the term "sunlight" will be used to include the total solar radiation reaching the earth (see pp. 24-25).

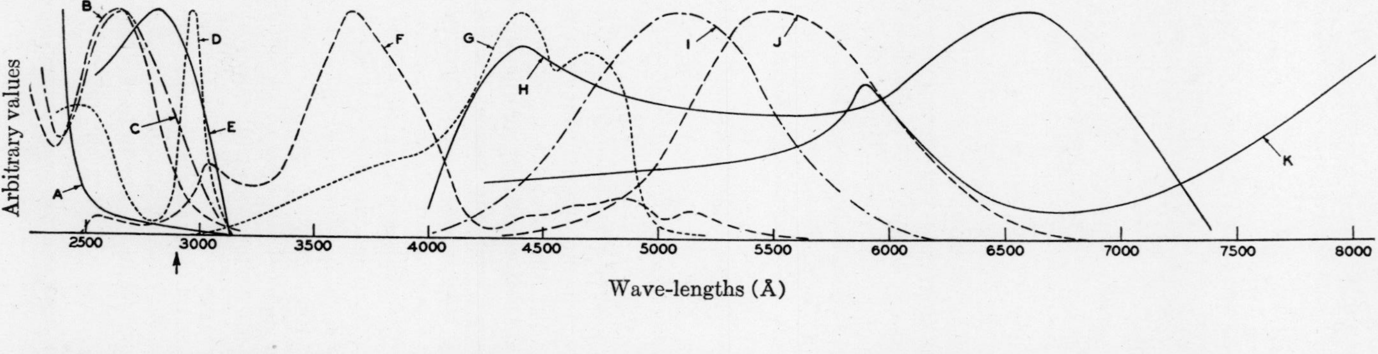

FIGURE 8. Action spectra for some photobiological processes. The arrow at 2900 Å indicates the short wave-length limit of sunlight. A, Hemolysis of erythrocytes (Sonne, 1927). B, Killing of a bacillus, *B. Coli* (Gates, 1930). C, Killing of a yeast, *Saccharomyces cervisiae* (Oster, 1934). D, Erythema of human skin (Coblenz and Stair, 1934). E, Vesiculation of a protozoan, *Paramecium multimicronucleata* (Giese and Leighton, 1935). F, "Vision" of an insect, *Drosophila* (Bertholf, 1933). G, Tropic bending of the oat coleoptile (Composite from Johnston, 1934, and Sonne, 1929). H, Photosynthesis of the wheat plant (Hoover, 1937). I, Scotopic vision (rods) of the human eye (Hecht and Williams, 1922). J, Photic vision (cones) of the human eye (Hyde, Forsythe and Cady, 1918). K, Photosynthesis of a purple sulfur bacteria, *Spirillum rubrum* (French, 1937). The curve reaches a maximum about 9000 Å.

biology; roughly this is a region extending from about 2900 to 18,500 Å. At least two factors are of major importance in the limitation of the effects of sunlight on living organisms. First of these is the penetration of light into the organisms; this varies, of course, with the absorption spectra of the substances found at the surface of the particular organism. The absorption of protoplasm increases very rapidly from about 3300 Å toward the shorter wave-lengths, so that penetration at the lower wave-length limit of sunlight, about 2900 Å, is very slight. Thus direct effects of this short wave-length radiation must be quite superficial. The absorption spectrum of water sets the long wave-length limit for the penetration of radiation into living systems in the near infrared, so that the region of greatest penetration lies, in general, in the visible. Since light absorption varies logarithmically (see p. 21) with the thickness of the absorbing substance, it is impossible to set exact limits for the penetration of light. A general idea may be gained from figures for penetration of human skin. At the region of maximum penetration, 7500 Å, light is 99 per cent absorbed after penetrating 2.5 mm. (see Bachem and Reed, 1931); this should be about the maximum depth at which radiation may have its action. Most radiation effects are much more superficial than this, those occurring in the region of 3000 Å being limited to within about 0.1 mm. of the surface. Thus the effects on micro-organisms, in which the radiation may penetrate the major part of the organism, may be quite different from the effects on larger animals where only a very superficial layer is reached.

A second factor is the variation of the energy of the quantum with respect to wave-length. The energy varies inversely with the wave-length, so that the quantum corresponding to 6000 Å is only one-half as great as that corresponding to 3000 Å. If a chemical reaction is to result from the absorption of a quantum of radiation, the quantum must supply energy of activation, and also energy of reaction in the case of a reaction taking place with an increase in free energy (see p. 16). Thus, other things being equal, more kinds of reaction should be possible at short than at long wave-lengths. Moreover, quanta corresponding to pure rotation and vibration do not seem to be effective in producing photochemical reaction, which property is restricted to quanta that produce changes in electronic energy levels. The incidence of the latter decreases with wave-length whereas the former increases, so that the incidence of photochemical reactions should fall off toward the infra-red.

The Importance of Absorption Spectra

While the factors last described may be of great importance in a general way in limiting biological effects of radiation, it seems probable that the spectral delimitation of a given photobiological process depends most often upon the limits of the absorption bands of the specific photo-active substance which serves as the light absorber in that process. From what has been said in the preceding chapter, it is obvious that the first act in any photobiological process must be the capture of a quantum of radiation by a particular kind of molecule which can thereafter promote a chemical reaction in the system. The ability of molecules of a particular kind to capture quanta must be limited to the region of the absorption spectrum of the substance, so that the limits for the particular photobiological process cannot lie outside this absorption spectrum. Absorption spectra of some organic compounds are shown in Figure 18, (p. 73). The curve relating wave-length and effectiveness in producing a given biological response, which may be referred to as the *action spectrum,* might be expected to follow rather closely the absorption spectrum of the substance whose molecules serve as the light absorber for that particular process.

Indeed it is possible in a few instances to show a remarkable degree of agreement between the action spectrum for a photobiological process and the absorption spectrum of a photoactive substance extracted from the particular biological system. In making such a comparison, it is important to recognize the fact that, while absorption spectra are ordinarily measured as absorbed energies, the action spectrum must depend upon the number of absorbed quanta. It is a simple matter to translate energies into relative number of quanta (see p. 57), and this correction should be made if a reasonable degree of accuracy in the comparison is desired. Exact agreement between absorption and action spectra, after making the above correction, can be expected only if the photochemical reaction is independent of intensity, which need not be true in all cases. Exact agreement is closely approached in Warburg and Negelein's (1928 a, b) studies of respiratory ferments, and in the case of visual purple and human rod vision (see Dartnall and Goodeve, 1937).

There are numerous factors which might cause important deviations from this agreement. The presence of absorbing, but inactive, substances in the system may reduce the action spectrum in certain regions. The energy required for a given reaction limits the action spectrum to quanta large enough to supply this energy, and this might correspond to a wave-length shorter than the long wave-length limit of absorption. In this case the longer wave-lengths of the absorption spectrum would not be represented in the action spectrum. The quantum efficiency might vary for different wave-lengths, which would tend to modify the action spectrum relative to the absorption spectrum; or there is always the possibility that only a certain structure within the molecule is capable of producing the particular photochemical process, so that only the absorption spectrum corresponding to this structure is active, and the rest inactive (rotation and vibration spectra may be considered as inactive, so that it is not necessary to explore absorption spectra far into the infrared). Absorption spectra may be somewhat different in different media, and therefore the action spectrum cannot be expected in all cases to fit exactly with the absorption spectrum of the light-absorbing substance which has been extracted and dissolved in some homogeneous medium. Absorption spectra of the same substance in different solvents usually resemble each other, however, particularly in the case of large molecules such as are concerned in photobiological processes, so that there should be some resemblance.

In spite of all the possibilities of deviation, a certain amount of agreement between action spectrum and absorption spectrum is to be expected. At least certain very useful postulates may be formulated which should hold in a general way:

(1) The limits of an action spectrum probably correspond to the limits of an absorption band of the light-absorbing substance.

(2) Some degree of correspondence between absorption spectrum and action spectrum is to be anticipated.

(3) Although action spectra and absorption spectra may not agree completely, a maximum of action is not to be expected in a region of minimum absorption. An interesting exception is discussed in Chapter 17.

(4) Two or more photobiological processes having similar action spectra may have the same substance or similar substances as their light absorbers. There is good probability of this because of the limitation of important biological compounds to relatively few types as compared to organic compounds in general.

Justification for applying such postulates will appear in this and the following chapters, and they should be kept in mind in any attempt to get at the mechanism of a photobiological process, or in comparing such processes.

Kinds of Photobiological Effect

Destructive Effect of Short Wave-Lengths. Although no definite wave-length limits may be set for different types of effect produced in biological systems by light, one general category may be distinguished from the rest. This comprises a group of destructive effects produced by radiation shorter than about 3300 Å. From the high incidence of such effects, a loose generalization may be made that such radiation is, as a rule, destructive to living systems. It is lethal to microscopic organisms, it causes sunburn in man, and in other large animals it produces effects varying from mild irritation to severe surface lesions. In some cases the effects of such radiation are stimulative rather than destructive, as with many destructive agents when acting in mild degree (see Sperti, Loofbourow and Dwyer, 1937 a, b, and Loofbourow, Dwyer and Morgan, 1938).

Action spectra for a number of such processes are shown in Figure 8 (A to E). In all cases the destructive activity is limited to the same region, *i.e.*, wave-lengths shorter than about 3300 Å. It is impossible to set this long wave-length limit exactly, as somewhat longer wave-lengths may bring about the same effects if the dosage is sufficiently great, but the curves indicate that there is virtually no effect beyond this region. This suggests that longer wave-lengths are incapable of producing such destructive effects, and the hypothesis has been advanced that the energy of the quantum sets this limit, *i.e.*, that the photoprocess underlying these effects requires energy equal to quanta corresponding to 3300 Å. Hausser (1928) offers this explanation for the long wave-length limit of sunburn. However, the definite maxima displayed by the action spectra shown in Figure 8 point toward the entrance of some other factor.

It will be noted that all the action spectra shown in Figure 8 for killing of unicellular organisms—a bacterium, a yeast and a protozoan —are similar in shape. A number of additional action spectra obtained by other investigators for other unicellular organisms are in general agreement. All indicate long wave-length limits short of 3300 Å, and maxima of destructive effect in the region of 2800 to 2500 Å; there is also a marked similarity in shape. Direct comparison is difficult because different criteria of destructive effect, and different methods of irradiation and measurement have been employed. The general resemblance is striking, however, as the reader may assure himself by consulting the following references in addition to those cited above: for bacteria, Ehrismann and Noethling (1932), Hollaender and Claus (1936); for yeast, Ehrismann and Noethling (1932); for dermatophytes, Hollaender and Emmons (1939); for protozoans, Sonne (1927), Weinstein (1930); Gates (1934b) obtains a similar curve for bacteriophage; and Henri (1934) compiles somewhat similar curves for a number of destructive phenomena studied earlier in his laboratories. This resemblance between the action spectra for the destruction of various unicellular organisms suggests that the photoactive substances are the same or similar.

Victor Henri (1912) pointed out the similarity between the action spectra for such destructive processes and the absorption spectra of white of egg and of serum, substances which are essentially protein. Reference to the absorption spectra of various proteins, Figure 9, shows that there is certainly some resemblance to the action spectra for the destruction of unicellular organisms, in that maxima occur in the same spectral region in both. However, this resemblance is rendered less important by the fact that many substances absorb characteristically in this spectral region, particularly benzenoid ring compounds. But since proteins are very important constituents of the cell and are readily altered by ultraviolet radiation, they must be suspected as light absorbers in such destructive photobiological processes.

Another important group of cellular constituents absorbs in this same spectral region, namely, the nucleic acids. In Figure 10, the absorption spectrum of thymo-nucleic acid is shown in comparison with the absorption spectrum of a typical protein. There is a great quantitative difference in absorption at the maxima, but this may be disregarded in making comparison with action spectra, since the latter provide only relative values. Important differences are shown, however, when only the shape of the curves is considered. The maxima occur at different wave-lengths, about 2800 Å for protein and about 2600 Å for nucleic acid, and protein absorption increases much more rapidly toward shorter wave-lengths.

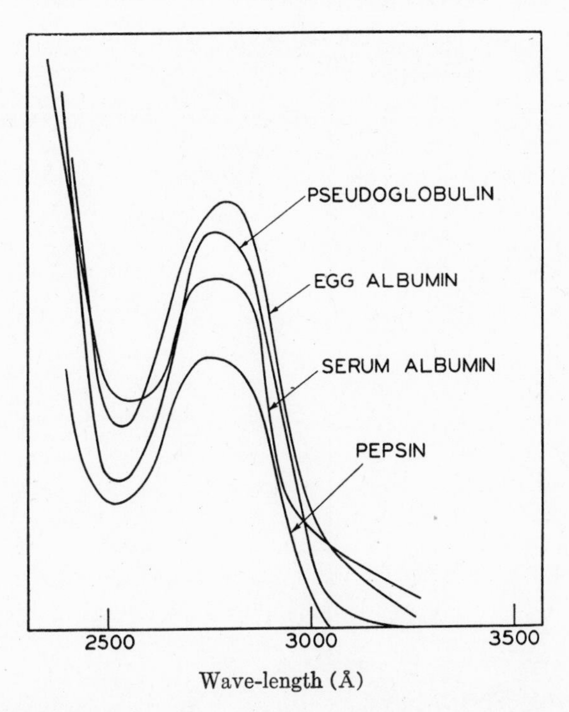

FIGURE 9. Absorption spectra of proteins. Ordinates: absorption coefficients, scales arbitrarily chosen to bring the curves into proximity.

Pseudoglobulin, egg albumin and serum albumin from Hicks and Holden (1934). Pepsin from Gates (1934a). The shape of the absorption spectrum curve of egg white (Henri, Henri, and Wurmser, 1912) is in close agreement with that for egg albumin.

Hollaender has recently pointed out that action spectra for some processes resemble nucleic acid absorption spectra, *i.e.,* killing of bacteria and fungus spores (dermatophytes) (Hollaender and Emmons 1939), and mutation production in dermatophytes (Emmons and Hollaender 1939), while others resemble the absorption spectra of proteins, *i.e.,* artificial parthenogenesis of echinoderm eggs (Hollaender 1938), and inactivation of pin worm eggs (Hollaender, Jones and Jacobs 1940).

Sonne's (1927) action spectrum for hemolysis of red blood cells (Figure 8A, p. 30), does not show the characteristic maximum, and he suggests that the lipoid component of the cell determines the action spectrum in this instance.

It is thus probable that radiation shorter than 3300 Å may cause injury to cells by action on the nucleic acid constituent of the nucleus in some instances, on the protein component of the cells in others, and possibly at times on other substances absorbing in this region. From the point of view of this book, it is only important to recognize that cells,

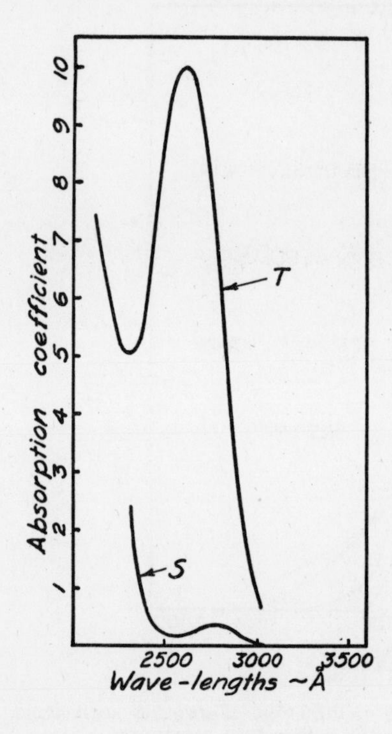

FIGURE 10. Absorption spectra of a nucleic acid and a protein. T, sodium-thymo-nucleate. S, serum albumin. (After Caspersson, 1936).

in general, are damaged by such radiation, probably because of the fact that important photolabile compounds common to all cells absorb strongly in this spectral region but not appreciably at longer wavelengths.

The action spectrum for sunburn (Figure 8D, p. 30), differs from those for the killing of unicellular organisms, although showing a remote resemblance. In Chapter 17 the possibility that sunburn is a manifestation of the same fundamental mechanism involved in the killing of unicellular organisms by radiation in this spectral region will be considered, together with reasons for the divergence of the action spectra in the two cases.

From the foregoing discussion it appears that there is no reason for assigning a specific destructive power to the "ultraviolet ray," as is too frequently done. The generally destructive action of "ultraviolet" radiation may be most reasonably attributed to the fact that important, photolabile components of protoplasm show specific absorption below wave-length 3300 Å, so that most cells are vulnerable to such radiation. When the term "actinic" is employed today it seems usually to refer to these destructive processes.

A point of practical and experimental importance is that common window glass absorbs virtually all the wave-lengths which produce these destructive effects (see Figure 50, p. 241), and hence may be used as a filter to eliminate such processes when it is desired to study effects produced by longer wave-lengths.

Destructive Effects of Longer Wave-Lengths. Wave-lengths longer than 3000 Å may produce destructive effects when their incident intensity is very high. This seems to be a different phenomenon from that just discussed. It will be taken up in a later chapter (p. 117).

Antirachitic Action. This important non-destructive process which depends upon the synthesis of vitamin D or other antirachitic substance from precursors has about the same long wave-length limit as the destructive effects described above. The action spectrum differs however (see Bunker and Harris, 1937), and must be determined by the absorption spectrum of the precursor substances.

Vision, Oriented Movements, and Other Photosensory Processes. The spectral range of human vision extends from about 3900 to about 7400 Å. This includes both rod (scotopic) and cone (photopic) vision; the curves relating visual sensitivity to wave-length for these processes are shown in Figure 8, I and J (p. 30). These are really examples of action spectra, and in the case of rod vision there is good agreement between the absorption spectrum of visual purple, a substance found in the rods, and the spectrum of rod sensitivity (see Dartnall and Goodeve, 1937). The spectral sensitivity of the eyes of other vertebrates is generally similar to that of the human eye (see Honigman, 1921; Laurens, 1923; Chafee and Hampson, 1924; Grundfest, 1932a and b; Graham and Riggs, 1935; and Brown, 1936).

Among the invertebrates oriented movements provoked by light and other photosensory processes have been studied as regards their action spectra. Among such studies may be mentioned those of Hecht (1920, 1928), White (1924), Crozier (1924), Moore (1926), Visscher and Luce (1928), Graham and Hartline (1935), and Waterman (1937). The studies cover several phyla, and the action spectra display wide variation. All are subject to the criticism that the measurements cover only the same spectral range as human vision. The action spectra obtained by Bertolf for the honey bee (1931a and b) and for *Drosophila* (1933), the latter of which is shown in Figure 8F., (p. 30) indicate that these insects are far more sensitive in the ultraviolet than in the visible region of the spectrum, and therefore measurements which do not include the ultraviolet are of questionable value. Fardon, Carroll and Sullivan

(1937) find minor variation between different genetic strains of *Drosophila,* but uniformly greater stimulation by the ultraviolet than by the visible. Such measurements show that vision is not limited to any very definite region of the spectrum, and that the assumption that all animals see approximately the same wave-lengths as man is entirely unwarranted.

Action spectra for photo-oriented movement among the chlorophyll-bearing flagellates have been studied by Mast (1917) and Laurens and Hooker (1920). There is a great similarity between these action spectra, but there are definite variations. All appear to lie in the violet and blue, and the measurements of Blum and Fox (1933) indicate no important sensitivity in the ultraviolet.

Among plants, the action spectrum for the phototropic bending of the oat seedling (etiolated coleoptile) has been measured by a number of workers, whose values show general agreement, but certain variations. The data of two of these (Sonne, 1929; and Johnston, 1934) have been used to draw the composite curve in Figure 8G (p. 30). The spectral sensitivity of the combined processes is confined chiefly to the blue and violet, with some sensitivity in the ultraviolet. Action spectra for the phototropic bending of the sporangiophores of certain fungi have been studied, and found also to lie in the blue and violet; the ultraviolet has not been fully explored (see Parr, 1918, and Castle, 1931).

Assuming that action spectra must correspond to the absorption spectra of the photoactive pigments, it is interesting to speculate as to the possibility that some one molecular species is concerned in all photosensory processes. The carotenoids suggest themselves as such substances. Wald has presented convincing evidence that visual purple (*rhodopsin*) is a conjugated protein in which the prosthetic group is a carotenoid, and has shown the relationship between this substance and vitamin A (1935a, b and c). Thus the carotenoid participation in the rod vision of vertebrates is established. No photoactive substance has been isolated from the retina which might be suspected of being that of the cones, except that recently found by Wald (1937) in the chicken's eye; the chemical nature of this pigment is not yet established. Wald and DuBuy (1936) isolated carotene and xanthophyll from the oat seedling, and pointed out the similarity between the absorption spectra of these substances and the action spectrum for phototropic bending. This similarity appears obvious when Figure 8G and Figure 44 (p. 206) are compared, since the absorption spectra of the uncombined carotenoids are limited to the same general region, and show characteristically two large maxima separated by a minimum. The data of Haig (1935) present a somewhat different picture, which may be more difficult to harmonize with this concept. The absorption spectra of the carotenoids and carotenoid proteins cover a wide spectral range (see Verne, 1930, and Zechmeister, 1934), and it may be that the photoactive substances involved in the photosensory processes of many invertebrates belong to this group. On the other hand, the action spectra for the honey bee and for *Drosophila* certainly do not resemble the absorption spectra of carotenoids, but might fit well the characteristic absorption of porphyrins (see Figure 16, p. 56). An answer to this interesting question can come only from determination of the action spectra for photosensory responses in a number of representative plant and animal forms. The existing data are seldom satisfactory, the wave-lengths chosen for study being, as a rule, too restricted both in number and spectral range.

From the preceding discussion, it is clear that no particular portion of the spectrum may be set apart as a region of photosensory response. Such responses cover a wide range of the spectrum of sunlight, extending from its lower limit to the long wave-length limit of human vision at about 7400 Å, and possibly beyond. The attempt to interpret the vision of invertebrates in terms of human vision is obvious anthropomorphism.

Photosynthesis. The photosynthesis of the green plant, which may be referred to as chlorophyll photosynthesis, takes place almost exclusively in the visible spectrum. The action spectra for photosynthesis by the wheat plant, as measured by Hoover (1937) is shown in Figure 8H (p. 30); it displays the two principal maxima of chlorophyll, the photoactive substance concerned in this process. The resemblance between the action spectrum and the absorption spectra of chlorophyll *a* and *b* (see Figure 18, p. 73) is greater if energies are converted into a number of quanta; the differences may be due to the concentration of chlorophyll and the thickness of the absorbing layer (see Chapter 5 for a consideration of these factors). The action spectrum for photosynthesis by the purple sulfur bacteria, recently determined by French (1937), is also shown in Figure 8K; it differs strikingly from that of chlorophyll photosynthesis, and is in good agreement with the absorption spectrum of bacteriochlorophyll, a substance found in these bacteria.

Effects of Infrared Radiation. Absorption of radiation from the far infrared represents changes in rotation and vibration of the atoms within the molecule, not changes in the electronic energy levels. Only the latter produce photochemical reaction. In the near infrared, electronic energy level changes occur as in the visible, and photobiological effects produced by such radiation are possible; but, on the other hand, their incidence should decrease as wave-length increases. Relatively few well-substantiated effects in the near infrared have been reported; the case of photosynthetic activity by the purple sulfur bacteria is an example. Many of the alleged effects are doubtful, and it cannot be assumed that this part of the spectrum is very active biologically; an example of a negative experiment with such radiation is that of Clark and Chapman (1932).

In all studies of radiation phenomena in biological systems, the possible effect of increase of temperature by the absorption of radiation which does not produce photochemical reaction must not be forgotten (see pp. 17-18). Examples are known where rates of biological processes are tremendously accelerated by increase of temperature, *e.g.*, the denaturation of proteins. Radiation of any wave-length may increase the temperature of the system which absorbs it; such effects are not limited to the infrared, and hence the term "heat radiation" has no real meaning, since any wave-lengths may "heat" the system exposed to them. The action spectrum of such a heating effect should not show sharp maxima and minima as do those of the true photobiological processes which have been discussed above, although some variation with wave-length might be expected because of differences in absorption by the system. Such a difference in action spectra should help to distinguish between these two types of effects.

Other Effects. Certain other effects of light have been described, for example, an effect on the growth and sexual cycle of birds, which may result from stimulation of the retina, and, if so, should be classed with vision (see Hemmingsen and Krarup, 1937).

"Mitogenetic Radiation." The topic of the so-called "mitogenetic radiation" is controversial at present. The existence of such radiations is not firmly established, and the evidence is conflicting. A discussion of the subject cannot be included here, but the review of Hollaender (Duggar, 1936) may be recommended as a fair and critical analysis of the problem. To invoke explanations based on the existence of such radiation at the present time only serves to confuse the general subject, so long as other explanations are available which are based on more firmly established principles. This concept cannot be expected to add to the clarity of the material to be discussed in the following pages.

"Antagonism" of Radiation of Different Wave-Length

Much has been published about the supposed antagonism of different wave-lengths, but the subject seems shot through with error and misunderstanding. There is no theoretical reason for believing that specific wave-lengths are antagonistic to other specific wave-lengths, yet the idea that infrared radiation tends to oppose visible or ultraviolet radiation is current in some places. It seems rather premature to invoke such explanations for effects observed in biological systems before the attempt has been made to explain them in terms of the absorption spectra of different photoactive substances which participate in independent photochemical reactions.

A nice example of the success of such an explanation is found in the experiments of Flint and McAlister. First experiments (Flint, 1934) indicated that red or yellow light increased the germination of dormant lettuce seeds, whereas blue light decreased their germination—an apparent case of "antagonism." Further analysis (Flint and McAlister, 1935, 1937) disclosed the existence of three distinct action spectra regions: (1) inhibition by wave-lengths at about 4000 to 5000 Å; (2) promotion of germination between 5000 and 7000 Å and (3) inhibition at about 7000 to 8000 Å. The results are represented in Figure 11. The action spectrum of inhibition in the blue and violet shows two maxima; it corresponds approximately to the absorption of a substance in the extract from lettuce seeds, to the curve for phototropic response of the oat seedling, and to the absorption spectra of carotene and xanthophyll. The action spectrum for increase of germination agrees very well with the absorption spectrum of chlorophyll in this region, and with that of an acetone extract of the seeds. Inhibition by longer wave-lengths is not accounted for as yet. This analysis indicates the existence of more than one separate photochemical process depending upon absorption by different molecules, and it is probable that many, if not all, cases of so-called "antagonism" will be explained eventually on this basis.

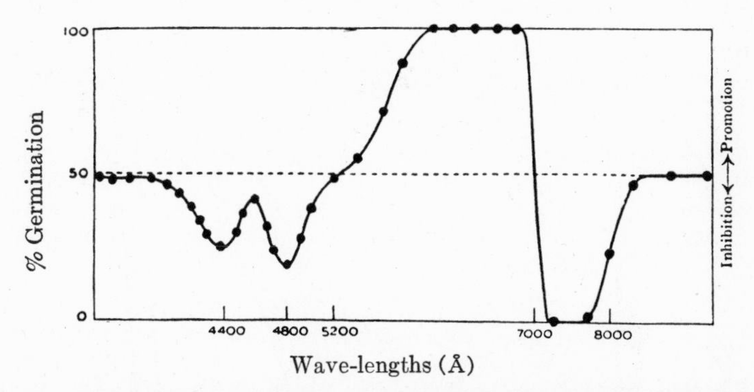

FIGURE 11. Inhibition and promotion of germination of lettuce seeds by light. The seeds are first irradiated with enough red light to produce fifty per cent germination; subsequent irradiation with the wave-lengths indicated either increases or decreases the percentage of germination according to the wave-length used. [From L. H. Flint and E. D. McAlister, *Smithsonian Miscellaneous Collections*, **96**, No. 2 (1937).]

Prát has recently (1936) reviewed the subject of "antagonism" of radiation; in very few of the studies he reports have energies and wave-lengths been sufficiently controlled to allow critical evaluation. Often color filters used to isolate spectral regions have been chosen according to the color which they present to the eye without proper spectroscopic control. The experience of Flint and McAlister provides an excellent example of the fallacies that may enter into such experiments. Flint (1934) found that light passing through some green filters inhibited germination of lettuce seeds, whereas that passing through other green filters promoted it. Flint and McAlister (1935) made a spectral analysis of the transmissions of the filters, some of which are shown in Figure

12. A comparison of these transmissions with the inhibition and promotion curves, Figure 11, will explain the discrepancy.

Radiation Outside the Limits of Sunlight

X-rays and Gamma Rays. Living organisms are normally subjected to an extremely small amount of such radiation, which may have some importance from a genetic and evolutionary standpoint since it may

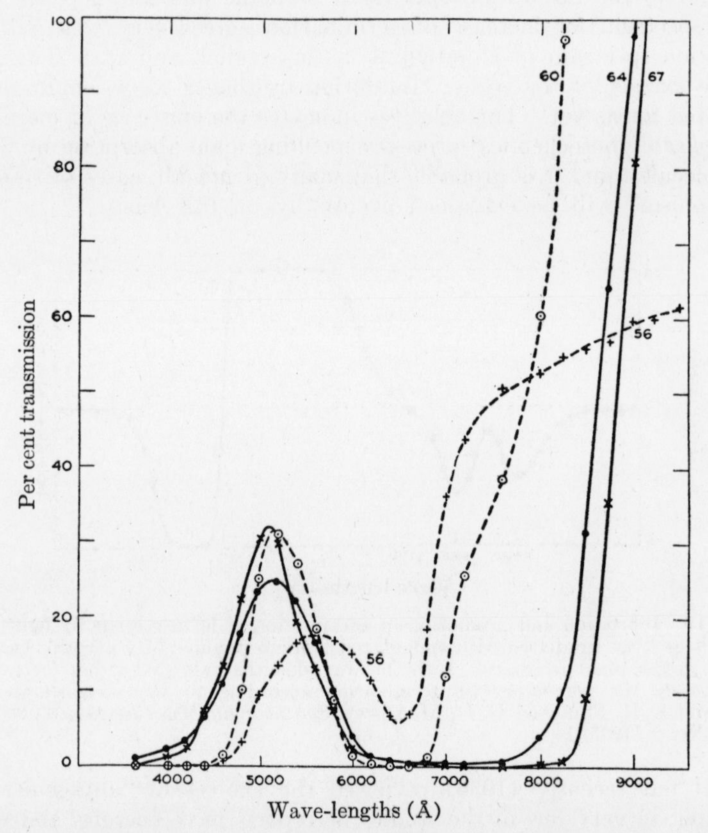

FIGURE 12. Transmission of light by green filters used by Flint (1934). Numbers 56 and 60 are filters which inhibit germination of lettuce seeds. Numbers 64 and 67 are filters which promote germination. All filters appear green to the eye. [From L. H. Flint and E. D. McAlister, *Smithsonian Miscellaneous Collections,* **94,** No. 5 (1935).]

produce mutation. Certain lethal and stimulating effects are caused by much larger doses than would normally be encountered. These effects are beyond the scope of this book. For treatment of this subject the reader is referred to numerous chapters in Duggar (1936).

Particle Bombardment. Modern physical theory finds little difference between waves and particles, and the effects of bombardment by alpha and beta particles are quite similar to those by gamma rays. High-momentum neutron and deuteron bombardment produces phenomena of yet another type, for the nucleus of the atom may be disturbed by such particles. Study of the effects of such bombardment is in its infancy, and the mechanism is so different from that which will be dealt with in the following pages that it merits no further discussion here. Possible cosmic ray effects also belong in this category.

Radio Frequency. While concentrated short radio waves may raise the temperature of living systems to dangerous limits, it is very doubtful if any specific effects are to be expected. Theoretically they are not probable, and none seem to have been experimentally established despite numerous claims. The subject has been recently reviewed by Curtis, Dickens and Evans (1936).

Part II

Photodynamic Action

Chapter 4

General Photochemical Aspects of Photodynamic Action

The present chapter and those which immediately follow are devoted to an analysis of the mechanism of photodynamic action, leading to the picture of the phenomenon which was briefly sketched in the introduction. In this analysis it will be necessary to distinguish between the photochemical part of the process, which is fundamental to the phenomenon in all its manifestations, and the events induced by this process, which vary with the nature of the biological material. Only certain systems lend themselves to such treatment, and the immediate discussion will therefore be limited to relatively few examples of photodynamic action, description of the multiformity of the phenomenon being postponed to Chapter 10. Consideration of the numerous factors which modify and lend variety to the general aspect of photodynamic action will also be delayed until after analysis of the photochemical mechanism, as will be the criticism of alternative hypotheses, since these have often been influenced by such modifying factors.

A Typical Photodynamic Process. Mammalian red blood cells suspended in salt solution may be exposed to sunlight for several hours without the occurrence of detectable damage, provided wave-lengths shorter than 3300 Å are excluded.* But if a small amount of any one of a large number of fluorescent dyes (*e.g.*, eosin†) is added to the cell suspension, hemolysis, *i.e.*, damage to the cell structure resulting in the release of hemoglobin, may occur in a few minutes if the suspension is exposed to light of those wave-lengths which are absorbed by the particular dye.

This process, which will be referred to as photodynamic hemolysis, will serve as an example of photodynamic action which is characteristically a destructive effect, often lysis.

The Course of Photodynamic Action. If the course of photodynamic hemolysis is followed under the microscope, it is found that the cells do not begin to hemolyze immediately. Only after a relatively long period of time do the first cells fade out due to the escape of hemoglobin; but once these first cells have hemolyzed the others follow quite rapidly, until in a short time almost all have disappeared. The progress of hemolysis

* See p. 33. Hemolysis does not occur after excessive exposure to longer wave-lengths (see p. 118).

† See pp. 72-73 for a discussion of the wide variety of these substances.

in thin layers of cell suspension may be followed more or less quantitatively by means of an appropriate photometric method, yielding a curve such as is shown in Figure 13, curve *a*, where the percentage of hemolyzed cells is plotted against the time after exposure to light. The curve indicates that no appreciable hemolysis occurs during a long initial period, after which hemolysis begins abruptly and proceeds quite rapidly.

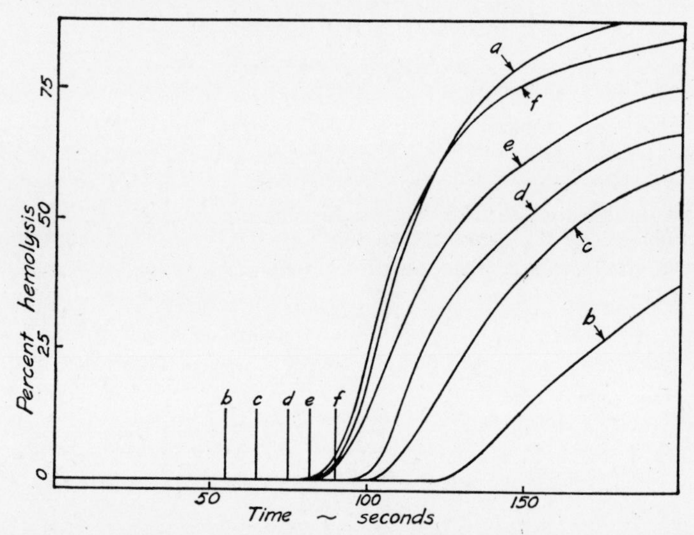

FIGURE 13. Curves showing the course of photodynamic hemolysis.* Curve *a* was obtained when light was continuous throughout the run. In curves *b* to *f* the light was discontinued after the point on the time ordinate indicated by the vertical line bearing the corresponding letter. (From Blum and Morgan, 1939.)

* These curves were obtained by photometry with "indifferent" light, *i.e.*, wavelengths not absorbed by the dye, and hence not producing hemolysis. They may not accurately represent percentage hemolysis throughout their courses, being subject to errors which are discussed by Blum and Gilbert (1940a).

This long initial period before detectable change is typical of photodynamic action. It is easily observable in photodynamic cytolysis of echinoderm eggs, and the quantitative curve presented by Moore (1928) is not unlike the percentage hemolysis curve for photodynamic hemolysis. Photodynamic stimulation of muscle also displays a very long latent period (see Figure 26, p. 109), and many more instances might be cited.

The nature of curve *a*, Figure 13, suggests that a photochemical reaction takes place during the initial period which causes damage to the cell, but that hemolysis actually begins only after a given amount of damage has been accomplished. This suggests, furthermore, that the photochemical reaction is necessary only to initiate the hemolytic process which is itself independent of light, and this is confirmed by the fact that hemolysis continues after the light is cut off, as illustrated by curves *b* to *f* in Figure 13. These curves also show that the earlier the light is cut off the more slowly does the percentage hemolysis curve rise, indicating that either hemolysis itself or the initiation of hemolysis is accelerated by photochemical damage beyond that necessary to initiate lysis.

Actual change in the red-cell membrane during the period of irradiation prior to the beginning of hemolysis has recently been shown by Davson and Ponder (1940), who find a progressive alteration of permeability to potassium during this period. How closely this follows the photochemical reaction has not been shown.

It seems clear, then, that photodynamic hemolysis may be divided into a photochemical process and a lytic process which are to a certain extent separable. It is obvious that the characteristics of the photochemical process can be determined only if it is separated in some way from the lytic process. Failure to do this has frequently resulted in misinterpretation.

Irreversibility of the Photochemical Process. If the cell suspension is subjected to short, alternate light and dark periods, hemolysis will occur only after the total period of irradiation is equal to that which would be required if the light were continuous, as is illustrated in Table 1. This must mean that the photochemical process is irreversible and

Table 1. Effect of Intermittent Light on Photodynamic Hemolysis.

(After Blum and Morgan, 1939)

Time (in seconds) to 50 per cent hemolysis.

light	dark	light	dark	light	dark	light	total light
60	60	60	60	60	60	150	330
60	60	60	60	60	60	160	340
60	60	60	60	60	60	187	367
continuous	..	..	..	..	..	...	350–380

additive. Obviously, if the light and dark periods were too long this would not be true, because hemolysis, once started, would continue through the light period.

A similar result was reported by Efimoff (1923) for the photodynamic killing of paramecia, indicating that this is a general characteristic of photodynamic action.

The Reciprocity Law. As was pointed out in Chapter 2, the reciprocity law should hold for the primary reaction in any photochemical process, and deviations shown by the overall process result from subsequent complicating reactions. If measurements to test this relationship for photodynamic hemolysis are made in such a way as to include the lytic part of the process in the measurement of the time of irradiation (t in equation 3, chapter 2) deviation from the reciprocity law is observed. This is shown in Table 2, the data for which were obtained by treating percentage hemolysis curves after the manner common for rate curves of

Table 2.

(From Blum and Gilbert, 1940a)

Experiment	Q Incident energy in ergs sec.$^{-1}$ mm.$^{-2}$ λ = 5461Å	t Time to "50%" hemolysis (seconds)	$Q \times t \times 10^{-5}$
6–20a	310	830	2.57
	208	1060	2.21
	165	1280	2.11
	75.8	1860	1.41
6–23a	276	800	2.21
	204	960	1.96
	50	2860	1.46

chemical reactions, *i.e.*, by measuring t as the time to a given degree of hemolysis. When t is measured in this way it must include a period required for hemolysis, during which the light need not act (compare curves a and f in Figure 13).

Table 3.

(From Blum and Gilbert, 1940a)

Experiment*	Q Incident energy in ergs sec.$^{-1}$ mm.$^{-2}$; λ = 5461Å	t Period of irradiation (seconds) required to produce hemolysis in 24 hrs.	$Q \times t$ ergs mm^{-2} $\times 10^{-4}$	Maximum deviation from mean of $Q \times t$	Uncertainty in measuring t
6–20	326	37.5	1.22		± 20%
	230	50	1.15		± 20%
	173	75	1.30		± 20%
	101	100	1.01	14%	± 20%
6–23	274	62	1.70		± 11%
	181	90	1.63		± 11%
	50	34.3	1.71	3%	± 13%
6–30	280	18.5	.52		± 80%
	142	36.5	.52		± 80%
	52	94.5	.49	4%	± 9%
7– 6	261	15	.39		± 33%
	261	30	.78		± 33%
	88.5	88	.78		± 33%
	88.5	88	.78	27%	± 33%
7–19	318	143	4.55		± 9%
	174	261.5	4.55		± 9%
	111	422	4.68	2%	± 11%

* Different concentrations of dye were used in the various experiments so that the results are comparable only within a given experiment.

On the other hand, good agreement with the reciprocity law is obtained when the lytic part of the process is eliminated from the measurement of t, as is illustrated in Table 3. In this instance, t is the minimum period of irradiation required to produce complete hemolysis twenty-four hours after the irradiation; this method eliminates the rate of the lytic process from direct participation in the measurement of t. The details of the method, and its apparent accuracy, are discussed by Blum and Gilbert (1940a). Such close adherence to the reciprocity law indicates that the fundamental photochemical reaction underlying photodynamic hemolysis follows this relationship, and hence is independent of the intensity of radiation. This speaks for a relatively simple photochemical mechanism.

All attempts to demonstrate the reciprocity law for photodynamic action previous to that which yielded the data of Table 3, have indicated deviation from this relationship; but in all instances measurement of t seems to have been influenced by processes other than the fundamental photochemical mechanism. The studies of photodynamic hemolysis carried out by Blum and Hyman (1939a) employed a method similar to that with which the data of Table 2 were obtained, and are subject to the same error although the deviation is less apparent because of the fact that higher intensities were used. Dognon (1928b) measured t as the time required for killing paramecia, finding approximate agreement with the reciprocity law. His curves indicate deviation similar to that found by Blum and Hyman, however; and since his measurement of t included the time required for death of the animals, his data are subject to the same criticism. The effect of light intensity on photodynamic stimulation of muscle was studied by Lillie, Hinrichs, and Kosman (1935) who found, over a limited range, fair agreement with Schwartzchild's law; an exponential relationship which holds for the photographic plate, but has no theoretical significance as applied to other systems. Their measurement of t was based on the amplitude of contraction of the muscle, which is probably not a true index of the photochemical process, since experiments of Kosman and Lillie (1935) indicate that muscle contraction may be modified without altering the rate of photo-oxidation.

Since the only experiment in which such complicating factors were ruled out shows good agreement with the reciprocity law, it is reasonable to assume that the underlying photochemical mechanism follows this simple relationship.

The Effect of Temperature. The effect of temperature on chemical reactions was discussed in Chapter 2, where it was pointed out that photochemical processes not complicated by thermal reactions should be independent of temperature, *i.e.*, should have a Q_{10} equal to unity. This seems to be virtually true for photodynamic action as far as such studies have been made. For photodynamic hemolysis, Blum, Pace, and Garrett (1937) found the value 1.2 for the Q_{10} of the overall process, including the time necessary for lysis (see p. 85), and recently Davson and Ponder (1939) have found a Q_{10} of unity. The disagreement is not significant, and may be due to the fact that red cells from different species were used, which might effect the lytic part of the process, although it is unlikely that the photochemical mechanism is different in the two cases. Hannes and Jodlbauer (1909) found 1.12 for the Q_{10} of photodynamic inactivation of invertase, and Wohlgemuth and Szörény (1933a) found the uptake of oxygen by photosensitized tissues to be independent of temperature. Some of the foregoing measurements include secondary processes, but nevertheless have very low Q_{10} values; and there can be little doubt that the photochemical mechanism itself is independent of temperature, or virtually so.

Characteristics of the Photochemical Reaction. The findings given above throw considerable light on the photochemical mechanism underlying photodynamic action, indicating its relative simplicity. Irreversibility and independence of light intensity show that this is not a steady state reaction, where a photochemical process is in dynamic equilibrium with some other reaction. These same factors, together with independence of temperature, rule out any long-chain reaction and the existence of dominant secondary thermal reactions. The whole picture presented by these findings is that of a relatively simple photochemical mechanism which would probably have a quantum yield of unity or near unity. We may now proceed to the examination of other important aspects of this photochemical mechanism.

Chapter 5

Light Absorption in Photodynamic Action

By a more or less crude experiment, Raab (1900) showed that the wave-lengths which produce photodynamic action are those which are absorbed by the photosensitizing dye. He found that photodynamic killing of paramecia was prevented by interposing a layer of the dye between the photosensitized protozoa and the source of light—a qualitative demonstration that the wave-lengths absorbed by the dye, when acting as a filter, are the same as those which initiate the photodynamic effect. This is of fundamental importance because it shows that the absorption of a quantum of radiation by the dye molecule is the initial reaction which sets off the events leading to the observable photodynamic effect.

Numerous investigators have confirmed Raab's finding by one method or another with various sensitizers and biological systems (*e.g.*, Hertel, 1905; Metzner, 1924; Passow and Rimpau, 1924; Lippay 1929; Blum and Scott, 1933; Dreyer and Cambell-Renton, 1936); but there are occasional statements to the contrary, and it is distressing to find that photodynamic action is frequently attributed to "ultraviolet" radiation, as this indicates a fallacious association with the destructive effects of wave-lengths shorter than 3300 Å (see pp. 33 and 114).

Action Spectra and Absorption Spectra

If the dye is the light absorber in photodynamic action, its absorption spectrum should agree on a quantitative basis with the action spectrum of that process, subject to certain limitations. It seems advisable to discuss at some length the conditions necessary for such agreement in order to understand these limitations and to evaluate the agreement which may be anticipated. In the course of this discussion it will become clear that certain reported instances of disagreement between action spectra and absorption spectra, represent only such deviations as may be expected under the experimental conditions. Although designed to apply only to the immediate problem of photodynamic action, the discussion will constitute an extension of that in Chapter 3 on action spectra and absorption spectra in photobiological processes in general.

Measurement of Absorption Spectra. The nature of absorption spectra may be made more clear by a brief discussion of their measurement, which is rendered the more important because errors in this regard have contributed a certain degree of confusion. It is, of course, beyond the scope of this book to describe in detail methods of absorption spectroscopy, and only a few fundamental principles will be discussed.*

There are numerous ways of determining absorption spectra. Direct measurement may be made by using a monochromator to separate the wave-lengths, with a non-selective instrument such as a thermocouple to measure the intensities. Other methods are usually more convenient, however. A spectrophotometer, which makes use of visual photometry, may be employed for the visible spectrum, and a quartz spectrograph is commonly used for the ultraviolet. In the latter instrument the spectrum is made to fall upon a photographic plate which is subsequently developed, the density of the blackening being measured for each wave-length. By proper correction to relate blackening to light intensity, a quantitative value for absorption is obtained. A discussion of this method is given by Harrison (1929, 1934). Whatever method is used, the absorption of the solvent must always be taken into consideration and proper correction made. Unfortunately the photographic method has sometimes been used without correction, and this has led to serious errors in interpretation. This has been true for some of the measurements of absorption spectra of the porphyrins, substances which will be frequently discussed in subsequent chapters; and as an illustration the determination of the absorption spectrum of a porphyrin will be described.

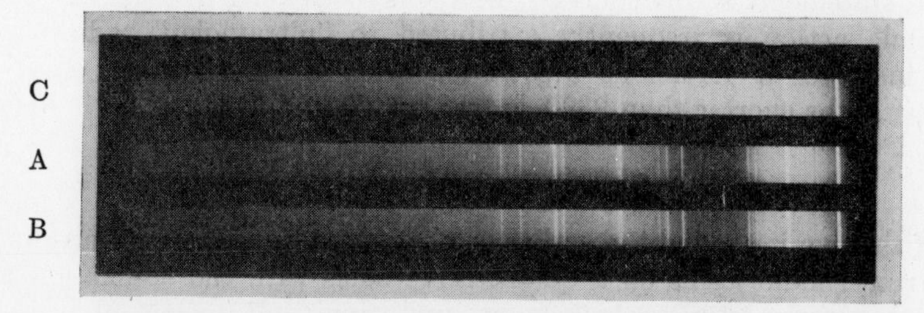

FIGURE 14. Positive of spectrogram of hematoporphyrin absorption. The light and dark regions are reversed from those in the original spectrogram. C, spectrum of hydrogen discharge tube through water; A, through 1/1000; B, through 1/2000 solution of photodyn (hematoporphyrin hydrochloride). The bright lines, which serve as wave-length markers, are due to mercury vapor.

Figure 14 is the positive of a spectrogram† which was taken in the following way: The spectrum of a hydrogen discharge was photographed after passing through a given thickness of water (C), and through two different concentrations of hematoporphyrin in solution in water (A and B). The same quartz cell was used to contain the solvent or solution. The bright lines in the photograph are the mercury lines due to traces

* For a general treatment of methods of spectroscopy, see Twyman and Allsopp (1934).

† The spectrogram itself is a photographic negative.

of mercury in the hydrogen discharge tube; they serve as wave-length markers, and the hydrogen discharge itself forms the continuous background. Figure 15 is a tracing from another photographic negative obtained by means of a microphotometer from the spectrogram photographed as Figure 14. It represents the deflection of a string galvanometer attached to a photoelectric cell as the spectrogram is passed between a light source and the photoelectric cell, and is thus a measure of the blackening of the original photographic negative or spectrogram (Figure 14). The curves corresponding to the hematoporphyrin spectra obviously do not represent the absorption of hematoporphyrin, but the blackening of a photographic plate exposed to light passing through a solution of this substance. The quantitative blackening must depend on

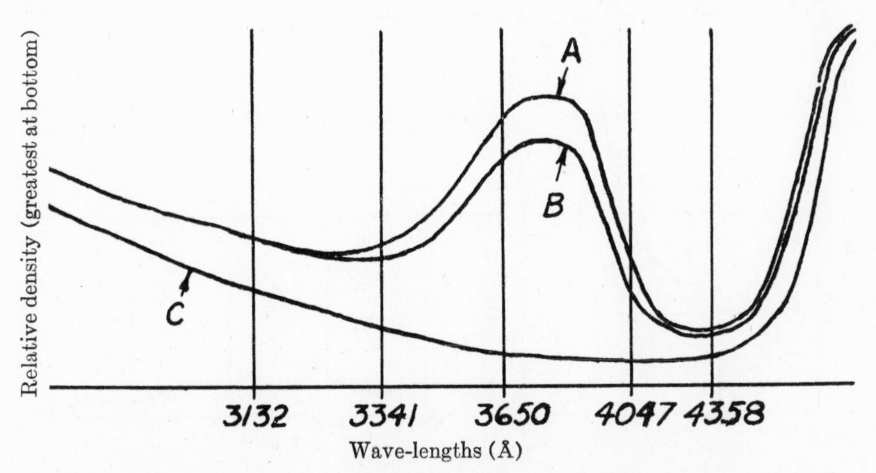

FIGURE 15. Tracing from the microphotometer curve obtained from the plate photographed as Figure 18. A, 1/1000 photodyn. B, 1/2000 photodyn. C, Water. The original curves show sharp spikes corresponding to the bright lines of Figure 14, which have been smoothed out in the tracing. The vertical lines represent the position of such spikes, corresponding to mercury lines, which are used for wavelength calibration.

three factors: the absorption of light by hematoporphyrin, the absorption of light by the solvent, and the blackening characteristics of the emulsion on the photographic plate. Unfortunately the last two factors have sometimes been neglected, and such curves have been taken as quantitative representations of the absorption of porphyrins. If the curve for water is subtracted from the curves for hematoporphyrin, something approaching the absorption spectrum of the latter may be obtained, since all the curves are subjected to the same errors. However, unless a correction is made for the relationship between the energy striking the plate and plate blackening, the curves are still in error. After both these corrections are made, the curve representing absorption by the hematoporphyrin in the solution is obtained; this is a true absorption spectrum.

The curves for hematoporphyrin at the two concentrations can be compared by applying the equation for Beer's law (equations 8-11, Chapter 2). The curve plotted in Figure 16 is the absorption spectrum represented as values of E in equation 10, Chapter 2. The values of E for the ultraviolet wave-lengths were obtained in the foregoing manner; for the visible wave-lengths a spectrophotometer was employed.

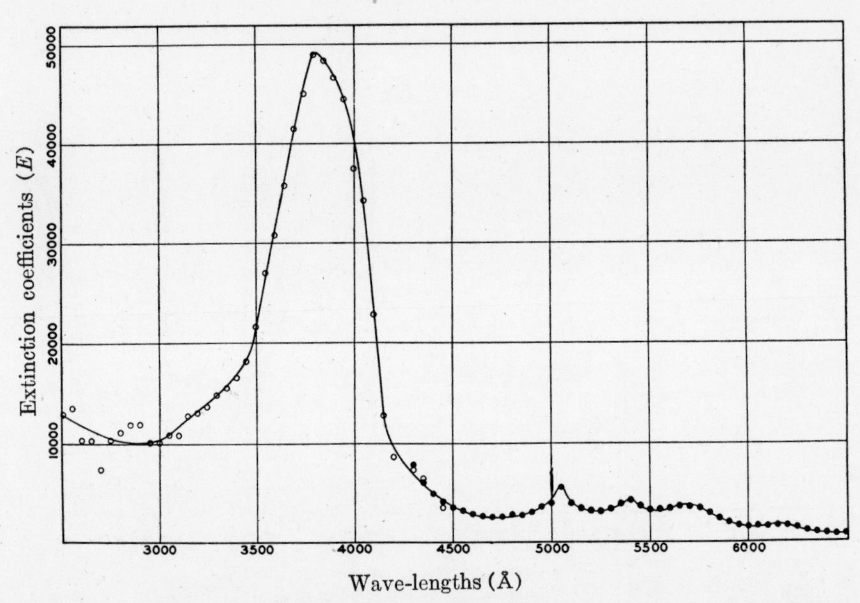

FIGURE 16. Absorption spectrum of hematoporphyrin-hydrochloride in solution in H_2O. Open circles, from corrected values obtained from Figure 15. Solid disks, measurements made with spectrophotometer.

Comparison of Action Spectra with Absorption Spectra. Assuming the system to obey Beer's law at all concentrations, it should be possible to calculate the absorption of any given thickness or concentration of solution from the absorption or extinction coefficient. However, it is usually impossible to know the concentration, let alone the thickness of the material, after it has been introduced into a given biological system where it is to act as a photosensitizer, so it would seem impossible to make a fair comparison of the absorption spectrum of a dye with the action spectrum for a biological system sensitized by that dye. There are conditions under which this comparison should be possible, however, namely, for dilute solutions and very thin absorbing systems. From Beer's law the absorption of a solution may be expressed as follows (equation 8, Chapter 2):

$$\frac{I}{I_0} = e^{-klc} \tag{1}$$

where I_0 is the intensity* of the incident light, I the intensity of the transmitted light, e the base of natural logarithms, k the absorption coefficient, l the thickness, and c the concentration of the dissolved substance.

If I_a is the radiant power per unit area absorbed, which will be called, for convenience, the absorbed intensity,

$$I_a = I_0 - I \tag{2}$$

Substituting and rearranging,

$$I_a = I_0 (1 - e^{-klc}) \tag{3}$$

Expanding the exponential,

$$I_a = I_0 (klc - \frac{k^2l^2c^2}{2!} + \frac{k^3l^3c^3}{3!} \ldots\ldots) \tag{4}$$

Neglecting all powers higher than unity,

$$I_a = I_0 klc \tag{5}$$

If the absorption is less than 10 per cent, no important error is introduced by neglecting powers higher than unity.

Thus if l and c are both constant, and of such values that the absorption is less than 10 per cent,

$$I_a \propto I_0 k \tag{6}$$

i.e., the product of the incident intensity and the absorption coefficient is proportional to the absorbed intensity.

Intensity, being the power of radiation per unit area, has dimensions of energy per unit time per unit area. Thus for unit time and unit area we may write,

$$Q_a \propto Q_0 k \tag{7}$$

where Q_a is the absorbed, and Q_0 the incident energy. If, then, the photosensitizing substance in a biological system is constant in amount and so dilute that it absorbs only a small fraction of the incident light, the product of its absorption coefficient with the incident energy provides a measure of the energy absorbed by the photosensitizer.

But when the effect of different wave-lengths is to be compared, the number of absorbed quanta should be used rather than the absorbed energies, since photochemical reactions depend upon the former. The number of quanta, N, is related to the energy, Q, as follows:

$$N = \frac{Q}{h\nu} = \frac{Q\lambda}{hc} \tag{8}$$ †

* Intensity is used in the sense of radiant power per unit area in an approximately parallel beam.

† See p. 9.

Since h and c are both constants

$$N \propto Q\lambda \tag{9}$$

and if in this specific case we write

$$N_a \propto Q_0\lambda\, k \tag{10}$$

N_a is a relative measure of the number of absorbed quanta, which may be used in comparing the action of different wave-lengths. The extinction coefficient (equations 10, 11, Chapter 2) may be used in place of k when only relative values are to be considered.

Action spectra are ordinarily measured by determining the incident energy, Q_0, necessary to produce a given response, and it is customary to plot the reciprocal of the energy; $\frac{1}{Q_0}$; against wave-length. To compare action and absorption spectra, the basic assumption may be made that the number of absorbed quanta, N_a, required to produce this given response is the same at all wave-lengths. Thus, taking N_a as constant, expression (10) may be written:

$$k \propto \frac{1}{Q_0\lambda} \tag{11}$$

or

$$k\lambda \propto \frac{1}{Q_0} \tag{12}$$

If agreement between action and absorption spectrum is to be tested at various wave-lengths, it is obvious that either expression (11) or (12) may be used according to which is the more convenient.

In the preceding discussion numerous conditions have been assigned which must be fulfilled if there is to be strict agreement between action and absorption spectrum. The conditions which may be expected in actual experiment will now be considered.

Full agreement between action and absorption spectra demands that Beer's law hold at all concentrations, but there are numerous departures from this law which may cause divergence.

It has already been pointed out that the absorption spectrum of a substance within the organism may be somewhat different from its absorption spectrum when extracted from the organism and in simple solution (p. 32), and needless to say, the absorption spectrum to be used in such a comparison should be obtained under conditions approaching those in the cell, *e.g.*, hydrogen-ion concentration should be taken into consideration, since some of the photodynamic dyes are indicators. It has been assumed that dyes are absorbed on cell surfaces, and that this may shift the absorption spectrum. A nice example of the effect of adsorption is given by Natanson (1937), who compares the absorption spectrum of the sensitized photographic plate, the sensitizing dye in solution, and the spectral sensitivity of the plate. His values, shown in Table 4, indicate the magnitude of the shift in absorption maxima due to adsorption and the agreement between the spectral sensitivity of the plate and the absorption spectrum of the adsorbed dye. It is doubtful, however, whether close analogy should be drawn between the manner in which a dye is taken up by the photographic plate and by the cell (see p. 80).

Table 4. Shift of Absorption Spectrum by Adsorption on the Photographic Plate. *(From Natanson, 1937)*

	Maxima		
Dye	Plate sensitivity (Å)	Absorption of adsorbed dye (Å)	Absorption of dye in solution (Å)
Erythrosine	5580	5580	5230
Phloxine	5630	5630	5240

For agreement between action and absorption spectra the photosensitizing substance must be so dilute that it absorbs only a small fraction of the incident light, a condition which should be adhered to in all quantitative studies of photodynamic action. A common error arises from the use of too concentrated dye, or too thick irradiation chambers. If the outer layers of the solution absorb so much of the incident energy that the intensity reaching the inner layers is much reduced, the total photodynamic effect may be actually less than when the concentration is lower and total energy absorption less. This effect is shown in time dilution curves for photodynamic hemolysis, Figure 20 (p. 84) where the hemolysis time increases above a certain dye concentration. If adequate stirring could be provided, this effect should not appear, since only the total light absorbed would then be of importance. This is often impracticable or even impossible, as for example in layers of tissue such as skin. The effect has been frequently described, and often interpreted as demonstrating an "optimum" dye concentration for photodynamic action. Obviously, this "optimum" should shift according to numerous conditions, *e.g.*, stirring, thickness of the system, etc.

This factor may affect the action spectrum, since at the region of maximum absorption the amount of photodynamic action may be less per absorbed light unit than at wave-lengths which are less completely absorbed. This would tend to depress the action spectrum in the region of maximum absorption with respect to regions of less absorption, and may account for Dognon's (1928a) finding that the region of maximum action did not correspond to the region of maximum absorption.

In the foregoing treatment it was assumed that the quantum efficiency of the photochemical process underlying photodynamic action is the same at all wave-lengths. Studies of analogous systems (see p. 67) indicate that this is true throughout the spectral range studied. Were this not true another source of deviation would be introduced.

Measurement of Action Spectra. The preceding discussion should indicate that it is futile to attempt close comparison between action and absorption spectra unless a number of factors are carefully controlled. If this were possible, measurements with monochromatic radiation, using a monochromator to isolate spectral lines and a thermocouple to measure energies, would be necessary to test the agreement accurately. Such measurements have not been made, partly because the intensities required have been too high for available light sources and monochromators. With the development of more intense sources and large monochromators such measurements may be possible, but in the meantime information must be derived from qualitative and semi-quantitative methods.

Lacking more appropriate apparatus, a method has been used by the writer which allows a comparison of action and absorption spectra with reasonable accuracy. An example is the study of photodynamically induced tropisms (Blum and Scott, 1933). The roots of wheat seedlings were sensitized by immersing them in a solution of erythrosine with the result that when subsequently exposed to light from one side the roots were caused to bend toward the source. The roots were exposed to the radiation from a 500-watt tungsten filament lamp, either directly or after passing through certain filters of known spectral transmission. The intensity was varied by changing the distance between the roots and the source; and the threshold, *i.e.*, the lowest intensity which would just produce an observable bending of the roots in 24 hours, was determined for each filter.

The distribution of the energy from a tungsten filament lamp may be calculated with sufficient accuracy from Wien's equation (see p. 23) if the color temperature of the lamp is known; and values of the intensity at a given distance from the filament, I_s, may be calculated for the wave-lengths desired. At this given distance from the source the energy per unit time, per unit area, Q_0, for any given wave-length will be proportional to the intensity, I_s, at that wave-length.

$$Q_0 \propto I_s \tag{13}$$

If a filter is interposed between the source and the roots

$$Q_0 \propto I_s F \tag{14}$$

where F is that fraction of the incident radiation transmitted by the filter at the given wave-length.

Substituting in (10)

$$N_a \propto I_s F \lambda k \tag{15}$$

By substituting values of I_s, F and k for different wave-lengths, values of N_a may be obtained for all wave-lengths. When no filter is used, F is dropped from expression (14). The values of N_a at different wave-lengths calculated for the various filters are plotted in Figure 17*; the areas, A, under the curves represent integrations of N_a at all wave-lengths for the various filters, *i.e.*, the relative numbers of quanta absorbed by the dye at the surface of the root when the different filters are used, and when the source is at a given distance from the root. By changing the distance, d, between lamp and root the total number of absorbed quanta is varied according to the inverse square law (see p. 27), but not the distribution with respect to wave-length. Thus A/d^2 may be used as a measure of the number of quanta absorbed by the dye on the root for any given filter and position of the lamp. Thus,

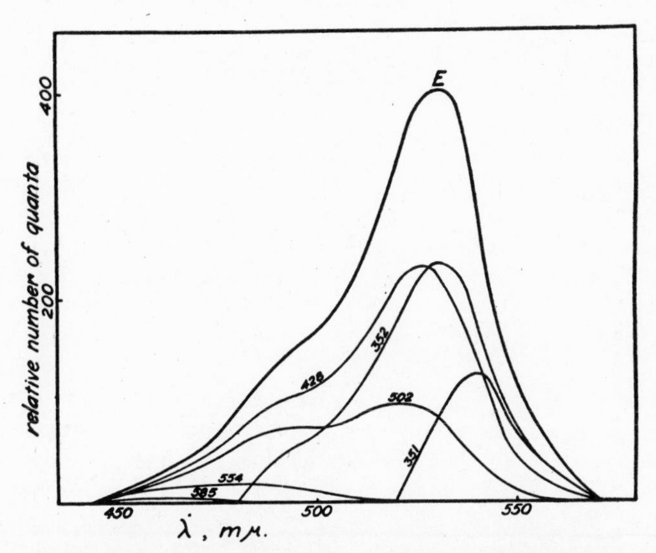

FIGURE 17. Estimated spectral distribution of quanta absorbed by the plant root sensitized with erythrosine. (See text for method of calculation.) E, without filter. 585, 554, 428, 352, 502, 351, with Corning glass filters designated by these numbers. (From H. F. Blum and K. G. Scott, *Plant Physiology,* **8,** 525, 1933.)

all values of A/d^2 for threshold position of the lamp should be equal, since threshold bending of the root should represent the absorption of a given number of quanta by the dye at the root surface.

The experimental values are given in Table 5. The agreement is not very exact; but when the differences in intensity of the incident light are considered, the data are convincing enough, and leave little doubt that the dye is the light absorber which initiates the changes leading to bending of the root. This approximate method of comparing action and absorption spectra has also been applied to the photosensitization of human skin by hematoporphyrin (see p. 215).

* A further correction was made in this particular experiment for the light absorbed by the erythrosine solution between the roots and the source.

Table 5.
(From Blum and Scott, 1933)

Filter	A	d (Meters)	$\frac{A}{d^2} \times 10^{-2}$	Tropism*
0	1.00	8.00	1.6	+
		8.00	1.6	0
351	0.16	3.32	1.4	+
		4.00	1.0	0
554	0.045	1.52	2.0	+
		1.50	2.0	0
352	0.48	6.17	1.3	+
		6.17	1.3	0
585	0.007	0.53	2.5	+
		0.60	1.9	0
502	0.30	3.62	2.3	+
		4.00	1.9	0
428	0.60	6.20	1.6	+
		6.50	1.4	0
348†	0.00	1.10		0
246‡	0.00	0.10		0
986§	0.00	0.25		0
			Average = 1.8	

* The threshold lies between the + and 0 values.
† Absorbs all wave-lengths shorter than 5700 Å.
‡ Absorbs all wave-lengths shorter than 5800 Å.
§ Absorbs wave-lengths between 4000 and 7000 Å.

Other attempts at such comparisons by various methods must be mentioned. The measurements of Metzner (1924) indicate that the dye molecule is the light absorber for photodynamic killing of paramecia. They seem to show that the absorption spectrum of a number of dyes is shifted toward the red by adsorption on the protoplasm, and that the action spectrum is likewise shifted. The estimates of Passow and Rimpau (1924) for the effectiveness of different wave-lengths in killing bacteria sensitized with various dyes indicate agreement between absorption and action spectra, although they are not quantitative. Eidinow (1930a, 1935) gives only the spectral limits of his filters and disregards the spectral distribution of the source, so that his measurements have little meaning; his statement that the maximum photosensitizing activity of hematoporphyrin is in the green and blue is erroneous, as will be shown in Chapter 19; the same criticism may be applied to Cuzin's experiments (1930a, b). Guerrini's statement (1934) that the photodynamic substances act independently of their absorption spectra cannot be seriously regarded, since his measurements (1930a, b, c; 1931a, b; 1934) are in no way quantitative. His finding that red light increased saponin hemolysis as well as hemolysis by most photodynamic substances suggests that heat may have been responsible for some of his results, since saponin does not seem to be photodynamically active (see Noguchi 1906b). The experiments of Karschulin (1929), who claims greater effectiveness of light through a green filter than for white light, should be repeated before they are accepted.

Numerous studies of the action spectrum for the photodynamic action of porphyrins have been made. These will be discussed in Chapter 19, where they find particular interest. Properly analyzed, they indicate that the porphyrin is the light absorber in these instances.

Alleged Sensitization to X-rays, Gamma Rays, Radio Frequencies, etc. Presumably, if a dye molecule captured a quantum of energy of magnitude corresponding to the x-ray region of the spectrum, it would be disrupted rather than activated in the photochemical sense. Thus sensitization to x-rays or gamma rays by photodynamic dyes is not to be expected, although it has been sought repeatedly. Jodlbauer (1904), Viale (1923) and Kammerer and Weisbecker (1926) have reported negative results. Podkaminsky (1930) claims to have sensitized white mice to x-rays by injecting hematoporphyrin, but this was not confirmed by Carrié (1933). The latter investigator found, however, that epilation occurred more rapidly in mice injected with porphyrin when subjected to Grenz-rays (long wave-length x-rays), than in mice which were not injected with this substance. Roffo (1932a) found that hemolysis by erythrosin was increased by x-rays or radium, and Nadson and Jolkewitz (1926) found that eosin and radium act additively to produce inhibition of plant growth.

There is some possibility that secondary radiation, *i.e.*, radiation of longer wave-lengths caused by fluorescence or other processes which result in the transformation of the large x-ray quanta into smaller quanta, might have produced the results reported by Nadson and Jolkewitch, Carrié, and Roffo. It is also possible that such "sensitization" is merely a summation of the action of x-rays on the biological system with the dark action of the dye (see Chapter 8). Baldwin (1920) has shown such a combined effect of dyes and x-rays on paramecia; he does not suggest the possibility of photodynamic action, and it is not probable that all the dyes he used were photodynamic. Thus, the possibility of sensitization to x-rays by photodynamic dyes must be treated skeptically for the present. Theoretically it is difficult to account for, and a combined action of x-ray and dark action of the dye explains better those results which have been reported.

The quanta of radiation of radio frequencies are much too small to produce activation of dye molecules in the photochemical sense, *i.e.*, changes in electronic energy levels. However, Roffo (1932a, 1933a) found hemolysis by erythrosin to be increased by the action of such radiation. A glance at Figures 21 and 22 (pp. 85, 86) will show how great, and how complex, is the effect of temperature on hemolysis of red blood cells by a similar dye; and it is possible that the effects described by

Roffo were due to the heating of the system by such radiation, rather than to photochemical action.

The Primary Act in Photodynamic Action. Viewing the evidence, there can be little doubt that the light absorber in photodynamic action is the photosensitizing dye, and not some other component of the living system. The instances of apparent lack of agreement between action and absorption spectra cannot be interpreted as evidence to the contrary, since they represent minor deviations at the most, and are explainable when the factors upon which this agreement depends are taken into account.

We may write, then, without further ado, the equation for the primary reaction in photodynamic action:

$$D + h\nu \longrightarrow D'$$

where D is the photosensitizer molecule, and D' the same molecule after its activation by acquiring a quantum $h\nu$. It is now time to consider the reactions which follow this primary act.

Chapter 6

The Role of Oxygen in Photodynamic Action

One of the outstanding characteristics of photodynamic processes is that they occur only in the presence of molecular oxygen. Table 6 summarizes data which show this to be true for such a wide variety of photosensitizers and of living systems that there is scarcely room for doubt that this is a fundamental requirement. There are few experiments which might be cited to the contrary, *e.g.*, Passow (1924), and it is probable that these are accounted for by failure to reduce the O_2 pressure sufficiently. Numerous experiments, less conclusive because of the conditions than those listed in Table 6 might be added to swell the evidence that O_2 is necessary for photodynamic action, including those of Ledoux-Lebard (1902) who first called attention to this characteristic of the phenomenon.

The actual uptake of O_2 during photodynamic action has been demonstrated by Harris (1926b), Gaffron (1926), Wohlgemuth and Szörényi (1933a, b), Kosman and Lillie (1935), Smetana (1938a), and others. Further evidence of the oxidative nature of photodynamic action is provided by experiments which show that it is inhibited by reducing agents (Sacharoff and Sachs, 1905; Noack 1920; and see pp. 85 and 91).

Contrast between Photodynamic Action and Normal O_2 Metabolism. The O_2 requirement suggests a relationship to the normal aerobic metabolism of cells, but the resemblance is entirely superficial. Experimental evidence provided by Wohlgemuth and Szörényi, and others makes it quite clear that the oxidations which are fundamental to photodynamic action are entirely distinct from those of normal cellular metabolism. This evidence may be summarized as follows:

(1) The respiratory quotient, *i.e.*, the ratio between CO_2 production and O_2 uptake $\left(\frac{CO_2}{O_2}\right)$, is near unity for normal aerobic metabolism, whereas Wohlgemuth and Szörényi (1933b) found 0.047 for this ratio during photodynamic action.

(2) Normal O_2 metabolism is virtually abolished if the structure of the cell is destroyed, but Wohlgemuth and Szörényi (1933b) found that the O_2 uptake during photodynamic action was the same for both intact or hemolyzed red blood cells.

(3) Cyanide inhibits normal O_2 metabolism, but increases the uptake of O_2 in photodynamic action (Wohlgemuth and Szörényi 1933a; Kosman and Lillie 1935; Rocha e Silva 1937d). (See p. 96 for further discussion).

Table 6. Photosensitized Effects in Presence and Absence of Molecular Oxygen.

(Wave-lengths longer than 3300 Å)

System	Photosensitizer	O_2 present Effect	O_2 absent Effect	O_2 absent Conditions	Investigators
Bacteria *B. proteus*	Rose Bengal Methylene blue Phenosafranine	Killed	None		Jodlbauer and Tappeiner (1905a)
Protozoa Spirostomum	Eosin	Killed	None	Vacuum	Straub (1904b)
Erythrocytes (mammalian)	Eosin Erythrosin Rose Bengal Methylene blue Sodium dichlor-anthracene disulfonic acid	Hemolyzed	None	Vacuum	Hasselbalch (1909)
	Eosin	Hemolyzed	None	Vacuum, illuminating gas, H_2	Schmidt and Norman (1922)
	Eosin Erythrosin Rose Bengal Hematoporphyrin	Hemolyzed	None	?	Eidinow (1930b)
	Eosin	Hemolyzed	None	CO*	Blum and McBride (1931)
	Lactoflavin Chlorophyll	Hemolyzed	None	Vacuum after CO	Kuen and Blum (unpublished)
	Hypericin "Y"	Hemolyzed	None	Vacuum	Blum and Pace (unpublished)
	Neutral red	Hemolyzed	None	Vacuum after CO	Blum (unpublished)
	Methylcholanthrene	Hemolyzed	None	Vacuum after CO	
Skeletal muscle (frog)	Rose Bengal	Stimulation to contraction	None	H_2, N_2	Lippay (1930)
	Eosin	Stimulation to contraction	None	N_2, vacuum	Spealman and Blum (1933)
	Eosin	Stimulation to contraction	None	N_2	Lillie, Hinrichs and Kosman (1935)
Whole frogs	Rose Bengal Hematoporphyrin Eosin	"Sensory stimulation"	None	N_2	Blum and Spealman (1934b)
Human skin	Rose Bengal Hematoporphyrin	Whealing, etc.	None	Circulation occluded	Blum, Watrous and West (1935)
Blood clot Fibrinogen, calcium and prothrombin mixture	Methylene blue	Clotting prevented	Normal clotting	Vacuum	Baumberger, Bigotti and Bardwell (1929)
Bacteriophage	Methylene blue	Inactivated	None	Vacuum	Clifton (1931)
	Methylene blue	Inactivated	None	N_2, vacuum	Perdrau and Todd (1933a)
Enzymes Invertase	Erythrosine	Destroyed	None	H_2	Jodlbauer and Tappeiner (1905a) (1906a)
Peroxidase	Eosin	Inactivated	None	Vacuum	Jamada and Jodlbauer (1907)
Catalase	Eosin	Inactivated	None	Vacuum	Zeller and Jodlbauer (1907)
Toxins Ricin	Erythrosine	Inactivated	None	H_2	Jodlbauer and Tappeiner (1905a)
Viruses Herpes	Methylene blue	Inactivated	None	Vacuum	Perdrau and Todd (1933b)
Proteins Egg Albumin	Rose Bengal	Denatured	None	H_2	Mirsky (personal communication)

* CO does not inhibit photodynamic hemolysis when O_2 is present.

(4) Normal O_2 metabolism is abolished by heat, but boiling does not abolish the photosensitized uptake of O_2 (Kosman and Lillie, 1935).

There are other possible distinctions, but those listed should be sufficient to show the important differences which exist between the two processes, and that the fundamental chemical reactions are not the same in the two cases. This distinction will become more apparent as the mechanism is further considered.

Photosensitized Oxidations

It would be difficult to go further with the analysis of the mechanism of photodynamic action were it not for knowledge of a type of photochemical reaction which seems analogous in every respect. The same dyes which bring about photodynamic action photosensitize the oxidation of many organic and inorganic substances by O_2; and Straub (1904a, b) early called attention to the possible identity of the two processes.

In these photosensitized oxidations which occur characteristically in solutions, the substrate undergoes oxidation and O_2 is used up; but the photosensitizer itself, although the light absorber, remains unchanged at the end of the reaction. Each dye molecule, after having absorbed its quantum of radiation and participated in the reaction, returns to its original state and may again take part in the reaction, this process being repeated over and over again. This is clear from the fact that the dye may bring about the oxidation of many times its equivalent of oxidizable substrate. In the absence of other oxidizable material, the dye itself may be oxidized so that the solution loses its color. These essential points are well established by experiment, and give a general picture of the reaction.

The general characteristics of these photosensitized oxidations correspond to those of photodynamic action. They are independent of temperature, are irreversible, and independent of the intensity of radiation (Spealman and Blum, 1937), as was shown to be true for photodynamic action in Chapter 4. They are also virtually independent of wave-length (Gaffron, 1927a, Spealman and Blum, 1937), which appears to be the case for photodynamic action, although the evidence presented in Chapter 3 is somewhat less definite in this respect. They are also virtually independent of concentration of the photosensitizer (Spealman and Blum, 1937), which the following chapter will show to be the case for photodynamic hemolysis.

Quantum yields (O_2 molecules reacting per quantum) of unity or approaching unity have been obtained for photosensitized oxidations by Gaffron (1927a, 1933), Bowen and Steadman (1934), Schneider (1935), and Koblitz and Schumacher (1937). Such values are found at higher concentrations of substrate and O_2; but at lower concentrations the quantum yield falls off (Gaffron 1927a, Spealman and Blum, 1937), indicating that all the dye molecules which are activated by capture of quanta are not able to reach other components of the reaction before losing their energy of activation, and hence do not bring about chemical reaction. Quantum yields near unity are to be expected for photochemical reactions displaying such characteristics as have been mentioned above.

Reaction Mechanisms. A number of hypothetical reaction schemes have been suggested for these oxidations, each supported by certain experimental evidence, but it is difficult to fit all the evidence with any one scheme. All the hypotheses and experiments cannot be discussed here, and only the more salient points will be touched upon. One of the simpler schemes is based on the transfer of activation from the sensitizer molecule to an oxygen molecule by collision. This view has been supported chiefly by Kautsky; it may be represented in a general way as follows:

$$\begin{aligned} D + h\nu &\longrightarrow D' \\ D' + O_2 &\longrightarrow D + O_2' \\ O_2' + X &\longrightarrow X\text{ox} \end{aligned} \qquad \text{(I)*}$$

The dye molecule D captures a quantum, $h\nu$, and becomes an activated molecule D'; D' then transfers its energy of activation to an oxygen molecule by collision, resulting in the formation of an activated (or metastable) oxygen molecule, O_2' and the return of the activated dye molecule to its original state. The activated oxygen molecule then reacts with a substrate molecule X, resulting in oxidation products Xox.

Kautsky has been the chief proponent of transfer of activation from the sensitizer to O_2, basing his argument primarily on fluorescence studies. Measurements by Gaviola (1927) indicated activated lifetimes for dye molecules in solution, comparable with those occurring in gases, *i.e.*, of the order of 10^{-9} to 10^{-8} second. These were measured by determining the time during which fluorescence persists after the exciting radiation is removed, which measures the interval between receiving and re-emitting the quantum of radiation. Kautsky *et al.* presented evidence that the activated lifetime is much greater than estimated by Gaviola. They found (1935) that fluorescence of these dyes in solution is diminished, and that the persistence of fluorescence after withdrawal of the light is shortened by the presence of O_2; hence Gaviola's measurement should be in error, since O_2 was not excluded from his solutions. Kautsky estimated lifetimes as great as 10^{-3} to 10^{-2} second when O_2 was not present. Moreover, the fact that O_2 decreases the activated lifetime, as indicated by the quenching of fluorescence, is presented by Kautsky as evidence that the activated dye molecules give up their energy of activation to oxygen molecules instead of emitting it as fluorescence.

Kautsky *et al.* (1933) suggest that the oxygen molecule is raised in this way to a high-energy metastable state. The $^1\Sigma$ state of the oxygen molecule, which corresponds to wave-length 7623 Å, seems to be the only state to which it might be raised by the energy available in this way; it is a metastable state which might have a relatively long lifetime. Important evidence against Kautsky's mechanism is presented by Gaffron (1935, 1936a), who found photosensitized oxidation by wave-lengths 8000 Å

* In this and the following schemes, which are to be regarded only as roughly descriptive, no distinction is made between "activated" and "metastable" states.

or longer when bacterio-phaeophytin was used as the sensitizer. This should not occur if it is necessary to activate the $^1\Sigma$ state of oxygen, since these wave-lengths do not represent sufficient energy.

Kautsky *et al.* (1933) presented an experiment which seems to support this theory rather nicely, at least for the conditions of the experiment. They used trypaflavine as the sensitizer, and the leucobase of malachite green as the substrate, sensitizer and substrate being absorbed separately on small particles of silica gel. Particles of malachite green silica gel were mixed with particles of trypaflavine silica gel in vacuum in the absence of light; O_2 was then admitted and the system exposed to light. Oxidation could be detected by the change from the colorless leucobase of malachite green to the colored dye. At low O_2 pressure there was no change in color; at a certain optimum range of O_2 pressure oxidation took place, but at higher O_2 pressure there was no change. Since sensitizer and substrate were apparently not in physical contact, it would seem that the gaseous oxygen molecules must receive energy from the activated dye molecules, and carry it to the substrate molecules. The optimum O_2 pressure is explainable in the following way: At low pressure, when O_2 molecules are too few, most of the dye molecules lose their energy by fluorescence before they can be reached by O_2 molecules. If the O_2 pressure is too high, the oxygen molecules which have been raised to the metastable $^1\Sigma$ state by collision with dye molecules, lose their energy by collision with other oxygen molecules before they can reach substrate molecules. Thus there should be an optimum O_2 pressure at which neither of these effects predominates. If sensitizer and substrate were brought into closer contact by adsorbing them on the same silica gel particles, oxidation took place more rapidly, but if sensitizer and substrate were too widely separated oxidation did not occur. Gaffron (1936a) objects to Kautsky's experiment on the ground that sensitizer and substrate might not actually be separated after the silica gel particles are mixed.

Activation of Substrate Molecules. Franck and Levi (1934) discuss the mechanism of quenching of fluorescence, pointing out that substances other than O_2 are very effective in suppressing the fluorescence of dye solutions. They suggest that the activated sensitizer may, without the intermediation of O_2, transfer its energy to the substrate by collision, resulting in chemical reaction, the product of which may then react with O_2. Schneider (1935) shows that KI is more effective than O_2 in quenching the fluorescence of dyes in aqueous solution, and that, parallel with this, the photosensitized oxidation of iodide is more dependent upon KI concentration than upon O_2. Schneider presents a mechanism which fits the suggestion of Franck and Levi, making the specific requirement of direct activation of O_2 superfluous.

Gaffron (1933) had previously proposed that the activated sensitizer transfers its energy to the substrate molecule, thus raising it to a metastable state, in which state it may react with O_2. A generalized scheme for such a reaction may be written:

$$\begin{aligned} D + h\nu &\longrightarrow D' \\ D' + X &\longrightarrow X' + D \\ X' + O_2 &\longrightarrow X\text{ox} \end{aligned} \qquad \text{(II)}$$

Gaffron offers several lines of evidence to support his conclusion, particularly studies of the photo-oxidation of rubrene. This hydrocarbon forms a peroxide with O_2 in the presence of light, thus acting both as photosensitizer and substrate. Gaffron claims that the quantum yield for this reaction is relatively independent of the concentration of O_2, and hence that the substrate itself must be activated. To fit this scheme, he assumes (1933, 1937) that the substrate enters a metastable state which has a long lifetime, 10^{-2} to 10^{-1} second. On the other hand, Bowen and Steadman (1934) contend that the quantum yield falls off rapidly with decrease in O_2 partial pressure, and that the reaction may be adequately accounted for on the basis of a trimolecular collision between an activated rubrene molecule, a rubrene molecule acting as substrate, and an oxygen molecule. Koblitz and Schumacher (1937) claim that the reaction is more complex, but maintain that the quantum yield is dependent upon the concentration of O_2; and Schumacher (1937) contends that the activated molecules need have lifetimes no longer than 10^{-7} seconds.

Without denying the possibility that Kautsky's mechanism may function under certain conditions, it seems probable that activation energy is transferred from the sensitizer molecule to the substrate molecule in most photosensitized oxidation by O_2. This point of view may be accepted without regard to the disagreement of the various writers regarding the exact steps in the mechanism. Development of this general hypothesis with relation to photodynamic action will be presented after some further reaction schemes for photosensitized oxidation by O_2, and some other photosensitized reactions have been discussed.

Sensitizer Peroxide Schemes. It has been repeatedly suggested that the first step in photosensitized oxidation is the formation of a peroxide of the sensitizer, which peroxide then reacts with the substrate. This may be represented as follows:

$$\begin{aligned} D + h\nu &\longrightarrow D' \\ D' + O_2 &\longrightarrow DO_2 \\ DO_2 + X &\longrightarrow X\text{ox} + D \end{aligned} \qquad \text{(III)}$$

where DO_2 is a peroxide of D.

Although there have been numerous attempts to isolate an intermediate peroxide, this has not been accomplished. A stable peroxide of rubrene is formed, but this apparently does not act as an intermediate to oxidize other substrate. Gaffron (1927b) found that when amylamine is irradiated together with chlorophyll, a peroxide is formed, but that this is a peroxide of the substrate, not of the sensitizer. If fluorescein dyes are irradiated in the absence of oxidizable substrate, a certain small amount of peroxide is formed accompanied by bleaching of the dye; but Blum and Spealman (1933a) found this to be hydrogen peroxide, not a peroxide of the dye. When porphyrins are irradiated (Fischer and Herrle, 1938) peroxides are formed, but this involves the splitting of the porphyrin ring structure, and these peroxides cannot be considered as intermediates in the sense of the scheme shown above.

The possible role of hydrogen peroxide as an intermediate in photodynamic action will be discussed in Chapter 9.

The Dissociation of O_2. The dissociation of the O_2 molecule after receiving energy from the photosensitizer is ruled out on energetic grounds. The energy required for this dissociation is about 128,000 cals. per mol, corresponding to wave-length 2200 Å, which would place the long wave-length limit in the short ultraviolet, whereas photosensitized oxidations occur in the visible and in the very near infrared.

The Dye as a Hydrogen Acceptor. If the photodynamic dyes are irradiated in the absence of O_2 there is no evidence of bleaching or other chemical change. However, when appropriate substances are present which may serve as hydrogen donors, these may be dehydrogenated by the transfer of hydrogen to the dye with the formation of the colorless leucobase. The following is a scheme for this type of reaction:

$$\begin{aligned} &D + h\nu \longrightarrow D' \\ &D' + H_nX \longrightarrow DH_n + X \end{aligned} \qquad \text{(IV)}$$

where DH_n is the leucobase of D. Such reactions can be carried only to the point where an amount of the substrate, H_nX, equivalent to the quantity of dye present, is dehydrogenated; at that instant the dye will be entirely reduced to the leucobase (Levaillant, 1923, Windaus and Borgeaud, 1928). In this case the bleaching of the dye is reversible, or nearly so; if oxygen is readmitted to the system after the color has disappeared, the color will return. This is in contrast to bleaching of the dye when exposed to light in the presence of O_2 and the absence of oxidizable substrate, in which case the dye is irreversibly oxidized.

It is possible to formulate schemes which will fit both this type of reaction and oxidations by molecular oxygen, for example:

$$\begin{aligned} &D + h\nu \longrightarrow D' \\ &D' + H_nX \longrightarrow DH_n + X \\ &DH_n + n/4O_2 \longrightarrow D + nH_2O \end{aligned} \qquad \text{(V)}$$

Here oxygen enters only in the return of the hydrogenated molecule to its original form, accompanied by formation of water. From this scheme, it might be expected that the products formed from the oxidation of the substrate would be the same when the reaction takes place in oxygen as in the absence of oxygen, or that the products formed in the latter case would proceed rapidly to those resulting from the former upon admission of O_2 to the system after irradiation. This certainly does not occur in the case of ergosterol, where Windaus and Brunken (1928) found a peroxide of ergosterol formed in the presence of O_2, while Windaus and Borgeaud (1928) and Inhoffen (1932) found "ergopinacol" to be formed in the absence of O_2. The latter compound does not change to the former when standing in the presence of O_2.

In such a scheme the effectiveness of the photosensitizer must be related to the oxidation-reduction potential of the dye-leucobase reduction, but so far as it is possible to make a comparison of the photosensitizing effectiveness of the various dyes, this does not seem to be the case.

Other Types of Reaction. Still other photochemical reactions are participated in by at least some of the dyes which bring about photodynamic action; many have been shown to photosensitize the decomposition of Eder's solution, a mixture of $HgCl_2$ and $(NH_4)_2C_2O_4$ from which Hg_2Cl_2 is precipitated. The photographic plate is sensitized to those wave-lengths absorbed by the dye. Chlorophyll may act as a photosensitizer for the photographic plate, or for photodynamic action, but, when in the green plant, serves as the light absorber for the endothermic photosynthetic process.

The fact that the same dye may act as the light absorber for such apparently dissimilar reactions as have been cited suggests that the environment in which the activated dye molecule finds itself is more important in determining the type of reaction which will occur than the chemical nature of the dye molecule itself. On theoretical grounds, Franck and Levi (1934) point out that the ability of an activated dye molecule to transfer its energy by collision, and any reaction produced thereafter, must depend on the characteristics of the recipient molecule. Thus schemes such as (V) seem unnecessary, since the existence of a common basic chemical reaction for all the photosensitized reactions in which a given dye may participate is not required.

It is, then, not unreasonable to assume that the activated sensitizer molecule generally, if not always, transfers its activation energy, by collision, to some other molecule which is available; and that subsequent participation of the latter in chemical reaction is determined by its particular chemical properties, and those of other molecules in the environment. In such case, the action of the light-absorbing sensitizer molecule would be, in a sense, non-specific.

Structure of the Photosensitizer Molecule. This concept of the non-specific role of the photosensitizing dyes is supported by the great diversity of chemical structure which they display. Figure 18 shows something of this diversity among compounds which cause photodynamic action; the list of dyes which have actually been employed as photosensitizers for oxidations by O_2 in simpler systems would show no less diversity. Other types of dyes which act as photodynamic sensitizers are listed in Table 6 (p. 66). Metzner (1927) presents a list of photodynamic dyes which adds representatives of the acridine, phenoxazine, quinoline, and xanthone groups, phycoerythrin and bacteriofluorescein—all of which are fluorescent. He lists the following non-fluorescent dyes as photodynamically inactive: naphol green B, picric acid, several azo-dyes, the diphenyl methane dye auramin, and several triphenylmethane dyes including malachite green, some dyes of the xanthone group, indigo-carmine and hematoxlyn. Review of the literature adds a number of photodynamic substances not included in these lists. One searches in vain for a common structure which characterizes these photodynamic substances.

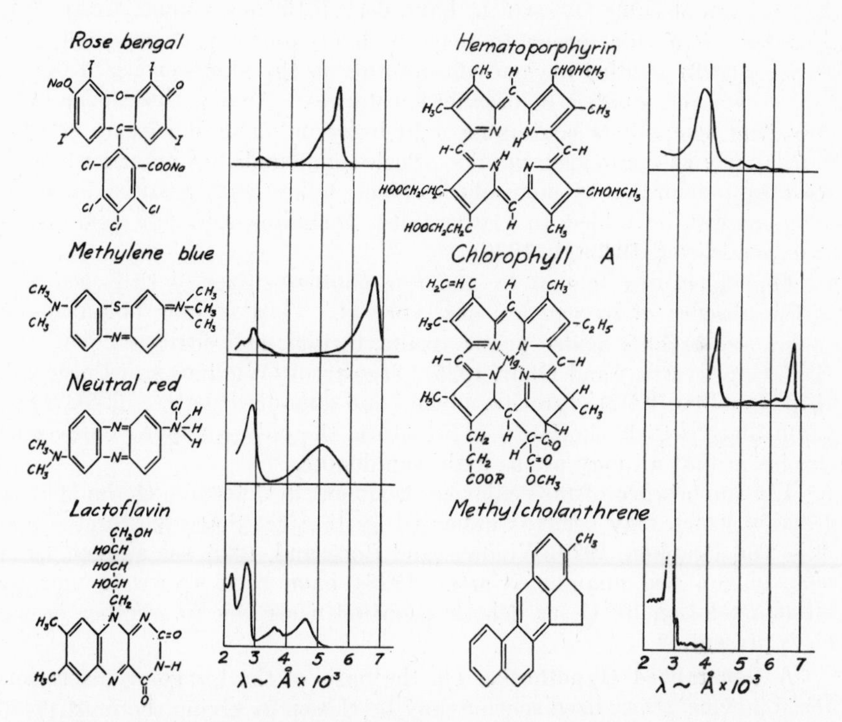

FIGURE 18. Structural formula and absorption spectra of some photodynamic substances. The ordinates of the absorption spectra are extinction coefficients, but are numerically different for the different spectra. They are intended to give an idea of the shape and position of the various spectra, but not their relative values.

On the other hand, all photodynamically active substances possess a distinguishing physical characteristic, namely the ability to fluoresce. Although it has been claimed that some of the photodynamic dyes are not fluorescent, it is probable that fluorescence would be observed in all cases if careful observation were made, and the quenching action of various substances, particularly O_2, were taken into consideration. Fluorescence means that the molecule has the ability to hold the quantum of energy which it has captured until it is released at one jump as fluoresced radiation, rather than frittered away as radiation of small quanta, or in collisions. Such a molecule has a better chance of retaining its energy of activation until it can meet another molecule to which it may transfer this energy more or less intact.

Variety of Substrate. If molecular environment is as important as the sensitizer in determining the nature of the chemical reaction which occurs, the nature of the substrates which participate in these reactions is of considerable interest. The substrates which undergo photosensitized oxidation by molecular O_2 seem to have very little in common except that they are, as a rule, prone to react with O_2 under various conditions. Carter (1928) studied oxygen consumption in the photosensitized oxidation of a large number of organic substances. Among those examined, only ring compounds containing a hydroxy or an amino group showed particularly active oxygen uptake. Such compounds are notoriously susceptible to oxidation by molecular oxygen. Other readily oxidizable compounds must be added to Carter's list, including non-ring compounds (*e.g.*, see Meyer, 1933a, b, 1935).

The substrates known to undergo photosensitized dehydrogenation in the absence of oxygen are also various. They include polyalcohols, sugars and organic acids—acetic, oxalic, tartaric and citric—(Levaillant, 1923, Chackvararti and Dhar, 1925); ergosterol (Windaus and Borgeaud, 1928, Inhoffen, 1932); tyrosine, phenol and guanine (Carter, 1928). Here again there is little similarity in structure, the only common characteristics being that all may act as hydrogen donors.

The importance of molecular environment in determining the type of reaction which may occur is indicated by the fact that ergosterol (Windaus and Brunken, 1928, Windaus and Borgeaud, 1928, see above), tyrosine, phenol and guanine (Carter, 1928) have been shown to undergo either oxidation by O_2 or dehydrogenation according to whether or not O_2 is present.

A Generalized Hypothesis. On the basis of the foregoing discussion, the following generalized scheme may be chosen as giving the most probable explanation of the photosensitized oxidations by O_2, which interest us because of their analogy with photodynamic action:

$$\begin{aligned} D + h\nu &\longrightarrow D_r \\ D_r + X &\longrightarrow X_r + D \\ X_r + O_2 &\longrightarrow X\text{ox} \end{aligned} \qquad \text{(VI)}$$

This is actually scheme II written in the non-committal fashion suggested in Chapter 2 (p. 20). While the transfer of activation (or less specifically, "reactivity," according to the above scheme), to the substrate molecule seems most probable, it must be emphasized that some other of the schemes proposed above might fit the known evidence.

It may be well to formulate a rather generalized hypothesis upon the foregoing scheme, which will serve adequately for the subsequent discussion of photodynamic action. The essential parts of this hypothesis are: (1) Certain molecules, widely different in chemical constitution, possess a non-specific ability to photosensitize chemical reaction. The ability to photosensitize resides, principally, in the ability of the molecule to hold its quantum of absorbed energy of activation more or less intact for a finite, but very short, time. If during this time the sensitizer molecule enters into collision with an appropriate molecule of another kind, the energy of activation may be transferred to the second molecule. This property of the photosensitizer molecule is disclosed by its ability to fluoresce. (2) The molecule to which the energy of activation is transferred may participate in chemical reaction, the nature of which is determined by the chemical properties of that molecule, and those of other molecules in the environment. A very common type of reaction is that between an oxidizable substrate molecule which is activated in this way, and O_2. The application of this general hypothesis to photodynamic action will be discussed in the next chapter.

Chapter 7

The Mechanism of Photodynamic Action

In the last chapter the role of O_2 in photodynamic action was discussed, and an analogy drawn between this process and photosensitized oxidations which occur in more simple systems. More intimate details of the mechanism will now be dealt with, showing how they fit the theory that photodynamic action is no more nor less than such photosensitized oxidation taking place in living systems.

Quantum Requirements. No method is available for direct determination of the quantum yield of photodynamic action, but it is possible to estimate the quantum requirements for photodynamic hemolysis from experiments such as those used to obtain the data presented in Table 3 (p. 50). These give the number of ergs of monochromatic radiation necessary to produce hemolysis under conditions such that the measurement is a reliable index of the underlying photochemical mechanism. With certain additional data it is possible to calculate the number of quanta required to bring about hemolysis of a single red blood cell under these conditions, using the following equation:

$$R = \frac{N \times A}{n} \tag{1}$$

where R is the number of absorbed quanta required per red cell, N is the number of quanta incident per unit area, A is the fraction of incident radiation absorbed by that portion of the dye taken up by the cells, and n the number of red cells in the volume subjected to radiation upon unit area.

N is obtained by dividing the incident energy, measured in ergs per square millimeter ($Q \times t$ in Table 3), by the value of the quantum in ergs. This is 3.6×10^{-12} for the wave-length used, 5461 Å.*

Knowing the amount of dye taken up by the cells, estimated as mols per liter of the cell suspension,† the fraction of light absorbed, A, may be calculated from the equation:

$$A = 1 - \frac{I}{I_0} \tag{2}$$

and

$$\log \frac{I_0}{I} = Elc \tag{3}$$

* For calculation of the energy of the quantum see p. 9.

† The method for determining this will be discussed on pp. 79-80.

obtained from (9) and (10) Chapter 2. E is the extinction coefficient for 5461 Å, and l the depth of the cell suspension in the irradiation chamber (0.11 cm.).

Table 7 summarizes the data from several experiments with two dyes, erythrosin and rose bengal, covering a wide range of concentrations. In

Table 7.

(From data of Blum and Gilbert, 1940b)

Experiment	Dye	Dye taken up by cells as mols/liter $\times 10^{-5}$	N Number of quanta incident per mm² $\times 10^{15}$	A Fraction of incident quanta absorbed	n* Number of r.b.c. irradiated per mm² $\times 10^{4}$	R Quanta absorbed per r.b.c. $\times 10^{10}$	S Quanta absorbed per dye molecule
I	Rose Bengal	.74	1.7	.133	2.0	1.1	46.
		.37	3.6	.069		1.3	104.
		.077	11.1	.015		.83	323.
		.030	26.5	.006		.80	797.
	Erythrosine	.80	11.1	.032		1.8	67.
II	Erythrosine	2.0	2.9	.078	2.3	.97	17.
		1.0	4.5	.040		.77	27.
		.34	15.9	.014		.96	98.
	Rose Bengal	.43	2.8	.077		.93	75.
III	Rose Bengal	.81	0.87	.144	2.4	.51	23.
		.086	9.0	.018		.66	283.
	Erythrosine	3.4	2.7	.129		1.4	15.
IV	Rose Bengal	.074†	12.5	.014	1.9	.94	358.
			12.6			.93	
			13.0			.96	
V	Rose Bengal	.0037	21.5	.0007	1.4	1.04	6800.

* Number of red blood cells per mm.³ obtained by hemocytometer count, multiplied by the thickness of the irradiation chamber, 1.1 mm.

† Values for three different intensities at this concentration.

these data, R is close to 1×10^{10}, regardless of the concentration of the dye on the cells,* or which dye is the photosensitizer. This constancy of R is to be expected if photodynamic hemolysis results from photosensitized oxidation of a given quantity of substrate in the cell since, as pointed out in Chapter 6, the quantum yield for such processes is unity under favorable conditions, and is independent of concentration of the photosensitizer. In such case each quantum absorbed would result in the participation of one molecule of O_2 in reaction, so that R should be constant. Even if only some definite fraction of the absorbed quanta took part in the reaction, as is probably the case in this instance, R should be constant, since Spealman and Blum (1937) found that at concentrations of substrate and O_2 at which the quantum yield was less than unity, it was nevertheless virtually independent of dye concentration.

Moreover, column S in Table 7, where the number of quanta absorbed

* Independence of dye concentration was not indicated by earlier studies (Kawai, 1928; Dognon, 1928b; Blum and Hyman, 1939b), but all were subject to the type of error discussed in Chapter 4 with regard to the reciprocity law (p. 51).

per dye molecule is calculated, shows that the quanta absorbed by a single molecule may range from a few to thousands. This is difficult to explain unless the molecule goes through a cycle in which it always returns to its initial state, as is the case in photosensitized oxidation (see p. 67). It is probable that the values of R are much too high,* but this does not affect S, which is a relative value.

A reaction in which the dye molecule was itself changed in the initial act, and subsequent reactions did not complicate the picture would also be independent of dye concentration; but such a mechanism is incompatible with the fact that a single molecule may capture so many light quanta.

Thus the fundamental photochemical reaction of photodynamic action proves, from a quantum standpoint, to be a photosensitization comparable to the photosensitized oxidations which were discussed in the last chapter, further strengthening the evidence that the two processes are identical.

The Oxidation Process. The hypothesis adopted at the end of the last chapter assumes that the activated photosensitizer molecule transfers its activation to a substrate molecule, which is then able to react with an O_2 molecule (scheme VI, Chapter 6). It seems probable that living systems in general are vulnerable to this type of mechanism because they contain some common type of substance which may act as substrate. Whatever this substrate, it must be a substance whose oxidation is incompatible with the structural existence of the cell, and it is hence unlikely that the reaction is one which goes on to any marked extent under normal conditions. Since most, if not all, oxidations of organic compounds by O_2 go with a decrease of free energy, it is probable that the reaction is possible as a thermal reaction; but is not activated in the course of normal cell processes, and hence does not take place unless activated by an outside agency, such as light in photodynamic action. This indicates its essential difference from the oxidations which take place in normal O_2 metabolism, a distinction which has already been made on other grounds (see p. 65).

According to this reasoning, the universal occurrence of photodynamic action results from the presence in living systems of a common type of substance which is susceptible to a particular type of oxidation brought about by light through the agency of a photosensitizer.

In contrast to this, Kautsky's hypothesis would explain the ubiquity of photodynamic action in another way. The basis of his hypothesis is the activation of O_2 by transfer of energy from the activated photosensitizer molecule (scheme I, Chapter 6). Presumably the activated O_2 could produce oxidation of whatever oxidizable material might be present in the cell, and hence lead to destructive changes. Strong evidence against Kautsky's hypothesis has already been presented, but it may be considered as an alternative to the above.

It was pointed out in the last chapter that dyes which photosensitize

* Reasons for suspecting this are discussed by Blum and Gilbert (1940b).

oxidation by O_2 may also take part in other photochemical reactions, yet they seem to be limited to the former type of reaction when they act as photosensitizers in living systems. This may be less true than apparent, for it must be remembered that photodynamic action is observed as a destructive end result, and that other reactions may take place which do not lead to detectable change in the cell and hence go unremarked. For example, it is possible that, when photosensitized living systems are subjected to light in the absence of O_2, a certain amount of dehydrogenation takes place following scheme IV, Chapter 6; but if so this does not lead to the observable destructive effects characteristic of photodynamic action.

The Nature of the Substrate. There is little direct evidence to indicate the nature of the substrate which is oxidized in photodynamic action, but since it must be a type of substance common to all living systems and one upon which the structure of the cell is dependent, protein is first to be suspected. Some evidence is provided by studies of the uptake of O_2 by photosensitized tissue extracts, which seems due principally to the protein component of the extracts. Kosman and Lillie (1935) found that fat-free protein containing extracts from muscle took up O_2 when exposed to light much more rapidly than protein-free, fat-containing extracts; and similarly, Smetana (1938a) found much greater uptake of O_2 by the protein fraction of blood plasma than by the lipoid fraction. Carter (1928) found that fatty acids are not readily attacked by photosensitized oxidation, whereas hydroxy and amino ring compounds are readily oxidized, the latter being very easily oxidizable substances. Harris (1926c) also found that, among a number of amino acids, only tyrosine and tryptophane were susceptible to photosensitized oxidation, and that among proteins, gelatin, which does not contain these amino acids, was likewise not oxidized in this manner. Harris did not find that histidine was susceptible to photosensitized oxidation, and Carter found its O_2 uptake to be somewhat less than that of tyrosine and tryptophane. It seems probable from this that tyrosine and tryptophane in the protein molecule provide the principal point of attack for the photosensitized oxidation which underlies photodynamic action. This is in accord with the thesis of Schmidt and Norman (1922); but their experimental evidence, which has been regarded as substantiating this, cannot be accepted for reasons which will be dealt with in the following chapter.

Uptake of Dye by the Cell and Photodynamic Effectiveness. In Figure 19 the relationship between the amount of dye taken up by the cell and that in the original solution is shown in graphical form. The data are those which were used in calculating the values for Table 7, and were obtained as follows: washed red cells were suspended in dye solution, and allowed to stand until equilibrium was reached. The cells were then centrifuged from the suspension, and the dye concentration of the supernatant solution determined spectrophotometrically. The difference between this value and the dye concentration in the original solution before the cells were added represents the dye taken up by the cells.

A linear relationship between dye taken up and total dye is indicated. This means that a linear relationship must exist between the dye taken up by the cells and that in the solution with which they are in equilibrium,* which is more striking if these values are plotted against each other after making proper correction for the number of cells in the suspension. For purposes of the present discussion the plotting in Figure 19 is more convenient, however.

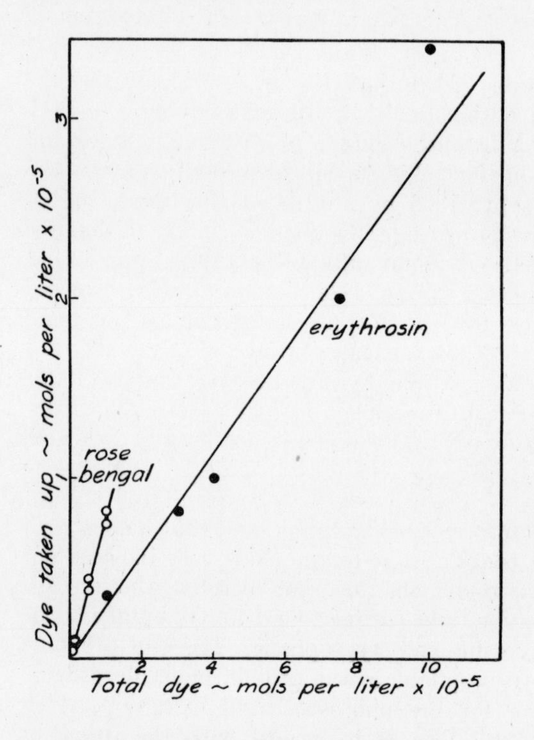

FIGURE 19. Uptake by rabbit red cells of fluorescein dyes from solution. These are data from several experiments in which the number of cells (about 10^{10} per liter) differs somewhat (see text).

Figure 19 shows that the uptake of rose bengal (80 per cent) is much greater than that of erythrosine (25 to 30 per cent). Thus, had the light absorption of the total dye been used in estimating the number of quanta required per red cell, rose bengal would have appeared two or three times more effective than erythrosine; whereas when the absorption of only that dye taken up by the cells is considered, they are equally effective, as shown by the constancy of R in Table 7. It will be shown, shortly, that it is the dye taken up by the cell which must be effective in

* This is a different relationship from that reported by Jodlbauer and Haffner (1921b), whose data indicate adherence to the adsorption isotherm. Further studies by H. W. Gilbert make it appear that the dye is not taken up by adsorption, employing that term in its usual sense.

photodynamic action; and thus distinction must be drawn between true and apparent photodynamic effectiveness.

From analogy with photosensitized oxidation in simpler systems, all photosensitizers might be expected to display the same effectiveness in producing photodynamic action. In all cases the quantum yield should be unity under the most favorable conditions. It is thus reasonable to expect that true photodynamic effectiveness is the same in all cases, and that apparent effectiveness is determined by the manner and extent of uptake of the dye by the cells. Rose bengal and erythrosin are closely related dyes, both belonging to the fluorescein series, and less direct evidence indicates that other dyes of this series have approximately the same true photodynamic effectiveness, although their apparent effectiveness varies greatly (Blum and Hyman, 1939b, and see p. 87). Among this series of dyes, only quantitative difference in uptake of dye is apparent among the different members of the series, whereas when photosensitizers of widely different chemical constitution are considered, qualitative differences also occur (*e.g.*, compare Commoner, 1938). This renders the comparison of true photodynamic effectiveness difficult.

It is obvious that the uptake of the dye may be altered by numerous factors, which should likewise affect the apparent photodynamic effectiveness. A number of these will be considered in the following chapter.

Implicit throughout the preceding discussion and that regarding quantum requirements is the assumption that only dye which is taken up by the cell takes part in photodynamic action. By a simple calculation, using the same data from which Table 7 and Figure 19 were prepared, it is possible to show that the number of activated molecules from the surrounding solution which strike the cell is insignificant in comparison to those which are in association with the membrane, and thus to justify this assumption.

From the number of mols of dye taken up by the cells it is possible to calculate the number of molecules per μ^2 of cell surface, accepting $80\mu^2$ as the surface area of the cell; such values calculated for the extreme

Table 8.

	Dye taken up by cells		Dye in solution		
				molecules striking μ^2 cell surface	
Dye	mols/liter $\times 10^{-5}$	molecules per μ^2 cell surface	mols/liter $\times 10^{-5}$	in 1 sec.	in 10^{-8} sec.
Erythrosine	3.4	1×10^7	6.6	1×10^{11}	1000.
Rose Bengal	0.81	3×10^6	0.19	3×10^9	30.
	0.0037	2×10^4	0.0013	2×10^7	0.2

concentrations represented in Table 7, appear in the third column of Table 8. In the fifth column of the latter table are values for the number of molecules from the solution striking per μ^2. To calculate the latter, the red cell surface is assumed to be stationary while the dye molecules travel at a definite average velocity.* Thus the number of dye molecules striking per second per μ^2 is equal to the number of molecules in that volume of solution which would be swept through by this area in one second if it traveled at the average velocity of the dye molecule.

To estimate the relative importance of the dye on the cells as compared to that in solution it may be assumed that all the dye molecules are activated at the same instant. The number of dye molecules which reach the red cell before deactivation is the number striking within the average lifetime, which Gaviola (1927) found to be 10^{-9} to 10^{-8} for dye solutions exposed to air. The number of molecules striking in the latter interval of time is shown in the last column of Table 8; it is insignificant compared to the number of molecules in association with the cell membrane. Thus it may be concluded that the dye molecules in the solution play a relatively unimportant part in the photochemical reaction leading to photodynamic hemolysis.†

The Mechanism of Photodynamic Action. The mechanism of photodynamic action indicated by the foregoing arguments may be sketched as follows: The dye is taken up by the cell so that it is in intimate association with oxidizable substances upon which the structure of the cell is dependent, probably protein. When the photosensitizer molecule is activated by capture of a quantum of radiant energy, it transfers its activation to this substrate which then reacts with O_2, a reaction which would not occur in the normal course of cell metabolism. Oxidation of the substrate results in damage to the cell structure. In this process the photosensitizer molecule is not altered, so that a single dye molecule may bring about the oxidation of many molecules of substrate. The fundamental photochemical efficiency for this process is always the same, but the apparent photodynamic effectiveness varies according to the photosensitizer and the conditions of experiment. Allowing for some uncertainty as regards minor points, this general picture seems the only one which will fit all the evidence presented.

* The dye molecules are assumed to travel at the velocity of gaseous molecules having the same molecular weight. This gives a theoretical maximum value, which is probably much too high, and hence the values in the fifth and sixth columns of Table 8 represent the greatest possible number of molecules striking the cell surface. The actual number is probably much lower.

† If reactive molecules were formed in the solution, which had considerably longer lifetimes than activated dye molecules, this argument might lose in importance. For example, if metastable O_2 molecules, as postulated by Kautsky (see p. 68), or dye peroxide molecules (see p. 70) were formed in the solution these might have a much greater chance of reaching the cell membrane. However, the dye molecules at or near the surface of the cell must always have a better opportunity to react, whatever the mechanism of the reaction.

Chapter 8

Factors Determining Apparent Photodynamic Effectiveness

Regarded only in terms of observed end effects, photodynamic action displays a variety which might suggest a corresponding multiplicity of mechanism. However, it is probable that this variety results from quantitative and qualitative differences in uptake of different photosensitizers by cells, and the action of numerous factors which modify this uptake. Such differences are nicely illustrated by the series of studies carried out by Tennent (1935, 1936, 1937, 1938a, b) on echinoderm eggs, which suggests numerous parallelisms between uptake of dye and photodynamic effect. A number of subjects will be taken up in this chapter which all have to do with this general question, although their relationship may be rather indirect in some instances.

Dark Action. Many of the photodynamic dyes produce effects in the dark which may ape those which occur upon irradiation. This is the dark action (*Dunkelwirkung*) of the Munich school. While the end results may be indistinguishable, the mechanisms of dark action and photodynamic action are essentially different.

To demonstrate this, hemolysis of the mammalian red cell by rose bengal will again be used as illustration. In concentrations about ten times greater than the strongest used to produce photodynamic hemolysis in the experiments previously discussed, this dye brings about hemolysis in the absence of light. The curve in Figure 20 shows in a general way the effect of dye concentration on the rate of hemolysis in the light and in the dark; the dye concentration is plotted in logarithmic increments as the ordinate with the greatest concentration at the bottom of the figure, and the time to complete hemolysis is plotted as abscissa. This gives what Ponder (1934) calls a time-dilution curve. The time to complete hemolysis has no truly quantitative meaning as regards photodynamic action, for reasons discussed in Chapter 4, but it serves as a convenient semi-quantitative means of comparison with the dark action.*

The curve shows clearly that there is a minimum concentration of dye which will produce hemolysis in the dark, and that at concentrations higher than this, light has little effect on hemolysis. In the range of dye concentrations at which hemolysis occurs only in the light, the curve shows a sigmoid inflection, which is due to the filtering effect of the outer

* Estimation of complete hemolysis was made in "indifferent" light by a method described by Blum, Pace and Garrett (1937).

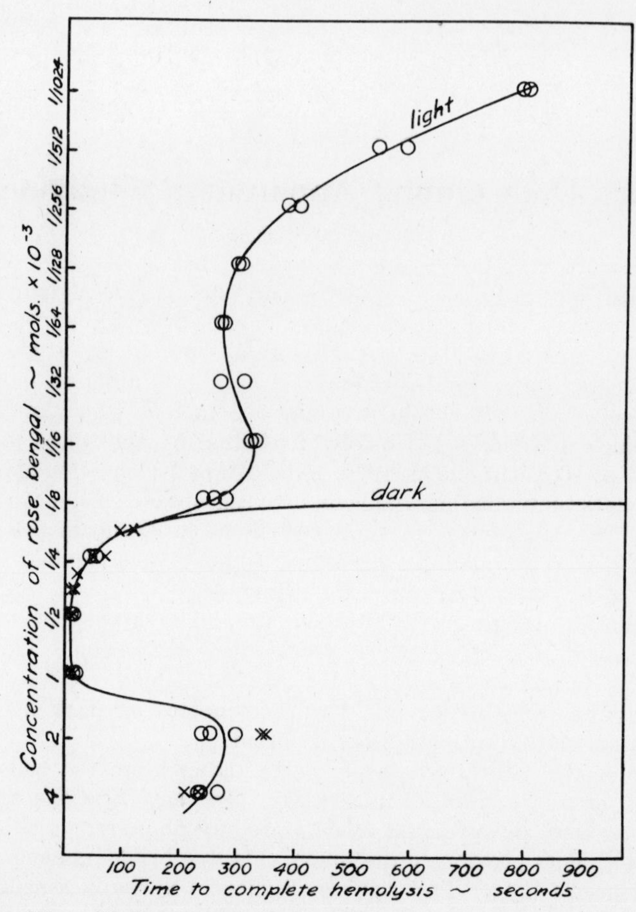

FIGURE 20. Time dilution curved for hemolysis by rose bengal at 20 °C. Circles, hemolysis times in light. Crosses, hemolysis times in dark. Human red blood cells. [From H. F. Blum, N. Pace, and R. L. Garrett, *Journal of Cellular and Comparative Physiology,* **9**, 217 (1937).]

layers of the system at high concentrations (see p. 59). This is the type of effect which has often been cited to show an optimum concentration for photodynamic action.

In the range of concentrations which produce hemolysis in the dark, the curve has again a sigmoid inflection which is similar to that observed in time-dilution curves for a number of typical lysins (see Ponder, 1934); the significance of this is not clear. The shapes of these curves need not concern us particularly as they may be typical for only this dye or others which are quite similar; but regarded in a general way, the curves are useful in illustrating the effects of different factors on dark action and photodynamic action.

It might be assumed that the mechanisms for these two processes are essentially the same, for example, dark action might be dependent upon a chemical reaction which is accelerated by the light to give the photodynamic effect. This would not be compatible with a good deal of the evidence which has been previously presented, and distinct differences in the mechanisms of the two effects can be demonstrated.

The response to temperature is very different for the two effects, as is illustrated by the curves shown in Figures 21 and 22. The Q_{10} of dark hemolysis varies in strange fashion from a slightly negative value to as high as 15 (see Figure 21). On the other hand, temperature has only a slight effect on hemolysis in the light, being uniform at all dye concentrations; the Q_{10} is approximately 1.2 (see also p. 51). The lower part of the curve in Figure 22 obviously represents dark action.

Difference between the two mechanisms is also shown if the red cell suspension is deprived of O_2; photodynamic hemolysis does not take place, but hemolysis in the dark is not inhibited (Blum and McBride, 1931). Obviously the dark action is not an oxidation by O_2, as is photo-

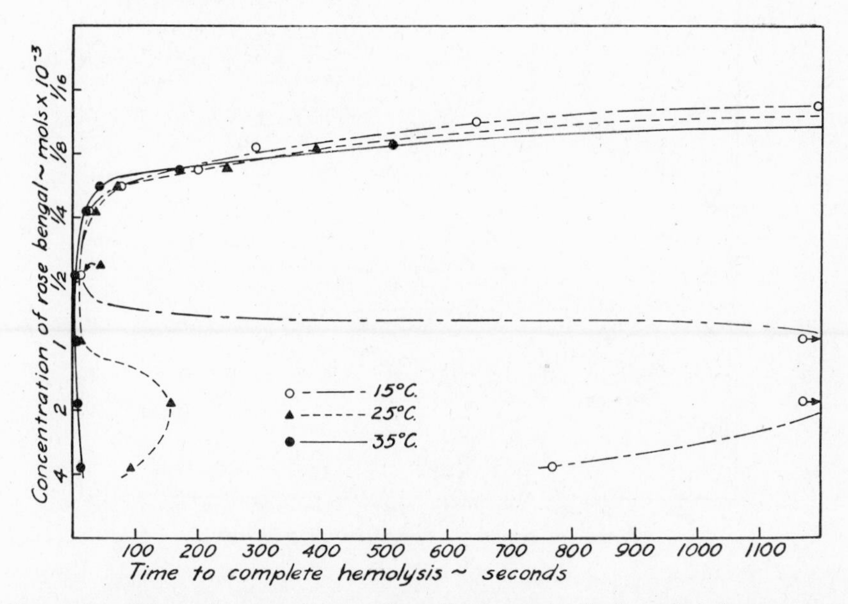

FIGURE 21. Effect of temperature on time dilution curve for hemolysis by rose bengal in the dark. Human red blood cells. [From H. F. Blum, N. Pace, and R. L. Garrett, *Journal of Cellular and Comparative Physiology,* **9,** 217 (1937).]

dynamic action. This is further demonstrated by the fact that inorganic reducing agents such as Na_2SO_3 and $Na_2S_2O_3$ inhibit photodynamic action but not dark action, as is illustrated by the time dilution curves in Figure 23.

Jodlbauer and Haffner's Correlation. As a rule, those dyes which have the greatest apparent photodynamic effectiveness are most effective as regards dark action. This was pointed out by Jodlbauer and Haffner (1921a) who suggested that it is to be explained by the fact that adsorption of the dye by the cell is prerequisite for both effects, and hence the

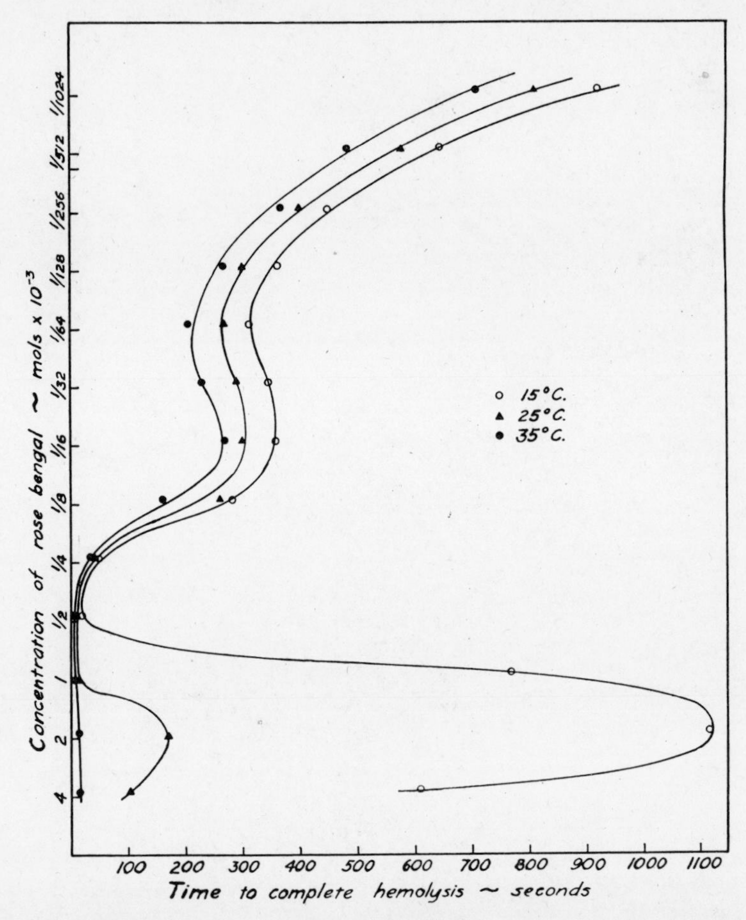

FIGURE 22. Effect of temperature on time dilution curve for hemolysis by rose bengal in light. Human red blood cells. [From H. F. Blum, N. Pace, and R. L. Garrett, *Journal of Cellular and Comparative Physiology*, **9**, 217 (1937).]

greater the uptake of a given dye, the greater its effectiveness in bringing about both photodynamic and dark action. Effectiveness is usually judged by the minimum concentration of dye which will produce an observable effect—a crude method which serves, however, as a means of comparison if properly employed.

An interesting illustration of the correlation is found in the action of various dyes of the fluorescein series on the mammalian red cell. In addition to photodynamic hemolysis and dark action hemolysis, these dyes cause a transformation of the cells from their common discoid shape to spheres (see Ponder, 1936). The latter effect occurs at concentrations of dye too weak to produce hemolysis in the dark, but does not require

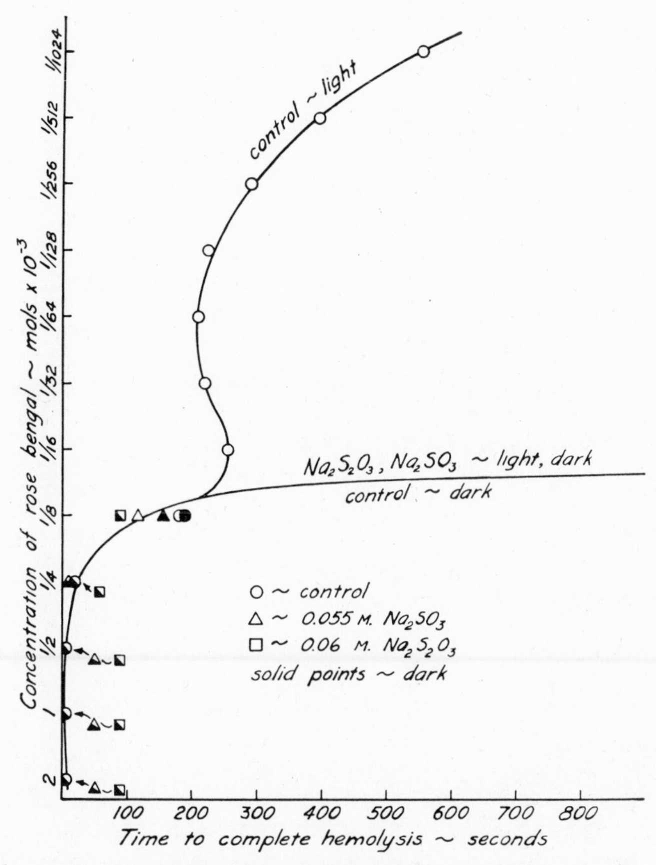

FIGURE 23. Effect of reducing agents on the time dilution curve for hemolysis by rose bengal. Human red blood cells. [From H. F. Blum, *Journal of Cellular and Comparative Physiology*, 9, 229 (1937).]

the action of light. A glance at Table 9, where the minimum dye concentration required to produce each effect is recorded, will serve to show that the same relationship holds for each dye of the series, *i.e.*, the ratio of the concentration producing the different effects is the same in each case. Following through the series, the effectiveness of the dyes in producing each of the three effects increases with increasing molecular weight.

Table 9.*

(Data from Blum, 1935a, and Spealman and Blum, 1937)

Dye	Minimum conc. for hemolysis in the dark (mols/liter)	Minimum conc. for disk-sphere transformation (mols/liter)	Minimum† conc. for photo-dynamic hemolysis in sunlight (mols/liter)	Quantum yield for photo-oxidation of I^-
Fluorescein Na_2R	insoluble at 5×10^{-2}	10^{-2}	10^{-5}	
Eosine $Na_2R\,Br_4$	5×10^{-3}	10^{-3}	10^{-6}	0.032 to 0.050
Erythrosine $Na_2R\,I_4$	5×10^{-4}	10^{-4}	10^{-7}	
Rose Bengal $Na_2R\,Cl_2I_4$	5×10^{-5}	10^{-5}	10^{-8}	0.063 to 0.10

* The hemolysis measurements are for human red cells.

† The absorption spectra of these dyes are very similar and fall so near the maximum of sunlight that equal concentrations may be assumed to absorb the same amount of light; and hence for low concentrations the absorbed radiation will be approximately proportional to the concentration. This would not be true for dyes of widely different absorption spectra.

All the dyes have the structure of fluorescein* except that different numbers and kinds of halogens are substituted. For all three effects, eosin (tetrabromfluorescein) is approximately 10 times, erythrosine (tetraiodofluorescein) 100 times, and rose bengal (dichlortetraiodofluorescein) 1000 times as effective as fluorescein. A single exception is the failure of fluorescein to produce "dark action," which is apparently prevented by the insolubility of this dye at the concentration at which this might be expected. It seems probable that the effectiveness of each dye depends upon the extent to which it is taken up by the cell,† and this is supported by evidence presented in the previous chapter regarding two of these dyes. This is in general accord with Jodlbauer and Haffner's explanation of the correlation.

The fifth column of Table 9 shows that the quantum yield for photosensitized oxidation of iodide ion, measured for corresponding concentrations of I^- and O_2, is not greatly different for rose bengal and eosin when compared to the difference in effectiveness of these two dyes in producing their effects on red cells. The difference in photochemical efficiency of the two dyes is in the same direction as the effects on red cells, however, suggesting that the ability of the dyes to "associate" with the substrate may influence their effectiveness as photosensitizers for oxidation in simple systems. It is interesting in this respect that Ruggli and Jensen (1935) find that chromatographic adsorption of fluorescein dyes occurs in the same order as their effectiveness in producing the phenomena listed in Table 7, *i.e.*, increases with molecular weight. Barnes (1939)

* Fluorescein itself is an acid and is not soluble in water. The form used in these studies is the water-soluble sodium salt, which is sometimes called uranine. The other dyes were also in the form of the soluble sodium salt.

† Blum and Gilbert (1940b) found that rose bengal is taken up more rapidly by the cell than erythrosine; rate of uptake may also play an important part in determining the effectiveness of these dyes in producing hemolysis in the dark.

also finds such a relationship with regard to the effect of fluorescein dyes on frog's skin.

Mottram and Doniach (1938) have recently shown a correlation between photodynamic and carcinogenic activity among the carcinogenic hydrocarbons. It seems probable that this correlation is to be explained in a manner similar to that for the fluorescein dyes (see p. 266). Numerous less quantitative but no less striking examples of this correlation are to be found.

Factors Affecting Dye Uptake and Photodynamic Effectiveness. The manner of uptake of dyes by the cell is rather uncertain, and, as an example, the studies of Tennent which were mentioned early in this chapter indicate a variety of detail. There are a number of obvious factors which might affect the uptake of the dye by the cells, only a few of which will be discussed here.

Ability to penetrate, and electrical charge. The fluorescein dyes which have been the subject of much of the preceding discussion, are taken up, principally at least, at the surface of the cell. This is nicely shown by watching an echinoderm egg or a paramecium during the course of photodynamic damage. The cell is not stained at first because the dye is taken up only at the surface, but after exposure to light for some time the cytoplasm takes on the color of the dye. This is usually the first sign of damage to the cell membrane, lysis following shortly after. As contrasted with the fluorescein dyes, methylene blue and neutral red both penetrate the undamaged cell.

Dognon (1927) found that the dyes which were most effective photodynamically, were taken up on the surface of the cell. Dyes which penetrate the cell were, in contrast, weakly effective, and irregular in their action. His data indicate that dyes which penetrate the cell have a positive charge, whereas negatively charged dyes are taken up at the membrane; hence the latter have greater apparent photodynamic effectiveness. The measurements are qualitative and require further study.

Hydrogen ion. Taking up of the fluorescein dyes is affected by hydrogen-ion activity (Jodlbauer and Haffner, 1921b), more dye being taken up at lower pH. Turner (1932) found apparent photodynamic effectiveness related to pH in the same way, as did Blum (1930b) for both photodynamic action and dark action. Beck and Nichols (1937) find that pH changes which tend to increase the penetration of the dye into the cell increase apparent photodynamic effectiveness. It is doubtful that penetration *per se* is important, but the taking up of the dye on the membrane is probably enhanced in the same way as penetration.

Fixation and precipitation. Blum (1930b) showed that at high dye concentrations, and particularly at low pH, red cells may be fixed by the action of light, so that they are refractory to hemolysis by low osmotic pressure. This fixation seems to be coupled with precipitation of cellular material, since the fixed cells are found in a mass of precipitated cellular debris. With acridine dyes, *e.g.*, trypaflavine, this may occur at moderately low dye concentrations at neutrality, to such an extent that one might doubt the occurrence of hemolysis upon superficial observation. The probable course of events in such instances is that photodynamic hemolysis of a certain fraction of the cells first occurs, and that this is followed by photosensitized oxidation of cellular material, presumably hemoglobin, with formation of a flocculent precipitate which carries down and fixes the remainder of the cells. Under proper conditions of dye concentration and pH, it can be shown that the dye alone does not precipitate the hemoglobin, nor does irradiation of the hemolyzed cells alone; but when dye and hemolyzed cells are irradiated together a flocculent precipitate is formed.

Obviously these effects depend upon the chemical and physical properties of the dye, pH, the kinds of cells, and probably the salt ions in the suspending solution. In the writer's experiments on photodynamic hemolysis, which have provided the basis of much of the discussion thus far, such effects were avoided by choosing a pH (6.7) at which they are not appreciable when the fluorescein dyes are used. In most cases the cells were suspended in a mixture of primary and secondary sodium phosphates.

Salt ions. Turner (1932) found a marked effect of certain salt ions on photodynamic hemolysis by fluorescein dyes. This is not surprising, since these dyes show a marked tendency to "salt out," and the uptake of the dye by the cell must be influenced by the aggregation of the molecules in the solution in which the cell is suspended. There are, of course, instances in which the salt affects the cell itself, but does not alter photodynamic action, *e.g.,* Lillie, Hinrichs and Kosman (1935) find that the response of photodynamically stimulated muscles is affected by various inorganic ions in much the same manner as is the response to electrical or other stimulation, whereas the photosensitized O_2 uptake by the muscle is not affected by these inorganic ions. The effects of hypertonic and hypotonic salts reported by Bier and Rocha e Silva (1935b) may belong in this category.

Retention in the cell. The degree of reversibility of the combination of dye with cell seems to vary greatly. The studies of Tappeiner (1908a, b), Harzbecker and Jodlbauer (1908), and Dognon (1928b) show that all the dye is not removed from red cells or paramecia by moderate washing, as they retain their sensitivity to light. In complex tissues, the dye may be retained for a long time in some cases in spite of the opportunity of sharing it with other cells or with body fluids. For example, hematoporphyrin injected into the skin renders a local area sensitive to light, and this sensitivity may persist for as much as two months; but if rose bengal is used in the same way, the photosensitivity lasts only two days (Blum and Pace, 1937; Blum and Templeton, 1937). Rask and Howell (1928) found, similarly, that the cold-blooded heart was rendered photosensitive after perfusion with hematoporphyrin, and that subsequent washing out with the perfusion fluid did not abolish the photosensitivity.

Water-insoluble photosensitizers. Hausmann devised a method for photosensitizing red blood cells with water-insoluble substances, by using methyl alcohol as a solvent, and adding the methyl alcohol solution to the red cell suspension in concentration not greater than ten per cent. In this concentration the methyl alcohol seems to have very little effect on the cells, and low concentrations of other solvents, *e.g.*, acetone, may be used in the same way. Even when the photosensitizing substance is rapidly precipitated when mixed with aqueous solution in these proportions, the cells may take up enough to become sensitive to light. Such technique is necessary with chlorophyll, for example. Obviously photodynamic effectiveness might be difficult to estimate in such cases.

Inhibition of photodynamic action. Certain hypotheses regarding the mechanism of photodynamic action have been built upon the results of experiments in which inhibition of photodynamic action was obtained in one manner or another. In order to avoid false interpretations it is important to recognize that inhibition may be accomplished in more than one way.

The principal types of inhibition of photodynamic action may be separated as follows:

(1) Inhibition by interference with the oxidative process:
   (a) by removal of molecular oxygen,
   (b) by reducing agents.
(2) Inhibition by agents which prevent the combination of the dye with the cell or other substrate.

Examples of both subdivisions of the first type have appeared in the foregoing chapters. An example of the second type is that described by Busck (1906), who showed that the proteins of blood serum may effectively inhibit photodynamic action. He explained this as due to combination of the dye with the serum proteins, which is equivalent to reducing the concentration of the dye in solution. Since the effective dye is that taken up by the cells, the serum may be said to inhibit by preventing the combination of dye with the cells. That inhibition is due to this rather than to any reducing action of the serum is shown by the fact that dark action as well as photodynamic action is inhibited by small quantities of serum or plasma. This is clearly illustrated in Figure 24.

That difference in effect may be expected for different types of cells and photosensitizers is shown by experiments of Jancsó (1931), who found that trypanosomes take up trypaflavine from the blood and are photosensitized in this medium. This is in distinct contrast to the case of red blood cells and fluorescein dyes, where plasma protects the cells from injury. Obviously no general rule can be drawn.

Schmidt and Norman (1922) suggested that the inhibiting action of serum is due to the reducing action of the amino acids, tyrosine and tryptophane, in the protein molecule, and cited experiments to show that these amino acids themselves could inhibit photodynamic hemolysis by eosin. This argument is not valid; tyrosine precipitates fluorescein dyes, and the inhibiting action of tryptophane must be due to combination

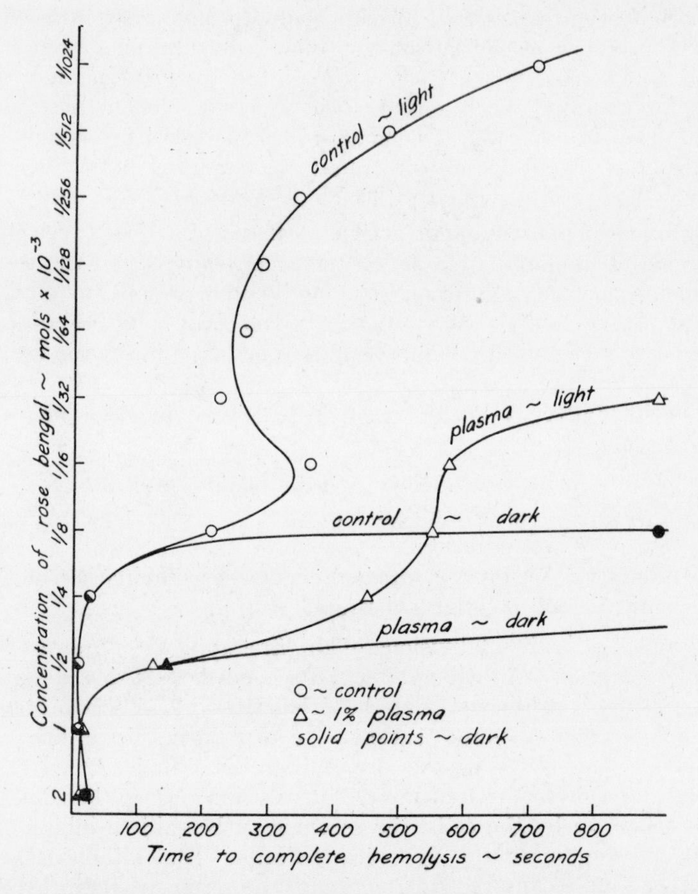

FIGURE 24. Effect of blood plasma on the time dilution curve for hemolysis by rose bengal. Human red blood cells. [From H. F. Blum, *Journal of Cellular and Comparative Physiology,* 9, 229 (1937).]

with the dye since, as shown in Figure 25, dark action as well as photodynamic action is affected. Jírovec and Ziegler (1933) also noted that serum and various proteins inhibit the dark as well as the light reaction. Rask and Howell (1928) showed that euglobulin, pseudoglobulin, egg albumin and peptone, all of which contain tyrosine, failed to inhibit the photosensitization of the turtle heart with hematoporphyrin. They found the inhibiting action of serum and plasma to be due principally to the serum albumin.

Another example of such errors in interpretation concerns the inhibition of photodynamic effects by fluorescein dyes when the dye phenosafranine, itself photodynamically active, is added to the system. Both

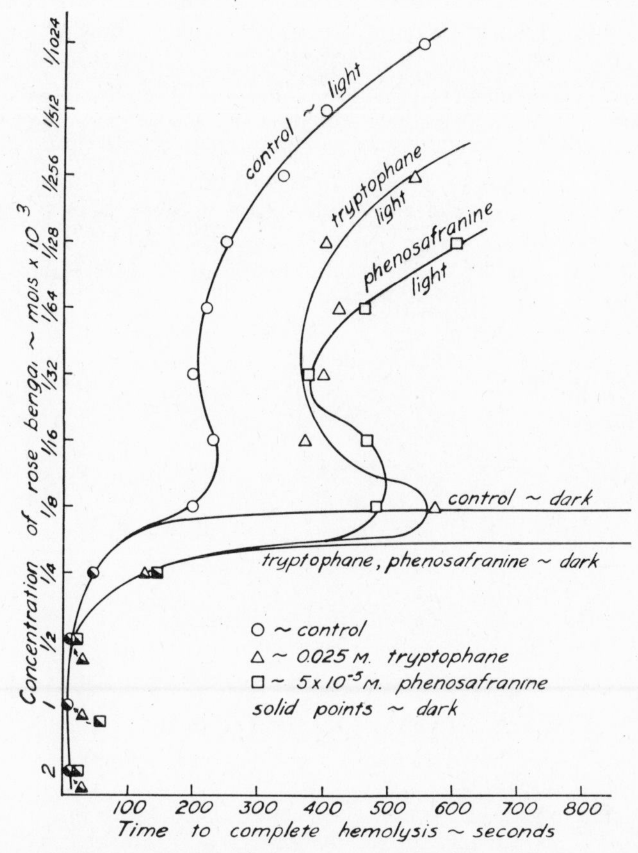

FIGURE 25. Effect of tryptophane, and of phenosafranine, on time dilution curve for hemolysis by rose bengal. Human red blood cells. [From H. F. Blum, *Journal of Cellular and Comparative Physiology*, **9**, 229 (1937).]

Dognon (1928c) and Szörényi (1932) explained this by analogy to "optical desensitization" of the photographic plate. Actually, the inhibition is due to combination of the two dyes, which renders both incapable of being taken up by the cell. Figure 25 shows that the addition of phenosafranine to rose bengal solution decreases both dark action and photodynamic hemolysis; in fairly strong solutions the two dyes form a precipitate.

In like manner, the well known tendency of the two photodynamically active dyes, eosin and methylene blue, to precipitate each other should render them mutually inhibitory. Both methylene blue and phenosafranine are basic dyes, whereas the fluorescein dyes are acid. The effect of different substances upon the combination of the dye and cell must differ with the nature of the dye and with other factors, so that the same substance will not affect all photodynamic processes uniformly. This may account for some apparent discrepancies, *e.g.*, Rask and Howell (1928) found that peptone does not inhibit photodynamic effects on the turtle heart sensitized by hemotoporphyrin; but Schmidt and Norman (1920) and Jírovec and Ziegler (1933) both found that this substance inhibits photodynamic hemolysis by fluorescein dyes. The amphoteric nature of the proteins may render them somewhat more universal in their inhibiting action; but other substances may be more specific, preventing combination of the dye and cell only in particular cases. Wide variation in such effects is shown in the studies of Testoni (1923), Viale (1924), Pennetti (1927), and Saeki (1932c).

Chapter 9

Other Theories of Photodynamic Action

In the forty years since its discovery, a considerable number of hypotheses have been put forth to explain the mechanism of photodynamic action. The more important of these will now be examined in the light of the existing experimental evidence, and in comparison with the mechanism which is proposed in this book.

Oxidation Hypotheses. Straub in 1904 (a, b) compared photodynamic action to photosensitized oxidation of iodide, and suggested that the processes are essentially the same, *i.e*,, that both are photosensitized oxidations by O_2. The experimental evidence presented in the foregoing chapters all fits this basic proposition, and the mechanism outlined at the end of Chapter 7 may be considered as an elaboration of this hypothesis. Unfortunately, Straub's ideas were overruled by Tappeiner, who proposed a more complicated hypothesis (see p. 98). Numerous subsequent workers have admitted the oxidative nature of the phenomenon, but have developed different theories for its mechanism.

Peroxide Hypotheses. In discussing the mechanism of photosensitized oxidations in Chapter 6, the lack of evidence for the existence of intermediate peroxides was pointed out However, the possibility that hydrogen peroxide is formed as an intermediate has also been suggested and requires discussion.

Blum (1930a) found that solutions of fluorescein dyes which have been exposed to sunlight contain a small proportion of peroxide, and that the ability of these solutions to produce hemolysis in the dark is enhanced. He presented this as evidence that peroxides of the dye are formed as intermediates in photodynamic action. However, Blum and Spealman (1933a) found that the peroxide was hydrogen peroxide, and Blum and Spealman (1934a) and Menke (1935) later showed that the hemolytic properties of such solutions are not appreciably diminished by destruction of the peroxide.

On the other hand, Liebert and Kaper (1937) found that the lethal factor for bacteria, which is produced when bacterial media containing erythrosine and methylene blue are irradiated, is hydrogen peroxide, since the lethal activity is abolished by brushing the surface of the media with a catalase solution.

Much larger quantities of hydrogen peroxide are required to produce hemolysis than are found in fluorescein dye solutions after irradiation, however (Blum 1930b, 1935a). Dark action and hydrogen peroxide are additive, *i.e.*, hemolysis is produced in the dark by lower concentrations of dye when hydrogen peroxide is present, and *vice versa;* but combined dark action and hydrogen peroxide does not approach the effectiveness of photodynamic action.

That hydrogen peroxide produced *in situ* might be more effective than hydrogen peroxide added to the solution, since it would run less risk of being destroyed by the action of catalase in the blood cells was suggested by Blum (1935a) who cited the experiments of Wohlgemuth and Szörényi (1933a, b) to indicate that hydrogen peroxide is formed in photodynamic action, and destroyed by catalase. The latter investigators had found that the uptake of O_2, measured manometrically, was actually increased by addition of cyanide to the system. If catalase constantly destroys hydrogen peroxide with the formation of O_2, the apparent uptake of O_2 will be reduced below the true value; if cyanide is added to such a system, the action of catalase is inhibited, and an increase in the apparent volume of O_2 taken up by the system results.

If it were true that photodynamic action proceeds by the formation of hydrogen peroxide which subsequently oxidizes components of the living system, the introduction of cyanide should accelerate photodynamic action by inhibiting the catalase. Earlier observations of the action of cyanide are in conflict (see Blum, 1935a). Bier and Rocha e Silva (1935a), and Rocha e Silva (1937c), find photodynamic hemolysis enhanced by the addition of cyanide in certain concentrations. The writer was unable to demonstrate any marked effect on NaCN on photodynamic hemolysis, when the time required to reach a given percentage hemolysis was used as an index. A number of cyanide concentrations were used and slight differences were found in some cases, but these were almost within experimental error.

Rocha e Silva (1936, 1937a, b) found that active catalase inhibited as much as 85 per cent of photosensitized oxidation of iodide, whereas catalase which had been inactivated by heat had much less effect. He interprets this to show that hydrogen peroxide is an intermediate, the removal of which by catalase inhibits the oxidation. Cyanide could not be used to inhibit the catalase in these experiments because of the formation of a complex with the iodine. Mr. Frederick Shelden has carried out studies at the writer's suggestion which indicate that the inhibiting action of catalase on this reaction is due principally to the removal of dye from solution by combination with catalase, catalase which has been inactivated by heat taking up the dye to a much less extent. This would account for the inhibition of photodynamic hemolysis by catalase which Rocha e Silva reports (1937c).

Experiments of Rocha e Silva (1937d) on the photosensitized oxidation of serum provide a confirmation and extension of the findings of Wohlgemuth and Szörényi. He found that fresh serum took up only half as much O_2 as old serum, or serum which had been "inactivated" by heating to 60 °C; but if cyanide was added to the fresh serum it showed the same O_2 uptake as "inactivated" serum. These results are to be expected if the catalase contained in the fresh serum destroys hydrogen peroxide, since heat, age and cyanide should all inactivate the catalase.

It is highly probable that hydrogen peroxide is formed in the photosensitized oxidation of organic substances in living systems, and the experiments of Wohlgemuth and Szörényi (1933a, b) and Rocha e Silva (1937d) show that the formation of this substance might account for a large fraction of the total O_2 uptake. However, it may be calculated from the data of the former investigators (see Blum, 1935a) that if all the O_2 taken up in the photosensitized oxidation of red blood cells formed hydrogen peroxide, it would not be enough to account for an appreciable part of photodynamic hemolysis. If the action of catalase were minimized when the hydrogen peroxide is formed *in situ,* it is possible that enough would be formed to account for any accelerating action of cyanide which has been demonstrated, but it is hardly possible that it could account for the total photodynamic process. Probably hydrogen peroxide is not a true intermediate in the photosensitized oxidation, but a product which may act as an oxidizing agent to increase somewhat the total destructive effect in photodynamic action; its importance in the total process would not appear to be great.

Oxidation Products of the Dye as Intermediates. It has been demonstrated by a number of workers (Ledoux-Lebard, 1902; Sacharoff and Sachs, 1905; Moore, 1928; Blum, 1930a; Menke, 1935a) that fluorescein dyes, which have been irradiated for fairly long periods and subsequently added to living systems in the dark, produce hemolysis and other effects comparable to photodynamic action at lower concentrations of dye than those which produce dark action. Apparently the irradiation changes the dye in some way, rendering it more destructive. Danielli* has found that irradiated fluorescein dyes cause the disruption of serum films, and Rideal (1939) reports a similar finding for some carcinogenic hydrocarbons.

Menke (1935a) put forward a hypothesis which involves the formation of lytic substances. He credited the lytic action of irradiated fluorescein dyes to "photo-compounds" which Wood (1922) had shown by spectroscopic means to be formed under such conditions. Menke isolated the "photo-compound" from irradiated fluorescein and found it to be more lytic than fluorescein which had not been irradiated. He suggested that the formation of this compound *in situ* may account for the destructive effects which characterize photodynamic action. These compounds are formed only in the presence of O_2 (Blum and Spealman, 1933a), and must be oxidation products of the dye. Readily oxidizable material in living systems must compete with the dye for oxidation (see p. 67) making it doubtful that any great amount of the dye would be oxidized to the "photo-compound."

* Personal communication.

Much of the evidence will fit such an hypothesis, however (see Blum, 1935a) but the fact that many quanta per molecule may be required to produce hemolysis (see the values of *S*, Table 7) is conclusive evidence against it. Following Menke's hypothesis, each dye molecule would have to undergo a photochemical change by the absorption of a single quantum, and this would have to be sufficient to produce the lytic effect.

The relation of such hypotheses to recent ideas regarding the photodynamic action of the carcinogenic hydrocarbons will be discussed in Chapter 22.

Dehydrogenation. Photosensitized dehydrogenation of components of living systems probably occurs when O_2 is excluded (see p. 71). This might lead to destructive changes, but must always be limited by the concentration of the sensitizer which is reduced to the leucobase. The change produced in this way must be insignificant since photodynamic effects do not occur in the absence of O_2.

Other Hypotheses. *Tappeiner's hypothesis.* Tappeiner rejected the simple hypothesis of Straub (1904a, b), and finally arrived at a quite complicated one which he outlines in his review of 1909. This assumes an interference with normal metabolic processes. He believed that visible light produced destructive effects similar to those brought about by ultraviolet radiation (p. 117), but that ordinarily the reactions due to visible light are inhibited by their own products. When a photosensitizer is added to the system these products are removed by oxidation, thus accelerating the action of visible light. Subsequent findings provide no support for this hypothesis, so that it is of historical interest only. Evidence that photodynamic action is independent of normal cell metabolism has been presented earlier (see p. 65).

Fluorescence Hypotheses. The ability to fluoresce shows that the molecule, after receiving its quantum of energy, does not give it off as small quanta, but holds it for a brief period and then emits it as a quantum nearly the size of that absorbed. The fluoresced radiation is not different from other radiation, yet it has been credited, repeatedly, with bringing about photodynamic action. One of Raab's first experiments (1900) was designed to test this possibility by exposing unsensitized paramecia to fluorescence from a solution of eosin; the organisms were not affected.

An hypothesis of Viale (1920, 1934) is based directly upon fluorescence, and follows an early proposal of Jean Perrin (1918). Perrin assumed that the dye molecule is destroyed in the act of fluorescence, part of the energy of the absorbed quantum being used in the destruction of the molecule while the remainder appears as fluoresced radiation. Thus, in accordance with Stoke's rule, the fluoresced radiation is always of longer wave-length than the absorbed radiation. Perrin used fluorescein (uranine) in his experiments, observing that the bleaching of the dye was limited to that portion of the solution which absorbed the light. Numerous convincing objections to this hypothesis have been produced (see

Kistiakowski, 1928, p. 212), *e.g.*, it has been calculated that a fluorescein molecule in solution absorbs and emits many quanta without being destroyed. Viale's hypothesis assumes that the energy used in photodynamic action represents the difference in the size of the absorbed and radiated quanta, this energy being used in changes within the living system rather than in the destruction of the dye molecule itself. This hypothesis must suffer the same fate as Perrin's upon which it is based.

Photoelectric Effect Hypotheses. Schanz (1921) put forward the hypothesis that photodynamic action results from changes in cell constituents due to the action of electrons emitted by the sensitizer when it is irradiated, supporting his contention with studies of the photoelectric effect in egg albumin sensitized with photodynamic dyes. Actually, Jodlbauer and Tappeiner (Tappeiner, 1909, p. 723) had the same idea, and attempted to test it experimentally. They placed a hanging drop of paramecium culture as near as possible to a fluorescent solution and exposed it to sunlight for several days, but the paramecia remained unharmed. They took this as proof that if electrons were discharged from the fluorescent solution they did not injure the paramecia.

Clark (1922) summarizes a photoelectric effect theory of photodynamic action, and of photobiological effects in general, in the following way. "Light shorter than 300 $\mu\mu$ acts on the living cell by ionizing its photo-electric constitutents, and thereby leading to photochemical action. Light longer than 300 $\mu\mu$ acts in the same way in the presence of sensitizers, which so affect the surface conditions of these constituents that their photo-electric threshold is shifted into the visible, and they therefore become ionized, with resulting chemical action, when illuminated by visible or near ultraviolet light."

It is hard to harmonize these hypotheses with the fact that O_2 is a necessary component of photodynamic action, and Clark's hypothesis does not explain why O_2 is required in photodynamic action, but not for the destructive effects of ultraviolet light (see p. 114).

Chapter 10

Variety of Photodynamic Effect

It is scarcely possible to list here all the photodynamic effects which have been described, since very many kinds of living organisms, and biological systems have been subjected to such photosensitization. Examples will be discussed briefly, but the inclusion of a complete bibliography would be awkward and its value doubtful. The discussion in Chapter 8 should indicate the wide variety of effects that may be expected because of the number of factors which may influence photodynamic action. Only effects produced in the laboratory by the deliberate introduction of photodynamic dyes into biological systems will be discussed in this chapter, as the effects resulting from the natural presence of these substances in such systems will be dealt with in later parts of the book.

Photosensitized Oxidations in Non-cellular Biological Systems

Besides the oxidation of organic compounds which derive from living systems, many non-living and non-cellular biological systems have been subjected to the action of photodynamic dyes and light.

Blood Plasma and Serum. Numerous studies of the oxidation of serum and plasma have been made (Gaffron, 1926; Harris 1926, b, c; Wohlgemuth and Szörényi, 1933a, b). Most of the O_2 uptake is by the protein fraction, according to Smetana (1938a).

Howell (1921) found that the clotting of plasma was inhibited by the photodynamic action of hematoporphyrin or eosin. This he considers to be due to some change in the solubility of fibrinogen, since this substance loses its ability to be coagulated by thrombin or by heat after treatment with eosin or hematoporphyrin and exposure to light. Baumberger, Bigotti and Bardwell (1929) found, similarly, that the combined action of methylene blue and light inhibits the clotting of plasma.

Enzymes. Tappeiner and Jodlbauer (1907) and their students made early investigations of photodynamic action on enzymes; these and subsequent studies are tabulated by Metzner (1927), and those of Supniewski (1927a) and Kambayashi (1928) may be cited in addition. All the results indicate destructive or inhibitive effects. The protein nature of many enzymes has been established in recent years, and it is not surprising to find that such systems are susceptible to destruction by photodynamic action. Claus (1929) has shown an acceleration of enzyme action by certain photodynamic dyes acting in the absence of light.

Toxins, Antitoxins, Viruses, and Venoms. Such substances, many of which are known to be protein in nature, are susceptible to alteration or destruction by photodynamic action. This was the subject of early studies in Tappeneiner's laboratory (Jodlbauer and Tappeiner, 1906b), by Noguchi (1906a), and by Hausmann and Pribram (1909). Recently there has been a revival of such studies with attempts to produce attenuated viruses or toxins. It is hardly to be expected that more specific results may be obtained in this way than by the use of other oxidizing agents, as was indeed recognized by Noguchi, but more thorough studies of the photochemistry of such reactions might lead to their advantageous use.

Perdrau and Todd (1933b) found that methylene blue and light inactivated a number of viruses, including those of canine distemper and fowl plague. In addition, the inactivation of vaccinia virus (Hertzberg, 1933), tetanus toxin (Lippert, 1935), rabies virus (Shortt and Brooks, 1934), diphtheria toxin (Lin, 1935), and poliomyelitis virus (Rosenblum, Hoskwith and Kramer 1937) have been reported among others. Birkeland (1934) finds plant viruses more resistant than animal.

Perdrau and Todd (1933c) found that canine distemper virus inactivated in this way conferred immunity to the disease, which finding was confirmed by Dempsey and Mayer (1934). Galloway (1934) found the same for rabies virus, Lin (1936) for diptheria toxin, Li (1936a, b) for staphlococcal toxin, and T'Ung (1936) for pneumococcus vaccine; but Lippert (1935) found that tetanus toxin deactivated in this manner had no antigenic properties, as did Shortt and Mallick (1935) for cobra venom.

It seems probable that the antigenicity of substances so treated may depend upon the extent to which the photo-oxidation has been carried; if too complete the antigenicity may be lost. Studies of the destruction of antipneumacoccal serum and diphtheria antitoxin by methylene blue and light have been recently reported by Ross (1938 a, b). More careful control of the conditions of irradiation might lead to more consistent results.

Bacteriophage. This system, which may be difficult to class as either living or non-living, is detroyed by photodynamic action (Clifton, 1931, Perdrau and Todd, 1933a). It is not destroyed in the absence of O_2.

Hormones. Penetti (1926) found that epinephrine was destroyed by the combined action of fluorescein dyes and light, but insulin was not. The former substance belongs to the readily oxidizable hydroxy-phenyl ring compounds. It is probable that others of the hormones may be destroyed by photodynamic action, but no great significance is to be attached to this as regards normal physiological processes. Loewi and Navratil (1926) found that the inhibitory hormone of the heart (acetyl choline) was destroyed in this manner, and Haberlandt (1927) found the same for the accelerating hormone of the heart (sympathin), but no great significance can be attached to this as regards the physiology of these substances, nor does it offer any reliable clue to their chemical identity.

Vitamins. The interesting case of the destruction of vitamin C (ascorbic acid) in milk due to the photosensitizing action of vitamin B_2 (lactoflavin), will be discussed in the next chapter.

Cells and Unicellular Organisms

In most studies of photodynamic action on cells, only killing or inhibition of growth has been observed. In others, detailed studies of morphological or other changes have been reported (*i.e.*, see Menke, 1935b). On the whole, differences in the picture may be expected, depending on whether the dye is taken up on the surface of the cell or penetrates to the interior. This has been discussed in Chapter 8.

A great number of bacterial forms (Reitz, 1907; T'Ung 1935, 1936; T'Ung and Zia, 1937), protozoa and protophyta (Metzner, 1927), and higher fungi (Jodlbauer and Tappeiner, 1905b), have served as victims for such experiments. White blood cells (Salvendi, 1906; Menke, 1935b), the ova of Echinoderms (Lillie and Hinrichs, 1923; Periera, 1925; Moore, 1928; Tennent, 1938b), cells in tissue culture (Menke, 1935b) have also been used, and the list of references might be widely extended. Many such studies have been undertaken in the attempt to elucidate some particular phase of the photodynamic mechanism, but the majority merely record the destruction of the particular organism by the combined action of light and some dye.

There may be wide difference in the susceptibility of closely related organisms to photodynamic action, as shown by the studies of Jancsó (1931a, b, c), and Jadin (1932) on trypanosomes and spirochetes, and those of T'Ung on bacteria. These are probably due in most cases to differences in the extent to which the cell takes up the dye.

Excitation effects have been described, *e.g.*, photoorientation in protozoa (Metzner, 1919, 1921), and fertilization of echinoderm eggs (Lillie and Hinrichs, 1923). These will be discussed in the following chapter as regards their relationship to normal physiological processes.

Multicellular Organisms

The effects on the cells of larger, many-celled organisms are essentially the same as on unicellular forms, but in the former case the damage is restricted to those superficial layers of cells to which the light penetrates. This may result in somewhat more complicated pictures. On the other hand, in the case of multicellular organisms so small that light may penetrate to most cells of the organisms, *e.g.*, rotifers, hydra, amphibian larvae (Bohn and Drzewina, 1923; Gicklhorn, 1914; Hinrichs, 1924) the effects are more directly comparable with those observed in the unicellular forms. Thus mere size may be as important as morphological or physiological complexity in determining the picture produced by photodynamic action.

Effects on Mammals. In subsequent discussion of diseases produced by light, interest will center on photodynamic action in mammals. As Hausmann has shown (1910, 1914, 1923) the effects which follow injection of a dye into a mammal, with subsequent exposure to light, vary with the magnitude of dosage of sensitizer and the intensity of the light. He observed the following groups of symptoms in white mice injected with hematoporphyrin and exposed to sunlight. (*a*) With large quantities of dye, or (and) intense light, the animal may go into coma within a few minutes and die shortly thereafter. (*b*) With less intense light or smaller dosage of sensitizer an acute syndrome results in which the animal scratches vigorously at first and waltzes about; in a few minutes the ears become inflamed, the animal blinks the eyelids, and attempts to avoid light; shortly after, it becomes weak and dyspneic; in some cases agonal tetany may be observed; death may occur in a few hours. (*c*) With less severe dosage, subacute symptoms are observed, principally severe edema of the skin. (*d*) With still milder conditions very little immediate effect may be observed, but there may be prolonged chronic effects. As in the subacute cases, the ears may become necrotic and slough off. The hair often falls out, particularly in a ring about the eyes.

Numerous investigators have described effects in other animals, similar to those listed by Hausmann. Syndromes similar to at least the last three are seen in domestic animals which have become sensitized to light by feeding on certain plants, a subject which will be discussed at length in the following section of this book.

The general symptoms divide themselves into three groups: those due to sensory stimulation, those due to destructive effects in the skin which involve local vascular changes, and secondary changes in organs not reached by light.

Sensory stimulation. Sensory stimulation of nerve endings or sensory organs in the skin, resulting from destructive effects of photodynamic action, offers the best explanation of the scratching and running about which is observed in acute photosensitization. Human skin may be locally sensitized by photodynamic dyes, as will be described later, and in this case intense itching occurs when the sensitized part is exposed to light. Mice given sufficient dosage of photosensitizer and light to produce acute symptoms, but narcotized with urethane previous tc exposure to light, display no sensory symptoms although they die as readily as when the narcotic is not used (Awoki, 1925). This demonstrates that the sensory nerves are involved in the acute symptoms, but that these sensory effects play little or no part in the death of the animals.

Skin changes. The changes in the skin are primarily destructive. Histological examination (Levy, 1929, Videbech, 1931) shows effects not unlike those produced by sunburn radiation (see Chapter 17), *i.e.*, degeneration of epidermal cells, dilation of the vessels of the papillary layer, and a general picture of inflammation.* Levy found that the alterations produced by photodynamic action were entirely comparable with those produced by sunburn radiation, but he used a source which emitted much of this radiation, and probably studied a mixed effect. Videbech, who filtered out the sunburn radiation, found a greater involvement of the deeper tissues to result from photodynamic action than did Levy. This may have been due to the deeper penetration of the wave-lengths producing the photodynamic effect, since as a rule photodynamic effects are produced by visible, or near ultraviolet wave-lengths which penetrate to the corium to a considerable extent (see Figure 41, p. 202), and might act directly upon the minuet vessels in the papillary layer of the skin. The epidermis must also be involved, however, and damage at this point may play an important if not a dominant role. Although the changes caused by photodynamic action may appear similar to those produced by sunburn radiation, the mechanisms are fundamentally different (see p. 116).

The picture of skin changes is not inconsistent with the assumption that the vascular changes in the corium are secondary to injury to the epidermal cells or those of the corium itself. The vascular changes appear first, but they may be the result of changes in the epidermis or elsewhere, which cannot be demonstrated by histological technique until the cells have undergone degeneration.

The earliest outward evidence of vascular changes is the appearance of erythema and edema. An interesting example of this is shown in local photosensitization of human skin, which has been studied by Duke (1923), Frei (1926), and by Blum, Watrous and West (1935). Such sensitization is accomplished by the intradermal injection of a photodynamic dye in appropriate concentration, *e.g.*, 10^{-4} molar rose bengal solution. An immediate wheal appears which subsides in the course of an hour, leaving a small localized area faintly colored by the dye. If this sensitized area is now exposed to sunlight for a few minutes, a severely itching wheal appears. The response has all the characteristics of a "triple response," which name Lewis (1927) has applied to the events which occur when histamine is pricked into the skin, or in spontaneous urticaria (hives) or dermatographism. The changes are erythema (reddening) of the affected area, followed by edema (swelling) limited to the same area, with a "flare" of erythema extending outward from the region of the edema. Lewis explains these three parts of the triple response as due to dilation and increase permeability of the minute vessels of the skin which result from the production of histamine or a histamine-like "H" substance. The "H" substance is presumably released by the cells of the epidermis. Before applying Lewis' attractive hypothesis to the present case, it is necessary to point out one distinct difference between the triple response which he describes and the one which follows

* A section of skin demonstrating the different layers is shown as Figure 32 (p. 176).

photodynamic sensitization. The "triple response" itself seems identical in both cases, but whereas spontaneous urticaria, the response of dermatographism, and that following histamine pricks, disappear in the course of a few hours leaving no trace behind them, the photodynamic "triple response" is followed by a persistent pigmentation of the photosensitized area. Pigmentation is a common sequel to epidermal injury, and seems to indicate that the production of "H" substance in the photodynamic response results from damage to epidermal cells, whereas in the other instances the "H" substance is released by these cells without permanent injury. Dilator substance is generally released by injured cells. This argument will be further elaborated in Chapters 17 and 18.

Necrosis may follow the original skin injury. A characteristic sequel to photodynamic action in mice or other rodents is a parchment-like condition of the ears, which may eventually slough off. The degree of skin damage, like other effects, varies with the intensity of the dosage of sensitizer and light.

Circulatory Changes and the Cause of Death. Death may follow sufficiently severe photodynamic action (acute form). The cause appears to be a generalized circulatory collapse. Pfeiffer (1911a, b) found a fall in temperature to accompany the generalized weakness which rodents show after a short period of irradiation. The temperature continues to fall until the death of the animal. If the animal does not die the temperature gradually rises again to normal.

Supniewski (1927b) found that, in rabbits, a fall in blood pressure, diminished respiratory movements and a decrease in blood cells in circulation, precedes the death of the animal. Rask and Howell (1928) found, in the dog, an initial rise in blood pressure with an increased pulse rate accompanied by paroxysms of hyperpnea. This was followed by a rapid fall of blood pressure, but the pulse and respirations continued rapid until death; there was intense cyanosis.

It is probable that the circulatory changes result from the production of some substance in the skin which is carried throughout the body by the circulation. This is evidenced by the generalized hyperemia of the internal organs which has been reported by Pfeiffer (1911a, b), and Gassul (1920). Hemorrhage from the intestinal tract is a common finding (*e.g.*, see Horsley, 1934). Further support is given by the parabiosis experiments of Pfeiffer and Jarisch (1919). These investigators found that if a pair of rats with a common circulation were injected with a fluorescein dye, but only one animal was exposed to light, the unexposed mate showed the same fall in temperature as the irradiated animal. Smetana (1928) found that in such a parabolic pair injected with hematoporphyrin, the unexposed mate did not show any abnormal symptoms. Smetana cites only one experiment, whereas Pfeiffer reports results on several pairs of animals.

Since light may penetrate to the small vessels in the papillary layer of the corium, and since blood plasma may undergo changes due to photodynamic action, some of the generalized effects could be due to changes produced in the circulating blood. Smetana (1928) introduced glass cannulae into the circulation of animals (both cats and dogs were used), and arranged to expose the blood to sunlight as it passed through the glass tube, while the rest of the body was protected from light. No pronounced effects resulted, suggesting that photodynamic action on the blood itself does not play an important part in the generalized hyperemia observed when mammals are subjected to photodynamic action. However, Smetana later points out (1938b) that the amount of blood passing through the capillaries of the skin is very great, and that oxidation of the plasma proteins at this point might have an important role in these generalized effects. Smetana found (1928) that irradiation of the peritoneum alone produced shock, thus demonstrating that photodynamic damage to tissue other than skin may lead to generalized symptoms.

Pfeiffer (1911a, 1919) pointed out the similarity between the events observed after photodynamic action and those observed in anaphylaxis, or after scalding. He offered the general hypothesis that all these effects are due to the toxic action of protein breakdown products. It seems probable that there is nothing specific about the general symptoms which occur when mammals are subjected to severe photodynamic action, but that they result from extensive damage to the skin which is the primary site of action, and are not essentially different from the symptoms following extensive skin damage from other causes. Similar symptoms follow intensive irradiation with a mercury arc without sensitizer (Levy, 1919), due to sunburn radiation emitted by that source, but the fundamental mechanism is very different in the two cases (see Chapter 11). In general, the symptoms are those associated with shock.

It is probable that some of the experiments describing death of photosensitized rodents upon exposure to sunlight have been wrongly interpreted. It has been shown that rodents which have not been injected with a photodynamic substance are very sensitive to sunlight, being killed by relatively short exposures to such radiation. This is due to heat rather than to any specific photochemical effect (Remlinger and Bailly, 1932; Lumière and Sonnery, 1934; Pinner and Margulis, 1936). Levy (1933) reports that animals subjected to photodynamic action which was not lethal were subsequently sensitive to heat,* but Remlinger and Bailly (1932), who exposed their animals to very intense solar radiation in Algeria, found that animals injected with eosin died no more rapidly than animals which were not so treated. It seems possible that at least some of the instances of death of rodents reported as acute photodynamic action may have been due to combined photodynamic action and heat, or even to heat alone. A considerable degree of skepticism must, therefore, be permitted as regards conclusions from such experiments when care has not been taken to protect the animals from overheating.

* The writer has not been able to confirm this.

Blood cell count. Different accounts of changes in the blood-cell picture have been given. Penetti found for the guinea pig (1923), and Smetana (1928) for the cat, no change in the red cell count; but Supniewski (1927b) found a dimunition in the red cell count in the rabbit, and Guerrini (1932) reports anemia.

Changes in the white cell count are reported, but display little agreement. Pfeiffer (1911a, b) found an initial leucocytosis followed by leucopenia. Both Penetti and Smetana report a rise in total leucocyte count. Penetti describes changes in the differential leucocyte count.

It seems probable that this lack of agreement is due to differences in the dosage of sensitizer and of light, as well as to the fact that different animals were studied.

Blood and Urine Chemistry. The changes produced by photodynamic action might be expected to reflect themselves in the blood and urine chemistry, but findings are somewhat variable. Pincussen (or Pincussohn) reports various changes which he interprets as representing altered metabolism (1913, 1922, 1928). On the other hand, Kichiya (1924), Smetana (1928) and Rask and Howell (1928) found relatively unimportant changes, or none, in the phases of chemistry which they followed, with the exception of alterations in oxygenation and CO_2 combining power of the blood reported by Smetana.

Other Changes. Pfeiffer and Jarisch (1919) found marked changes in the adrenal glands in both members of parabiotic pairs of animals when only one animal was irradiated. Such changes did not occur in controls injected with dye, but kept in the dark. Jolly (1926) reports changes in the thymus gland. It is probable that effects on various internal organs are to be found, and that these may vary according to the dosage of sensitizer and dye.

Inhibition. Awoki (1925) could not inhibit photodynamic effects in mammals by injecting reducing substances, but Hausmann and Löhner (1926) found that lowering the O_2 tension by decreasing the total atmospheric pressure accomplished partial inhibition.

Effects on Lower Vertebrates. Frogs sensitized with photodynamic dyes (Blum and Spealman, 1934b) respond by general signs of excitation, jumping about and making movements of the legs to wipe the region of the eyes. The movements may continue intermittently as long as the animal is exposed to light, but cease upon removal to the dark. After long exposure the animals may show prostration and paralysis followed ultimately by death, even though removed from light. These effects are completely abolished if O_2 is removed. Destruction of the brain without destruction of the spinal cord diminishes, but does not stop the movements, whereas destruction of the spinal cord stops them. Thus, the movements must be reflex in nature, resulting from sensory stimulation in the skin.

Jodlbauer and Busck (1905) described photosensitization of fish by the addition of fluorescein dyes to the aquarium water. The animals first showed signs of distress, gulping air at the surface; then the scales and skin began to slough off, and the animals died.

Effects on Isolated Tissues and Organs of Higher Animals. Various excitatory and inhibitory effects have been described for isolated tissues and organs. All cannot be discussed in detail, but a few examples will be presented.

Isolated Heart. Numerous investigations have described stimulation, inhibition, and arrythmias of cold-blooded hearts, produced by photodynamic action (Amsler and Pick, 1918; Viale, 1921; Wastl, 1926; Cuzin, 1930c; Osawa, 1934; Djourno and Piffault, 1936). There seems to be nothing to distinguish between these effects and those produced by other destructive agents. All observations do not agree, and it is possible that different effects are to be obtained with different dyes, different concentrations of dye and different light intensities. Differences in behavior of cold-blooded hearts under apparently similar conditions are not uncommon. Wastl found that the arrythmias which she produced in this way were prevented by introducing serum into the perfusing fluid (see p. 91).

Supniewski (1927b) found that perfused strips of mammalian heart showed coronary contraction followed by rapid decrease in activity.

Smooth Muscle. Adler (1919) and Kolm and Pick (1920) found that the smooth muscle of frog's intestine could be stimulated photodynamically, and Supniewski (1927b) found that smooth muscle of various organs of frogs and mammals, both excised and *in situ,* was made to contract by exposure to light after sensitization with hematoporphyrin.

Skeletal Muscle. Skeletal muscle may be stimulated to contraction by photodynamic action. Lippay (1929) described two types of response: a series of rapid twitches, and slow, prolonged contracture. He concluded that the first type is due to the stimulation of nerve or nerve endings, and that the second is due to direct stimulation of the muscle fibers. He found (1929) that the first, but not the second type of response occurred in curarized muscle. However, he discovered later (Lippay and Wechsler, 1931) that curare precipitated the dye, and that tetramethylammonium bromide, which like curare prevents the transmission of the impulse from nerve to muscle but does not precipitate the dye, did not abolish the twitch response. He presents other evidence (1931) to indicate the difference between the two types of response, but the argument does not seem altogether conclusive. It should be recalled that chemical stimulation, *e.g.*, by the calcium precipitants, produces both twitch and contracture responses, and that the photodynamic response seems in all ways comparable to such stimulation. Figure 26 (p. 109) shows the type of response obtained by photodynamic stimulation of skeletal muscle. The photodynamic response of skeletal muscles ceases in the absence of O_2 (Lippay, 1930; Spealman and Blum, 1933, Kosman and Lillie, 1935). Serum also inhibits the reaction by combining with the dye, Lippay (1930) having pointed out that when serum is present the muscle is not stained by the dye, and is not stimulated by light. Lillie, Hinrichs and Kosman (1935) have studied the effect of various ions on this phenomenon.

Nerve. Lillie (1924) reports photodynamic stimulation of frog's nerve, and Auger and Fessard (1933) have accomplished this with crustacean nerve.

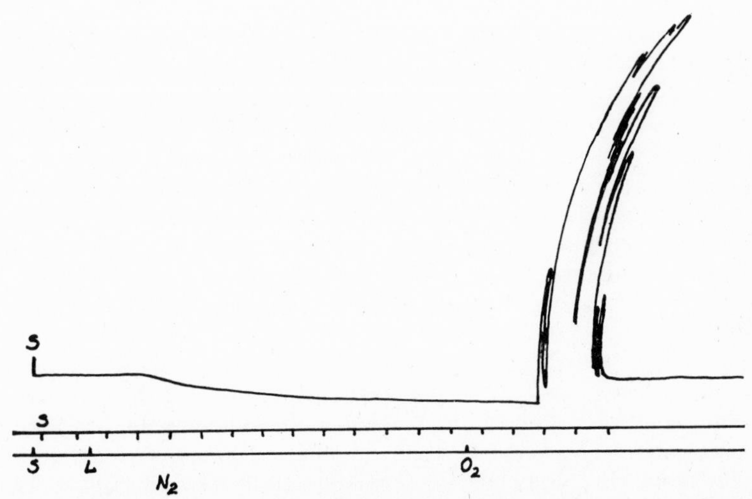

FIGURE 26. Response of a frog sartorius muscle sensitized with eosin and exposed to light from a 150-watt tungsten filament lamp. Tracing from a smoked drum record. Upper line, lever attached to muscle; middle line, time marker, minutes; lower line, signal. S, simultaneous ordinates.

The muscle was in nitrogen when the light was turned on at L. No response occurs until after the nitrogen is replaced with oxygen at O_2. In air or O_2 the response appears about two minutes after turning on the light. Contraction of muscle resulting from other stimuli is not inhibited by absence of O_2.

Ciliated Epithelium. Jacobson (1901) showed that the beating of the cilia of the membrane of the frog's esophagus is inhibited by the action of light and photodynamic dyes. Rocha e Silva (1935) has analyzed the effect of eosin and light on the bronchial cilia of the mussel. First, the beating of the cilia is depressed, and their rhythm is altered, finally the cilia become detached. Metzner (1921) observed that the cilia of paramecia become detached under similar conditions.

Mesentery. Stasis of the capillaries of the mesentery of frogs and of mice results from photodynamic action, according to Campbell and Hill (1924). The leucocytes collect and form thrombi without the formation of a clot.

Kidney. Singer (1936) describes changes in the kidney of the frog. Contraction of glomerular capillaries follows exposure of this tissue to light when the animal has been injected with photodynamic substances.

Tissue Metabolism. Wohlgemuth and Szörényi (1933a, b) studied photodynamic action on various mammalian tissues, using the Warburg technique for tissue slices. Some of the effects on O_2 uptake have been discussed elsewhere (pp. 65 and 97). As a rule, O_2 consumption and anaerobic glycolysis of tissue slices were decreased after irradiation of the slices. Tissue slices taken from animals killed immediately after having been subjected to irradiation of the whole body were found, on the other hand, to show increased O_2 uptake. Wohlgemuth and Szörényi (1933c) found that a comparable increase in O_2 uptake was produced by histamine. They suggest that this substance is elaborated during irradiation (apparently either by ultraviolet radiation or by visible radiation following the injection of photodynamic dyes), and is responsible for the stimulation of O_2 metabolism after irradiation ceases.

Skin. Barnes and Golubock have recently (1938) described changes in bioelectric potential across frog's skin treated with eosin and exposed to light, but in a later paper (Barnes, 1939) reports that light is not essential for this response. Kosman (1938) found that the irritability of frog skin, as manifested by contact simulus, is increased by photodynamic action.

Red blood cells. Photodynamic hemolysis has been discussed at length. Another change in these cells, which may accompany or even precede hemolysis, is the formation of methemoglobin by the oxidation of hemoglobin (Hasselbalch, 1909; Wohlgemuth and Szörényi, 1933b). Hemolysis is not related to methemoglobin formation (Blum, 1930b).

Effects on Larger Invertebrates. A number of effects on larger invertebrate animals have been described, and the list could no doubt be extended indefinitely. Hertel (1906) found that the atropinized retractor of the esophagus of *Sipunculus niger* could be stimulated by light if the animal was injected with eosin. Hinrichs (1924) studied the effect of light on hydra, planaria, and certain annelides as well as protozoa. Later (1926), she studied such effects in the echinoderm, *Arbacia,* at different stages of development. Her general finding was that the most active parts were the first to be injured, thus indicating the existence of an "axial gradient." Wide variety of effect is to be expected, depending upon the animal studied, the dye employed, etc. Welsh (1934) studied photodynamic action on the tentacles of a terebellid worm.

Effects on Plants

Effects on unicellular plant forms have been mentioned. Among those forms which are claimed by both the zoologist and the botanist, Gicklhorn (1914) and Jírovec and Vácha (1934) both found chlorophyll-bearing forms to be less susceptible to photodynamic damage than closely related forms which did not contain chlorophyll. Gicklhorn compared species of *Hydra, Paramecium,* and *Stentor;* Jírovec and Vácha compared species of *Euglena.* In addition, Gicklhorn obtained similar results when he compared etiolated and chlorophyll-bearing leaves of maize and of *Phaseolus multiforis.* This effect may be due to the action of chlorophyll as a light filter for the wave-lengths absorbed by the photosensitizer.

Gicklhorn also found that cells having a well developed cell wall were less susceptible than cells directly exposed to the sensitizer, another indication of the importance of contact of the dye with the cell substance.

Among larger plants, Gicklhorn (1914) found that leaves of cut shoots, which had been immersed in water containing photodynamic dyes, were injured by light; and Niethammer (1925) found that the breaking of dormancy was accelerated in plant stems treated with photodynamic dyes and exposed to light. Sellei (1935) found that growing plants may be either stimulated or inhibited by photodynamic action.* Niethammer also found that the percentage germination of seeds may be increased by soaking in solutions of photodynamic dyes followed by exposure to light.

Rebello (1920a, b) studied the inhibition of hyacinth roots by eosin and light, and Piskernic (1921) found that sprouting roots from seeds might be damaged by photodynamic action. Prescher (1932) made quantitative studies of root growth under such conditions, and found that inhibition was accompanied by suppression of cell division, and by the appearance of anomalies in mitosis. Metzner (1923) and Blum and Scott (1933) demonstrated that oriented bending of the roots of the oat or wheat seedling could be induced by photodynamic action; this is discussed elsewhere (p. 121).

Gicklhorn (1914) found that, in *Elodea, Vallisneria* and *Nitella,* photoplasmic streaming was first accelerated and then slowed by the action of photodynamic dyes and light.

Photodynamic Action in Biological Techniques

Since a number of the photodynamic dyes are commonly used as stains, and elsewhere in biological techniques, a brief note may be introduced regarding effects which may cause experimental errors or misinterpretations.

Action on Bacterial Media. Fruitman (1935), and Liebert and Kaper (1937) have called attention to the fact that agar media containing photodynamic dyes (Salle's medium) undergo changes when exposed to light, which cause inhibition of bacterial growth when the plates are subsequently inoculated. The latter investigators find that the inhibition is due to the production of hydrogen-peroxide (see p. 95). Mettler (1905) made extensive studies on the bacteriostatic action of agar and of gelatin media containing fluorescein dyes, when these are exposed to light. It is important to protect such media from light both before inoculating with bacteria and during incubation.

* This effect is undergoing extensive experiment at present, and judgment as to its importance must be withheld pending the results.

Ward (1928) reports that *Bacillus influenzae* grown on blood agar plates containing hematin shows no hemolysis surrounding the colonies after incubation, but if the plates are exposed to light a zone of hemolysis develops around each colony. This is probably due to the formation of a photodynamic sensitizer by the bacteria, perhaps a porphyrin arising from the hematin. Mallinckrodt-Haupt (1938) describes similar effects.

Vital Staining. Some vital stains, *e.g.*, methylene blue and neutral red, are strongly photodynamic, and their use in strong light, as on the microscope stage, may result in changes within the cell. Politzer (1924) found that neutral red has marked effects on cell division in salamander larva, and that these effects are enhanced by light. Efimoff and Efimoff (1925) found that many of the common vital stains cause death of the cell when exposed to strong light. The effects described by Haberlandt (1928) are probably of the same nature.

Methylene Blue. Methylene blue has been employed in a great number of studies on biological systems. It acts as a convenient hydrogen acceptor, and is useful in oxidation-reduction potential studies. Although it is generally recognized that the dye is photochemically active, and in some cases this property is taken advantage of, some of the effects reported may be due to the unrecognized action of light. The bleaching of the dye is generally accepted as an indication of the reversible reduction of the dye to the leucobase, but in appropriate systems the dye may be irreversibly bleached by oxidation to colorless end products due to the action of light (see p. 67).

Sterilization. Sterilization of water by photodynamic action has been considered by Welch and Perkins (1931), and it is possible that its use might be extended. This requires careful study, however, as regards appropriate dyes and sources.

Effects Similar to Photodynamic Action

Certain effects which resemble photodynamic action, but may have no relation to this phenomenon, have been reported and should be mentioned here. Clementi and Condorelli (1930) apply the term "falsa-azione fotodynamicii" to what appears to be merely an addition of effects. These authors found that the hemolysis caused by ethyl alcohol or sodium cholate is more intense if the system is exposed to sunlight, but since red blood cells which have not been treated with photosensitizing substances show some hemolysis if exposed to the sunlight for several hours, they conclude that this is an addition of effects similar to that obtained when the two lysins are employed together. This is not a true photodynamic action since the lysin is not acting as a photosensitizer. As a matter of fact, it will be pointed out in Chapter 11 that the hemolysis of unsensitized red blood cells by sunlight is probably an instance of photodynamic action; but this does not detract from the value of Clementi and Condorelli's argument, since the alcohol or cholate is not acting as a photodynamic sensitizer. It is possible that other effects which have been accepted as photodynamic action are of this type.

Other Types of Photosensitization. The categorical statement that all photosensitization of biological systems is photodynamic action, *i.e.*, photosensitized oxidation by O_2, may not be justified. Certainly, a large group of biological photosensitizations is of this type, whereas other types have not been clearly demonstrated. The possibility of different types of photosensitization must not be disregarded, however.

Oxidations may be photosensitized by salts of some of the heavy metals, *e.g.*, iron and uranium. However, Neuberg and Galambos (1914) found that very few of the same oxidations were brought about by the photodynamic dyestuffs. Thus it is questionable if there is a direct relationship between such photosensitization and the phenomenon which is described herein as photodynamic action.

Attempts to demonstrate effects similar to photodynamic action with such metals have varied in their outcome. Neither Noack (1920) nor Viale (1920) could produce such effects, but Henri and Henri (1912) found that colloidal selenium photosensitized paramecia if taken up in the vacuole. Löhner (1927) claims to have obtained hemolysis with arsenic oxide and light, although this substance alone inhibited hemolysis by other lysins. The writer, on the other hand, could not produce hemolysis with uranium compounds and light. Niethammer (1925) reports that uranium and iron salts act similarly to photodynamic dyes in accelerating the breaking of dormancy of plant stems. It is possible that the photoactive metals may produce photodynamic action when they are able to combine with or be taken into close association with the cell, as the finding of Henri and Henri indicates.

Chapter 11

Photodynamic Action and Other Photobiological Processes

Properly understood, photodynamic action may serve as a model process to help in the solution of some of the problems of general photobiology. Its importance in photopathology will become apparent in the following pages. But the hope that it might explain an important number of photobiological processes, which was entertained at the time of its discovery, is not justified by the evidence since accumulated. On the whole, it seems that photodynamic action is fundamentally a destructive process, which has been avoided in the evolution of living systems. Those wave-lengths shorter than 3300 Å which produce destructive effects in living organisms comprise a very small part of total sunlight, and the organism can rather readily find protection against them; but an organism which possessed a pigment capable of sensitizing it to longer wave-lengths could not easily survive except under quite restricted conditions. It is of interest in this respect that the photodynamic porphyrin compounds are closely related to, and derivable from, many important biological substances widely distributed throughout both kingdoms of living organisms (see Chapter 19); but they are seldom found free where they might act as photosensitizers (see p. 236). It is possible that some of the lethal effects produced by intense light from the visible spectrum are due to photodynamic action in which very slight quantities of these substances are the photosensitizers.

Distinction Between the Effects of Short Ultraviolet Radiation and Photodynamic Action. Relationship between photodynamic action and the effects of ultraviolet radiation has been sought more frequently than any other. As stated in Chapter 3, wave-lengths shorter than 3300 Å are generally destructive to living systems because of photolabile substances in the cell which absorb characteristically in this spectral region. The effects of these radiations are often very similar to those of photodynamic action, but such resemblances are misleading. For example, hemolysis is produced by both ultraviolet radiation and by photodynamic action, but this cannot be taken as evidence of the identity of the two processes, since hemolysis is caused in many different ways. Other apparent resemblances disappear on careful analysis.

An important point of distinction between photodynamic action and the destructive effects of ultraviolet radiation is that while the former take place only in the presence of O_2, the latter are independent of O_2.

Table 10. Effects of Ultraviolet Radiation (Wave-Lengths Shorter than 3300Å) in Presence and Absence of O_2.

System	Effect: In presence of O_2	Effect: In absence of O_2	Conditions	Investigators
Bacteria	Killed	Killed	H_2	Bie * (1905)
	Killed	Killed	H_2	Thiele and Wolf (1906a, b)
Erythrocytes	Hemolyzed	Hemolyzed	Vacuum	Hasselbalch (1909)
	Hemolyzed	Hemolyzed	?	Eidinow (1930b)
Enzymes (invertase)	Inactivated	Inactivated	H_2, CO_2, N_2	Jodlbauer and Tappeiner (1906c)
"Peroxidase"	Inactivated	Inactivated	Vacuum	Jamada and Jodlbauer (1907)
Catalase	Inactivated	Inactivated	Vacuum	Zeller and Jodlbauer (1907)
Skeletal muscle (frog)	Contracture	Contracture	N_2 vacuum	Spealman and Blum (1933)
Whole frogs	Scratching and rubbing	Scratching and rubbing	N_2	Blum and Spealman (1934b)
Human skin	Sunburn	Sunburn	Circulation occluded	Blum, Watrous and West (1935)

* See p. 117.

A comparison of Table 10, where evidence that ultraviolet effects are independent of O_2 is assembled, with Table 6 (p. 66) should serve to convince a reasonable skeptic. Some examples merit discussion at greater length.

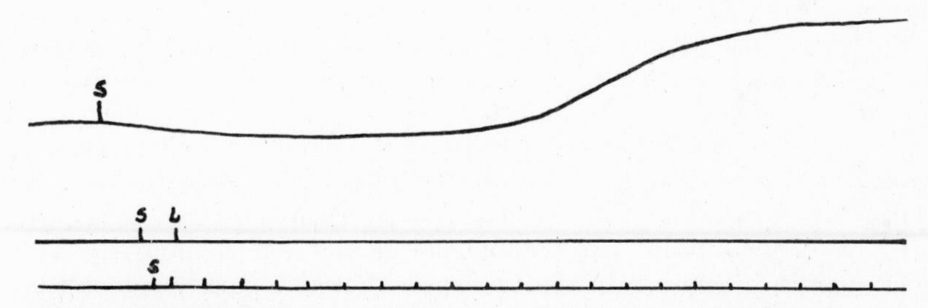

FIGURE 27. Response of frog's sartorius muscle when exposed to mercury-arc radiation. Tracing from a smoked drum record. Upper line, lever attached to muscle; middle line, signal; lower line, time, minutes. S, simultaneous ordinates. Irradiation begins at L.

This type of response is produced only by wave-lengths shorter than about 3200 Å. It occurs in either O_2 or N_2. A muscle photosensitized with eosin gives the same type of response to such radiation if O_2 is excluded, but the type of response illustrated in Figure 26 (p. 109) when O_2 is present.

Excitation of Skeletal Muscle. Lippay (1929) believed that he could demonstrate two types of response in frog skeletal muscle photosensitized with eosin or hematoporphyrin: an irreversible contracture, and a series of twitches. Arguments against the separation of these two types of response have been presented in the preceding chapter (p. 108). He found also (1930) that muscle which had not been sensitized responded to mercury-arc radiation by a slow, irreversible contracture which he thought to be the same as that sometimes occurring in photosensitized muscles. The photodynamic and ultraviolet responses are illustrated in Figures 26 and 27, respectively. Spealman and Blum (1933) found the latter to be caused by wave-lengths shorter than about 3200 Å, so that it belongs to the type of effect produced by ultraviolet radiation in unsensitized systems. They also found that muscles sensitized with eosin respond to mercury-arc radiation with the twitching response typical of photodynamic action as long as O_2 is present; but if O_2 is removed, only the slow contracture typical of the ultraviolet response is obtained. Thus there can be no doubt that the ultraviolet response—a slow, irreversible contracture—is, like other ultraviolet effects, independent of O_2, whereas the photodynamic response—a reversible series of twitches is dependent upon O_2. That the photodynamic response is entirely different from that produced by radiation shorter than 3200 Å is clearly shown in this instance.

The fact that other excitation phenomena are produced both by ultraviolet radiation and by photodynamic action is not evidence of similarity in the manner of action. For example, artificial parthenogenesis of echinoderm eggs may be accomplished by either ultraviolet radiation (Hollaender, 1938), or photodynamic action (Lillie and Hinrichs, 1923); but the same effect is elicited by various chemical or mechanical means as well.

Sunburn. The assumption that the skin changes produced by photodynamic action have basically the same mechanism as those produced by ultraviolet radiation (ordinary sunburn), has added much to the existing confusion of ideas about diseases produced by light in man and domestic animals. The outward signs of these two effects are somewhat similar: first erythema (with edema if the reaction is severe enough), followed by pigmentation. As pointed out in the last chapter, the histological picture is similar in the two cases, but the experiment of Blum, Watrous and West (1935) indicates a fundamental difference. These investigators found that depriving the skin of the forearm of O_2 by means of a sphygmomanometer cuff was sufficient to prevent whealing on a photosensitized area exposed to sunlight (see p. 104), but had no apparent effect on the erythema and pigmentation of sunburn. A further discussion of sunburn mechanism will be presented in Chapter 17.

Action of Radiation on Non-cellular Biological Systems. Ultraviolet radiation may bring about the uptake of O_2 by serum in the absence of a photosensitizer (*e.g.*, see Harris, 1926a, b, c). It is thus possible that the changes produced by ultraviolet radiation and by photodynamic action are sometimes essentially the same when one is dealing with noncellular systems, *i.e.*, both may be oxidations by O_2, for it would be unwise to state too rigidly that all effects of ultraviolet radiation which are produced in biological systems are independent of molecular oxygen.

However, the view that at least the great majority of effects on cellular systems which are produced by ultraviolet radiation are independent of O_2, and therefore basically different from photodynamic effects, is borne out by the fact that no exceptions to this rule have been demonstrated.

Photodynamic Action and the Destructive Effects of Longer Wave-lengths. *Killing of Bacteria.* About the turn of the century, there was considerable debate as to whether O_2 is necessary for the killing of bacteria by light. Bie (1905) undertook to examine this question and carried out a long series of careful studies. Using a carbon arc as source, he discovered that when the radiation passed through quartz the effectiveness in killing bacteria was nearly as great in an atmosphere of hydrogen as in air; but when the radiation passed through glass, the killing effect was much reduced in the absence of O_2; when the radiation passed through glass and a layer of agar medium, the presence of O_2 became still more important. Bie thought that the more "chemical rays" (see p. 29) present, the more necessary was O_2.

It seems safe to conclude that Bie studied a mixed effect made up of: (1) the destructive action of wave-lengths shorter than 3200 Å which pass through quartz, but not through glass to any great extent, and which are independent of O_2; and (2) an effect produced by longer wave-lengths which was dependent upon O_2. When the radiation passed through quartz, the predominant effect was due to (1); when it passed through glass (1) was eliminated or greatly reduced and (2) predominated.

Thiele and Wolf (1906a, b) made a more definite separation of the two effects, showing that the major killing effect was due to radiation which did not pass through window glass and was independent of O_2. Longer wave-lengths killed the bacteria, but this effect was much less intense, and required the presence of O_2. Realization of the existence of these two effects helps to explain many of the conflicting results which have been reported when mixed light was used.

It is possible that the lethal effect of longer wave-lengths, which is dependent upon the presence of O_2, is an instance of photodynamic action, in which some substance present in small quantity in the organism acts as the photosensitizer. Destructive effects of intense visible light on lower organisms are often reported, and it seems probable that these are instances of photodynamic action. Living systems contain substances which may act as photosensitizers, *e.g.*, porphyrins and lactoflavin, whose photodynamic action might be negligible with ordinary illumination, but destructive under intense light.

Inactivation of Enzymes. Jodlbauer and Tappeiner (1906b) found that inactivation of "invertase" by sunlight passing through glass required O_2, but later (1906c) found that inactivation by shorter wave-lengths which did not pass through glass took place in the absence of O_2. Jamada and Jodlbauer (1907) found a comparable condition when they irradiated "peroxidase." These enzyme systems may have contained impurities which acted as photosensitizers when O_2 was present.

Hemolysis. Hemolysis by wave-lengths shorter than 3300 Å takes place independently of O_2 (Hasselbalch, 1909; Eidinow, 1930b), but numerous investigators have shown that long exposure of red blood cells suspended in saline to the longer wave-lengths of sunlight also results in hemolysis (see Hausmann and Loewy, 1926). The latter is a slow process, so slow that it does not enter, as a rule, into quantitative experiments on photodynamic hemolysis, which are designed to take place within much shorter times. Lepeschkin and Davis (1933) studied the action spectrum, finding that it agreed rather well with the absorption spectrum of hemoglobin, and concluding that that substance is the light absorber. However, the action spectrum would agree quite well with the absorption spectra of porphyrins (see Figure 45, p. 216), and it has been shown that red blood cells contain small quantities of protoporphyrin (Van den Berg, Grotepass and Revers, 1932), so that it is possible that the hemolysis results from photodynamic action in which the small quantity of protoporphyrin in the cells acts as the photosensitizer, rather than from hemoglobin which is not known to display photosensitizing properties. I have recently found that this effect is prevented by the removal of O_2.

From the foregoing examples it seems probable that photodynamic action due to photosensitizing substances contained in normal living systems is not uncommon when these systems are exposed to intense light. This demands further investigation, less perhaps for its intrinsic importance under normal conditions of illumination than because of its possible confusion with other effects.

Vitamins.—*Vitamin D.* Antirachitic substances are formed by the irradiation of ergosterol and closely related sterols. The antirachitic substance found in fish oils (vitamin D) is identical with the product of irradiation of 7-dehydrocholesterol. The wave-lengths active in producing these antirachitic substances lie in the region of 3000 Å, in general agreement with the absorption spectra of the precursor compounds. Numerous investigators have been led to believe that the same reaction might be brought about by longer wave-lengths through the action of photodynamic dyes.

György and Gottlieb (1923) and Pilling (1924) claimed that the time required for the cure of rickets in children by mercury-arc therapy could be reduced by the administration of eosin before irradiation. Lakschewitz (1926) could not confirm this, however, in a series of careful studies. Strauch (1930) sensitized two rachitic children to light with hematoporphyrin, but he does not state that the treatment had any effect upon the rickets.

Van Leersum (1924) injected rachitic rats with hematoporphyrin, and reported indications of healing of the rickets. The rats were kept in a poorly lighted basement or sometimes in a cupboard, but nevertheless van Leersum believed that the small amount of light reaching the animals might have been responsible for the apparent cure, due to the photosensitizing activity of the hematoporphyrin. Van Leersum's evidence for the cure of rickets was based on the generally accepted "line test." In rachitic rats the calcification of the ends of the long bones either fails to develop or disappears, leaving a wide band of cartilagenous tissue between the epiphysis and diaphysis. If the rachitic animals are treated with vitamin D, a thin line of calcification appears in this cartilagenous tissue which is taken as the first sign of healing rickets. Marique points out (1938) that this line may also appear in rats which have remained on rickets-producing diet for a long time, being an indication of fatal termination of the disease rather than of healing, and he believes that it was this which van Leersum interpreted as healing rickets. Marique repeated van Leersum's experiments, but could find no difference between rachitic rats which received injections of hematoporphyrin, and rachitic rats which received none. The rats were exposed to diffuse light in the laboratory. Those which received hematoporphyrin showed definite signs of photosensitization.

Repling (1929, 1930) could not demonstrate curing of rickets by injections of hematoporphyrin or eosin, together with exposure to mercury-arc radiations through window glass or quinine bisulfate solution. His experiments are open to the objection that the mercury arc may not supply enough energy in wave-lengths which are strongly absorbed by hematoporphyrin, (see p. 219). Szczygiel and Clark (1935) injected eosin into rachitic rats and irradiated them with diffuse daylight, but obtained no evidence of healing rickets.

Activation of ergosterol by photosensitization *in vitro* has also been attempted. Windaus and Brunken at one time (1928) attributed antirachitic activity to the peroxide of ergosterol which they obtained by irradiating this substance together with eosin in the presence of O_2. However, Holtz (1928) states that while this was thought at first to be true, subsequent experiments in the Göttingen laboratories showed that the substance thus obtained did not have antirachitic properties. The ergopinacol, obtained by Windaus and Borgeaud (1928) by irradiating ergosterol together with eosin in the absence of O_2, was found to lack antirachitic properties. Meyer (1935) cites conflicting results when ergosterol was irradiated in the presence of O_2 and eosin or chlorophyll; some tests indicated antirachitic action, others none. Repling (1929, 1930) exposed linseed oil together with chlorophyll to mercury-arc radiation through window glass, and found the product was not antirachitic.

While some of the evidence suggests an antirachitic effect of photodynamic action, the balance is certainly on the negative side. Since there is no theoretical reason to expect such an effect, one must take the point of view that such action does not exist, unless it can be demonstrated by clearcut experiment; and thus far those experiments which seem to have been most carefully carried out have yielded negative results.

Other Vitamins. Lactoflavine, identified with the vitamin B complex and with Warburg's yellow respiratory pigment, acts as a photodynamic sensitizer. Blum and Kuen (unpublished) found that lactoflavine produces hemolysis of red blood cells in the presence of O_2; and various other photosensitizing activities have been recorded (Heiman, 1936).

The case of destruction of ascorbic acid (Vitamin C) in milk is an interesting example of the photosensitizing activity of lactoflavine. It was first shown by Mattick and Kon (1933) that vitamin C disappears from milk which is exposed to sunlight. Later Kon and Watson (1936) found that O_2 is necessary for this reaction, and that it is produced by wave-lengths in the short ultraviolet, violet and blue. Martini showed (1934) that lactoflavine brings about the oxidation of ascorbic acid. Hand, Guthrie and Sharp (1938) reasoned that since neither O_2 or ascorbic acid absorbs in the blue-violet region of the spectrum a sensitizer must be acting, and that lactoflavine which absorbs in this region and in the short ultraviolet as well would fit these requirements. These investigators destroyed both the lactoflavine and ascorbic acid in milk by long exposure to light. They then added ascorbic acid and found that this was not destroyed by further irradiation. They also found that the ascorbic acid in mare's milk, which contains little or no lactoflavine, is not appreciably destroyed by exposure to light. These experiments leave little doubt that lactoflavine acts as a photosensitizer for the oxidation of ascorbic acid in milk.

That this effect is an important one is shown by the observation of Kon and Watson (1936) that a pint of milk exposed to sunlight for one-half hour on a doorstep loses fully half its antiscorbutic property. It should certainly be taken into account where the attempt is made to increase the antirachitic activity of milk by irradiation.

The experiments of Buruiană (1937) indicate that photosensitized effects in milk are somewhat complex. Fats as well as ascorbic acid are oxidized, and there may be competition between the two reactions. Oxidation both by O_2 and by dehydrogenation may take place. Hand, Guthrie and Sharp (1938) found that removing O_2 from milk prevents the development of a typical flavor which is detected in milk exposed to sunlight in the presence of O_2. In the absence of O_2, however, another kind of flavor is produced. Perhaps lactoflavine acts in the dual manner characteristic of the photodynamic substances, to bring about either oxidation by O_2 or dehydrogenation, according to the conditions (see p. 71).

Hopkins (1938) has shown that lumichrome, an oxidation product of lactoflavine, is itself a photosensitizer, though somewhat less effective than the parent substance. Solutions of lactoflavine may preserve their photosensitizing activity after being decolorized by exposure to light, apparently because of the photosensitizing activity of oxidation products which are colorless but absorb in the ultraviolet. Hopkins found glutathione to inhibit the oxidation of ascorbic acid; apparently this is an example of the competition between oxidizable substances in photosensitized systems.

It is interesting that lactoflavin, itself a vitamin (B_2), acts as a photosensitizer to destroy another vitamin (C). Lactoflavine itself is destroyed by light, although probably not to a great extent so long as readily oxidizable substrate is present. It is doubtful that the photosensitizing properties of lactoflavine play any important part in the economy of living systems.

Vision and Photoörientation. It does not seem probable, *a priori*, that photodynamic action could serve as the basis for the process of vision. Such drastic irreversible changes are hardly compatible with the delicate mechanism found in that process. The bleaching of visual purple, which probably forms the basis for the photosensory process in all vertebrate animals, goes on independently of O_2 (Brunner and Kleinau, 1936), indicating that the process is not an example of photodynamic action.

Actually, orientations in light fields have been induced by photosensitization of organisms by dyes, which appear superficially analogous to the photoörientations (phototropisms and phototaxes) which can be demonstrated in many normal organisms. A striking example of this is the phototropism of plant roots induced by fluorescein dyes, which was studied by Metzner (1923), and by Blum and Scott (1933). The roots of the oat or wheat seedling growing in aqueous medium are ordinarily insensitive to light. If a small concentration of a fluorescein dye is added to the medium, the roots orient themselves and grow toward the light. This may be described as a positive phototropism, roughly analogous to the negative phototropism of the roots of the mustard seedling. Metzner 1919, 1921) has also described photoöriented movements, both positive and negative, when paramecia or spirilla are placed in media containing a small percentage of a photodynamic dye. These phenomena probably have little direct relationship to the photoörientations of normal organisms.

If light strikes a photosensitive organism asymmetrically, resulting in an asymmetrical effect on locomotion, the organism must tend to orient itself with respect to the light field in such a way that it will receive equal effective stimulus on both sides of the axis of symmetry. The organism need not be structurally symmetrical, but a functional axis may be assumed for the purpose of analysis. This was the thesis of Jaques Loeb, the consequences of which have been analyzed by Blum (1935b) who showed that such a mechanism may fit the different types of oriented movement found among animals when subjected to differently arranged light fields.

The same analysis must apply if local effects are produced by photodynamic action on one side of an organism, which in any way effect locomotion or growth of that side. In this case the organism must tend to orient in the light field so that the same amount of injury will be produced on both sides of the axis of symmetry. Such an explanation is easily applied to the induced phototropism of the plant root, for if growth on the side of the root toward the light be inhibited by photodynamic action, the growing root must bend toward the light source. This mechanism seems to be adequate to account for the phenomenon, and may bear little relationship to the normal phototropism of plants.

Similarly, if photodynamic action affects one area of a paramecium, so that movement of the cilia on that region is diminished, the normal movement of the cilia on the remainder of the body will tend to direct the forward end toward the light source if they are beating so as to move the organism in a forward direction. Only when all symmetrically effective areas receive equal illumination will the organism be free from orienting effect. Thus the organism must tend to move toward the light even though asymmetry of shape and movement, such as is characteristic of paramecia, cause it to move in what appears to be a random path. This explanation would seem to account for the positive photoörientation which Metzner observed. Orientation away from the light source would result, in an analogous fashion, if photodynamic action brought about a stimulation of ciliary movement on the area illuminated, and this will account for the negative orientation which Metzner describes as occurring under certain conditions. While affording an opportunity for analysis of the photoöriented movement in the case of paramecium, which may be analogous to orientation of protozoa to ultraviolet radiation shorter than 3300 Å (see Black, 1935), these photodynamically induced orientations probably have very little relationship to the true photoörientation found in organisms in general. Orientation is more complicated in the case of organisms in which the sensory and reflex processes are more complex. The writer has attempted to induce photoörientation in higher organisms by photosensitization, but without success.

Photosynthesis. It is not within the scope of this book to discuss the problem of photosynthesis by plants, nor the relationships between the chemistry of this process and the photosensitized oxidations discussed herein. References can be made to only a few papers which are concerned with the latter problem: Noack (1920, 1925, 1927), Kautsky (a series of articles in *Biochem. Ztschr.,* 1934-1936), Gaffron (1936b), Franck and Wood (1936), and Franck and Hertzfeld (1937a, b).

Hypersensitivity to Light in Normal Animals. Zielinska (1913) found the earthworm *Eisenia foetida,* and another of this genus, to be very sensitive to light. They became thin and elongated when exposed to light, eventually breaking up into segments. Members of other genera of earthworms, *Lumbricus terrestris* and *Helodrilus caliginosus,* were not found to be photosensitive. Zielinska extracted from *Eisenia,* an alcohol-soluble substance absorbing in the blue and ultraviolet, and displaying a bluish-green fluorescence. This was suspected of being a photodynamic sensitizer responsible for the effects of light on the animal. She also extracted a dark-brown pigment from the tegument of the earthworm, which Marschlewski thought to be a porphyrin, and Hausmann (1916) showed to be a photosensitizer for red blood cells. There has been considerable dispute as to whether porphyrins are really to be found in the tegument of earthworms, but Dhéré (1932) has recently demonstrated their presence in *Lumbricus terrestris.* He believes that the porphyrin is in colloidal form in the tissues of the animal, where it does not fluoresce and does not sensitize to light although it is an active photosensitizer when extracted from the animal.

Turner (1937) has called to attention the interesting fact that some rodents, the fox-squirrel *Sciurus niger* in particular, have a relatively high concentration of porphyrin in their blood, so much, in fact, that their bones are colored red by deposits of this material. Turner* could not show, however, that these animals are sensitive to light.

Rattlesnakes and other snakes are reputed to be killed by sunlight, but the experiments of Blum and Spealman (1933b) indicate that excessive heat is the cause of death in such instances.

Biological and Pharmacological Action of Photodynamic Substances. Some substances which have a specific pharmacological or biological action are also photodynamic, *e.g.,* quinine and lactoflavine; and it has been sometimes assumed that in such cases pharmacological or biological action may be increased by the action of light. The fact that a molecule can be activated by light in such a way that it brings about oxidation of cell components by molecular oxygen does not necessarily imply that such activation will accelerate other reactions in which the molecule may participate.

An example of the independence of photodynamic from other biological properties of a substance is given by studies of Wohlgemuth and Szörényi (1933b). They found that hematoporphyrin increased the O_2 uptake of mammalian red blood cells in the dark when the cells were intact, but not after the cell structure had been destroyed by lysis. Apparently hematoporphyrin possesses the property of stimulating the normal red cell metabolism, a property dependent upon the cell structure. In the light, the uptake of O_2 is further increased, and one might believe that light simply increases the stimulating action of hematoporphyrin, were it not for the fact that the increase in O_2 consumption by light occurs in hemolyzed as well as in intact cells. Furthermore, the O_2 uptake in light is increased by cyanide (see p. 65).

Many dyes, notably methylene blue (see Barron and Hoffman, 1929), display the ability to stimulate the O_2 metabolism of the mammalian

* Personal communication.

red blood cells; and Michaelis and Salomon (1931) have shown that this ability depends upon the oxidation-reduction potential of the dye. There seems, on the other hand, to be no relationship between oxidation-reduction potential and photodynamic effectiveness. As Wohlgemuth and Szörényi (1933b) point out, rose bengal is about as effective photodynamically as is hematoporphyrin, although it should, from its chemical structure, have a low oxidation-reduction potential, and does not increase the O_2 metabolism of mammalian red blood cells.

Part III

Diseases Produced by Light in Domestic Animals

Chapter 12

Diseases of Domestic Animals: General

Historical Discussion. In the *Philosophical Transactions of the Royal Society of London* for the year 1776 the following title appears: "*An Account of a very extraordinary Effect of Lightning on a Bullock,* at Swanborow, *in the Parish of* Iford *near* Lewes; *in* Sussex. *In sundry Letters, from Mr.* James Lambert, *Landscape-Painter* at Lewes; and *One from* William Green, *Esquire,* at Lewes, *to* William Henly, *F.R.S.*" Mr. Lambert's first letter begins, "I shall now inform you of a very extraordinary and singular effect of lightning on a bullock in this neighborhood, which happened about a fortnight since. The bullock is pyed, white and red. The lightning, as supposed, stripped off all the white hair from his back, but left the red hair without the least injury. . . . ." The letters go on to give a detailed description of the animal which does not appear to have been much the worse for its remarkable experience outside the loss of hair and subsequent skin damage. Similar cases were recalled to have occurred in the neighborhood, in which lesions developed on the white spots of piebald animals.

The account ends with some suggestions as to the mode of action of the lightning by Dr. A. Fothergill of Northampton, among others: ". . . . The phlogiston, or inflammable principle, is thought to be the foundation of colour in bodies, and to abound in proportion to the intensity of the colour. But phlogiston and the electric fluid are probably the same, or at least modifications of the same principle; therefore, red bodies are perhaps replete wtih electric matter, while white bodies may be destitute of it. A body saturated with it cannot receive more, and may escape, while a neighboring body, not calculated to receive it, may, on its admission, be destroyed. Or there may exist a chemical affinity between electricity and the different rays of light. . . ." It seems to have been difficult to explain how lightning could remove the white hair, but leave the red, even as it would be today.

The fact that the bullock was not burned nor the hair singed might lead one to suspect that some other factor than lightning removed the white hair, and a skeptic reading the letters will find that no one actually saw the bullock struck. Mr. Tooth, the local bullock leach, testifies to having seen similar cases, but does not state that he actually saw the animals struck by lightning.

What really happened to the bullock seen by Mr. Lambert will never be known, but other descriptions which occur in the literature offer a clue. Over sixty years later, Tierarzt Erdt (1840) at Cöslin, in Pomerania, recounts the case of a black and white cow which was stripped of her white hair and most of the underlying skin, the black parts remaining entirely normal. In this case the animal was observed over a longer period of time, so that the condition was not discovered with the dramatic suddenness of the case in Great Britain; there is no mention of lightning. Shortly after this, Tierärzte Steiner (1843), Schrebe (1843) and Burmeister (1844) describe a similar condition epidemic among horses. All these accounts are limited to north Germany in the districts of Pomerania and East Prussia; the last three associate the conditions with feeding on vetch and lucerne plants heavily infested with a species of aphis accompanied by a fungus. The lesions were limited to the white-haired or bare portions of the skin and to the eyes, which suffered severe damage, according to Schrebe; colored skin was not affected. Certainly lightning was not the cause of these epidemics. Mr. Lambert's bullock may have been suffering from a similar condition.

The predilection for white areas of skin and evoidance of pigmented regions suggests that light may have been a factor in all these cases, since the pigment of the epidermis and hair should serve as a protection of the skin against this agent, and the damage would thus be limited to the unpigmented areas. Without further evidence this explanation might be as difficult to accept as those of Dr. Fothergill; but the existence among domestic animals of diseases which are produced by light, and which present pictures quite similar to those described by these early writers, is established beyond doubt today, and lends support to the contention. Symptoms similar to those described by the German writers have since been reported in animals pastured on vetch or lucerne, and evidence presented which suggests that light is a factor in such cases (see Chapter 15). The belief that lightning produces such symptoms persisted for a long time; as late as 1932 Hendrickx expressed chagrin upon learning that light is responsible for symptoms which he had assumed, and taught, were caused by lightning.

An interesting anonymous account of a disease in which light has since been shown to play a dominant role appears in the *Atti del Real Istituto d'Incorragiamento alle Scienze Naturali di Napoli*, Volume II, published in Naples in 1818.* The name of the writer does not appear in the account, and his information is quoted from an earlier work by Domenico Cirillo, and from two correspondents, Manni di Lecce, and Marinosci di Martina. The quotation from Cirillo's "Fundamenta Botonicae," which was published in 1787, is as follow: "Hypericum crispum: a quick poison of white sheep; so that all which graze in the Tarentine fields are black; moreover, the wool among the rest is not so good as in the time of the Romans. Possibly this plant was then more rare. Cattle, on the other hand, feed upon *Hypericum crispum* with no harm; but if while they are eating this plant they, in licking with the

*I am indebted to the late Dr. Walther Hausmann for a transcript of this rare article and of others used in this discussion.

tongue, moisten any part of their own body the skin is quickly deprived of hair. It is commonly, by the inhabitants, called Fumulo. It grows also in the fields of Sicily."*

The accounts credited to Manni de Lecce and Marinosci di Martina confirm but add little to what is stated by Cirillo. According to them, the disease is usually fatal; it is accompanied by intestinal disturbances, and the eyes are often affected so that the animals go blind. This again is a condition in which there is a predilection for white skin, but in this case the skin lesions are explained as the result of external contact with a plant. To account for the difference in susceptibility to the action of the poison it is asserted that dark-colored animals are more robust than white.

Plants of the genus *Hypericum*, commonly called St. Johnswort, are widely distributed. Marsh and Clawson (1930) quote early accounts from the United States, in which skin lesions resulted from feeding on *Hypericum*. In these accounts it is recognized that the plant must be eaten, and does not act merely by contact with the skin. Subsequent experiments which will be discussed more fully in the following chapter have proved that feeding upon plants of this genus renders animals sensitive to sunlight, producing the same symptoms as those observed by the Italian writers.

The fact that animals fed on buckwheat become susceptible to light seems to have been recognized at a rather early date. Fuchs (1843) makes a statement to this effect, and according to Mangold (1932), Hertwig publishes an account in his "Veterinary Medicine" of 1833. Haubner, in his textbook written in 1867, also recognizes the role of light in this disease, but Wedding (1887) seems to have been the first to demonstrate this experimentally. Later experiments will be discussed in Chapter 13.

Within the last few years a serious disease of sheep, occurring in South Africa and known locally as geeldikkop, has been shown to result from abnormal sensitivity to light, although differing in certain important respects from St. Johnswort and buckwheat poisonings (see Chapter 14). More recently, abnormal sensitivity to light has been demonstrated among animals which feed on an agave growing on ranges in the southwestern United States (see p. 159). Thus the number of clearly established diseases among domestic animals in which light is a precipitating factor is increasing, and there are a large number of conditions in which its role is suspected (see Chapter 15).

In all instances of true sensitivity to light, susceptibility is greatest on least pigmented skin, and when lesions are limited to light-colored areas, as in the case of Mr. Lambert's bullock, abnormal photosensitivity must be suspected. There is a widespread belief, however, that light-colored skin is more susceptible to all kinds of irritation, which seems to

* This is the translation from the original Latin given by Marsh and Clawson (1930).

have had its origin in early misinterpretation. The idea was suggested by Heusinger in 1846 and has come down to us through Charles Darwin's "Origin of Species."* Heusinger cites, among others, the early reports which have been mentioned above, to show that such widely different factors as lightning, poisonous diet, and sunlight affect light-colored skin in preference to dark. He does not seem to have suspected that all the effects he mentions may have been due to the action of light. Darwin apparently accepted Heusinger's viewpoint.

Experimental Demonstration of Photosensitivity. It is possible that diseases produced in domestic animals by light are more common than is ordinarily recognized, but, on the other hand, it is probable that abnormal sensitivity to light is frequently suspected where it does not exist. It is therefore necessary to have a clear understanding of the general principles of photosensitization before proper diagnosis can be made, and before progress in the the elucidation of the etiologies of these diseases or their treatment and control can be anticipated.

Domestic animals, like man, are subject to sunburn, *i.e.*, skin changes which result from exposure to wave-lengths shorter than 3300 Å. This mechanism has received considerable study in the case of man and seems to be similar to the killing of microörganisms by radiation of these wave-lengths; it will be dealt with at length in Chapter 17. This can hardly be considered an abnormal response, but rather a normal response to excessive exposure. White areas of skin should be more susceptible to sunburn than pigmented areas, and perhaps a good many of the reports of lesions confined to white areas are simply instances of sunburn. On the other hand, there are a number of conditions, including St. Johnswort poisoning, buckwheat poisoning and geeldikkop, which certainly have different etiologies. As will be shown in Part IV of this book, it should be possible to distinguish between sunburn and photodynamic effects on the basis of the wave-lengths which produce them. The sunburn radiation is not transmitted through window glass, but the radiation producing the photodynamic effects is, probably in all cases, of longer wave-length, and hence will pass through this medium. It is important, therefore, to expose animals which are suspected of being light-sensitive to sunlight which has passed through window glass. If they develop symptoms of photosensitivity they are undoubtedly abnormally sensitive; if not, they have probably been suffering from sunburn. The discussion of abnormal sensitivity to light in man, to which Part IV is devoted, may contain a number of points which bear on the problem of photosensitivity in domestic animals, although the problem is, on the whole, a different one and is therefore considered separately.

Busck (1905) seems to have been first to suggest the relationship between such diseases and photodynamic action, when he assumed that

* This concept must actually be much older. In the article quoted on page 128, which was published in 1818, the fact that white animals are more susceptible to St. Johnswort poisoning is credited to their being less hardy than dark animals.

buckwheat poisoning results from the ingestion of a photodynamic substance existing in the plant, which substance is taken from the digestive tract into the blood stream and thence to the skin, where it acts as a sensitizer in the same manner as eosin or any other photodynamic substance. Following this reasoning, he attempted to isolate a photosensitizing substance from the buckwheat plant, with results which will be described in Chapter 15. Ray (1914) was apparently the first to recognize the relationship between St. Johnswort poisoning and photodynamic action. The more complex origin of the photodynamic substance in geeldikkop will be discussed in the chapter on that disease.

While there seems little doubt that St. Johnsworth poisoning, and geeldikkop are examples of photodynamic action, it must be admitted that certain steps in the proof are lacking in both cases. Critical demonstration of such an etiological factor must follow somewhat different lines from, for example, the demonstration of the etiology of a disease caused by a microörganism. A few simple postulates may be formulated which should be shown to be fulfilled before a given disease may be attributed with certainty to photodynamic action (Blum, 1938).

(1) The symptoms of light sensitivity must be elicited by exposure of the animals to sunlight, preferably to sunlight through window glass.

(2) A photodynamic substance must be isolated in pure form, which will produce the symptoms if injected into the experimental animals only when followed by exposure to light.

(3) It must be demonstrated that the wave-lengths which produce the sensitivity in postulates (1) and (2) are identical. The study of the absorption spectrum of the photosensitizing substance isolated in the development of postulate (2) should be of assistance in establishing postulate (3), since the action spectrum for the production of the symptoms of photosensitivity should resemble the absorption spectrum of the substance (see Chapter 5.)

Sunlight is specified in the first postulate because this is the radiation to which animals are exposed in the fields; artificial sources may be weak in the wave-lengths which elicit the symptoms of photosensitivity, and may be used in demonstrating postulate (3) only when some estimate of the limits of the action spectrum has been made. The reason for this must be evident from the discussion in Chapter 5. If the symptoms are abolished when sunlight is filtered through window glass, the condition is in all probability ordinary sunburn.

It is assumed that the photodynamic sensitizer which must be isolated to comply with the second postulate is derived in one way or another from plants. All green plants contain such a substance, chlorophyll, which may produce the symptoms of photosensitization when injected into animals, but it is improbable that chlorophyll is ever itself the causal factor in photosensitization of domestic animals, because as far as is known this substance is not taken into the blood stream. If chlorophyll passed through the lining of the digestive tract into the blood stream, we should all run the risk of becoming sensitive to sunlight after eating a lettuce salad. The possibility may be admitted that chlorophyll is taken through the digestive epithelium in some diseased conditions, but as far as the author knows this has never been shown to occur, and cannot be seriously proposed as an explanation of photosensitivity in domestic animals or man. Recognizing the photosensitizing properties of chlorophyll, it is obviously necessary to remove this substance when extracts of plants are made in the investigation of postulate (2). If this is not done photosensitivity is almost sure to result from the injection of the extracts, and erroneous conclusions may be reached. Unfortunately, this has not always been properly guarded against (see Chapter 15). Obviously the isolation from a plant of a fluorescent substance which produces photodynamic action in systems such as, for example, red blood cell suspensions, is not proof that the plant may cause photodynamic disease if animals feed upon it. Such a substance must pass through the digestive tract without being altered, and must be capable of being taken into the skin if it is to produce symptoms of photosensitivity in animals.

In the following chapters the experimental evidence, and to what extent it fulfills the foregoing postulates, will be considered for each of the diseases of domestic animals which has been attributed to photodynamic action. As in the case of Koch's postulates for demonstrating the etiology of a microbial disease, it may not always be possible to fulfill all the postulates, but the attempt should be made.

General Symptoms of Photosensitivity. The symptoms which are most often observed in these diseases are indicative of surface irritation. Evidence of sensory stimulation, such as scratching, rubbing and running about, are the immediate symptoms. These are usually followed by lesions of the skin, sometimes accompanied by blindness, and often complicated by secondary effects. These symptoms are entirely comparable to those displayed by mice which have been injected with photodynamic substances and exposed to light. The latter have been discussed in Chapter 10 (p. 103), where it is pointed out that the severity of the symptoms varies with the quantity and potency of the photodynamic substance which is injected, and with the dosage of radiation. Photodynamically produced diseases of domestic animals display similar variability. Often other symptoms are described which may be incidental or secondary to the effects of light, or obligatory accompaniments such as the icterus of geeldikkop.

Although it is usually reported that sensitization is strictly limited to white skin, there are occasional reports of colored areas being affected. This is to be expected, since the pigment of the colored areas serves only as a light filter, and may not remove all the active radiation, so that such areas are not exempt from the effects of light, but only subject to a less degree than white areas. Hausmann (1910) found that gray mice could be sensitized to light by the injection of hematoporphyrin, although he was not able to produce the acute, but only the subacute effects.

Chapter 13

Hypericism (St. Johnswort Poisoning)

It is convenient and proper to use the term hypericism, introduced by Hausmann (Hausmann and Haxthausen, 1929; Hausmann and Zaribnicky, 1929), to designate the disease which follows the ingestion of plants belonging to the genus *Hypericum*. Various common names have been applied to plants of this genus; in English-speaking countries they are usually called St. Johnswort, although a variety of local names are employed; in Germany, Hartheu, or Johanniskraut; in France, Millepertuis or Casse-diable; and by the Arabs of the Mediterranean region, Hamra. These terms are often used in connection with the disease, as for example; St. Johnswort poisoning, Hartheukrankheit, and Hamra.

The First Postulate. Numerous quotations from early literature might be cited to show that symptoms of photosensitivity are observed in animals which have fed on *Hypericum*. Such appear in the papers of Rogers (1914), and Marsh and Clawson (1930). The most significant observations are that the animals may show: (1) symptoms of sensory stimulation, such as running and dancing about wildly and scratching; (2) surface lesions, usually limited to the white areas of the skin. There have been, however, few direct attempts to prove that light precipitates these symptoms. Significant experiments have been performed in Australia by Dodd, Henry, Seddon and Belschner; and in South Africa by Quin. In the first of these studies (Dodd, 1920a, b), it was observed that the appearance of the sensory symptoms was related to the intensity of the sunlight; on cloudy days they did not appear, and at night they were replaced by extreme depression. Henry (1922) performed experiments which confirmed those of Dodd, and also succeeded in showing that external contact with the plant is not injurious. This was done by allowing the animals to run muzzled in a field heavily infested with *Hypericum* during the day, feeding them on clean lucerne at night only. No symptoms whatsoever were developed by these animals, although the experiment was continued for twenty days, thus providing a complete disproof of the contention of Cirillo (see p. 128). Seddon and Belschner (1929) and Quin (1933b) report similar results, the symptoms of sensory stimulation and surface lesions appearing only when the animals were exposed to sunlight. Dodd used dried plants in his experiments, and Henry pastured his animals in a field of mature flowering plants. Seddon and Belschner fed young immature plants, and Quin administered dried powdered plants by stomach tube. From this it appears that the age and condition of the plants is of much less importance than the intensity of the sunlight. The experiments of Marsh and Clawson (1930), in the United States, do not indicate any great sensitivity to light, but it is doubtful if these can be considered of great importance as the feeding period was not long in most cases. Dodd observed the symptoms only after thirteen days of feeding, although Quin's animals showed symptoms of photosensitivity within one or two days.

These positive experiments seem to leave little doubt that the first postulate listed in the preceding chapter is fulfilled in the case of hypericism; it seems certain that animals may become sensitive to sunlight after feeding upon this weed, with resultant sensory stimulation and skin lesions of varying severity.

The Second Postulate. Ray suggested in 1914 that the photosensitivity of hypericism is due to photodynamic action, and claimed to have isolated a red fluorescent substance from *Hypericum crispum* which produced the symptoms of photosensitization in rabbits and sheep. He gives no account of the mode of extraction or of administration, however.

Although Ray seems to have been the first to attempt the isolation of a toxic substance from Hypericum with a view to determining its relationship to photosensitivity in domestic animals, what was undoubtedly the same, or a similar, substance was isolated many years earlier by pharmacologists who were not concerned with photosensitivity. The pharmacologist's interest derived from the fact that *Hypericum* belongs to that legion of plants which have been used at one time or another as therapeutic agents, but subsequently abandoned. Extracts of the plant seem to have been popular ingredients of therapeutic decoctions in the eighteenth century (see Rogers, 1914). Its use antedates this by several centuries, as it was one of the sixty-one ingredients of the Theriac Andromachi Senioris attributed to Galen, a much-favored panacea during several centuries (see La Wall, 1927). The nineteenth edition of the "United States Dispensatory" (1907) states that the preparation known as *oleum hyperici,* or red oil, is still used as a domestic remedy, particularly for the treatment of bruises. The list of conditions in which it has been stated to be efficacious recalls the Spring of Nepenthe.

Dietrich in 1871 and Wolff in 1875 (quoted from Cěrný, 1911) both isolated a red fluorescent substance which was undoubtedly the same as that studied by Cěrný in 1911. The latter investigator applied the name *hypericin* which is more appropriate than *Hypericum-Roth* used by Wolff, since there are other red pigments in the plant. Apparently the same or similar substances were studied by Ray (1914), Fischer and Hess (1930), Sierch (1927), Mélas-Johannidès (1928, 1930a, b), Dhéré (1933) and Horsley, 1934). Recently, Mackinney and Pace (unpublished) have found that extracts from *Hypericum perforatum,* which resemble those of the previous investigators, contain several components which may be separated chromatographically. At least some of these components seem from their absorption spectra to be closely related. It is probable, therefore, that hypericin is not a pure substance but is made up of a number of related compounds; this leaves previous atttempts to define its chemical nature open to question.

All the extracted materials which have been studied by these workers have been brilliant red in color with an intense red fluorescence at neutral and acid reactions. In alkaline solution the color is a deep green. They have been reported as soluble in various organic solvents, but not in water, except for a component separated by Johannidès, which was sparingly soluble in water. Cĕrný determined the empirical formula as $C_{16}H_{10}O_5$, which was confirmed by Horsely; neither of these determinations was made on crystalline material. Cĕrný concluded that hypericin is a flavone. Siersch (1927) believes it to be an anthocyanine, but Mackinney and Pace could not reproduce her tests. The *Hypericum* plant contains large proportions of a yellow substance, quercetin (O'Neill and Perkin, 1918) closely related to the flavones. There is also a pigment, possibly of the logwood type obtained from the plant extracts after virtually all the hypericin has been removed. Either of these might be present as impurities in the extracts. Mackinney and Pace are still occupied with this question, which will not be discussed further at this time.*

Some confusion has arisen as to the chemical nature of hypericin, due to the unfortunate use of a name. In 1887 Reinke isolated a red fluorescent substance from a fungus, *Penicilliopsis clavariaeformis,* and measured the absorption spectrum for the visible wave-lengths with nice accuracy. He called the substance *Mykoporphyrin,* and believed it to be related to chlorophyll. Fischer and Niemann (1925) isolated this substance again, expecting from the position of the absorption bands to find a porphyrin; but examination of the spectrum in various solvents showed that it behaved differently from the compounds of the porphyrin group. Later, Fischer and Hess (1930) isolated hypericin from *Hypericum* leaves; finding the absorption maxima in the visible to be in agreement with those of the pigment from *Penicilliopsis clavariaeformis,* they concluded that the compounds are identical. Dhéré (1933) arrived at the same conclusion from a comparison of the fluorescence of the two compounds. A comparison of the absorption spectrum obtained by Reinke with those obtained by Mackinney and Pace for one component extracted from *Hypericum* is provided in Figure 28; the general agreement in position of maxima is clear. Comparison of these absorption spectra with that of hematoporphyrin, also shown in Figure 28, will indicate little similarity. The unfortunate application of the term

* Pace now reports the empirical formulae, $C_{30}H_{26}O_9$ and $C_{29}H_{22}O_8$ respectively, for hypericin "X" and "Y." He suggests that hypericin is a partially reduced methyl, polyhydroxy derivative of helianthrone. (Thesis, Univ. of Calif., 1940.)

Mykoporphyrin, and the sequence of events related, might lead to the belief that hypericin is a porphyrin, particularly since the latter substances are well-known photosensitizers. Whatever the chemical nature of hypericin, it is obviously not a porphyrin, since it contains no nitrogen, and nitrogen is a constituent of all porphyrins (see Chapter 19 for a discussion of these compounds).

That red fluorescent substances extracted from *Hypericum* act as photodynamic sensitizers has been shown by numerous investigators. The first of these was Ray (1914) whose experiments have been mentioned above. Photodynamic hemolysis was produced by Hausmann and Zaribnicky (1929) with extracts of *Hypericum,* and by Hausmann (1931) with hypericin. Mélas-Johannidès (1930a, b) and Horsley (1934) sensitized white rats to light by the injection of hypericin. Horsley made the interesting discovery that a "yellow waxy substance" which occurred in association with hypericin in the plant, increased the photosensitizing activity of the hypericin when mixed with it. The "yellow waxy substance," which was undoubtedly quercetin, did not sensitize the animals when injected alone. Whereas a dose of 4 mg. of hypericin was required to produce minimum symptoms of photosensitivity in a rat, 0.8 mg. of this substance mixed with four times this quantity of the yellow material would produce symptoms. Horsley suggests that this is due to increased solubility of the hypericin brought about by the presence of the other substance. Increase in the solubility of the photodynamic substance in the blood plasma, or the extent to which it is taken up by skin itself, should increase the effectiveness of the sensitizer. Blum and Pace (unpublished) found that at least one of the fractions from Hypericum extracts produces photodynamic hemolysis, which is characteristically inhibited by the absence of O_2.

Although the chemical identity is not firmly established, there can be no doubt that photodynamically active substances can be isolated from *Hypericum* plants, which produce symptoms of photosensitivity when injected into mammals. It is certain that all the extracts used in the experiments described above were chlorophyll-free, or virtually so. Thus the second postulate for the proof of a photodynamic etiology is established.

The Third Postulate. The third postulate has not yet been satisfied for hypericism, but there is a certain amount of pertinent evidence as to the character of the action spectrum which is to be expected. Figure 28 shows that the absorption spectrum for one of the hypericin fractions isolated by Mackinney and Pace absorbs less of the ultraviolet of sunlight than of the visible. The absorption spectrum for a second component is similar for the visible region, and from the similarity in chemical behavior of the two substances, further agreement in the ultraviolet is to be anticipated. Hausmann and Rosenfeld (1932) showed, by means of a method which will be discussed in Chapter 19, that red blood cells are sensitized in the ultraviolet by extracts of *Hypericum.*

Seddon and White (1928) found that window glass did not decrease the sensitivity to sunlight; window glass removes the sunburn-producing radiation from sunlight, but this represents only a tiny portion of the total energy (see Figure 36, p. 194). They also performed experiments in which the ears of sheep were painted with dyes, but these are of questionable significance.

Complete demonstration of the third postulate necessary for establishing a photodynamic etiology is lacking in the case of hypericism, but there seems little doubt that it will be possible to establish this last link in the chain of evidence.

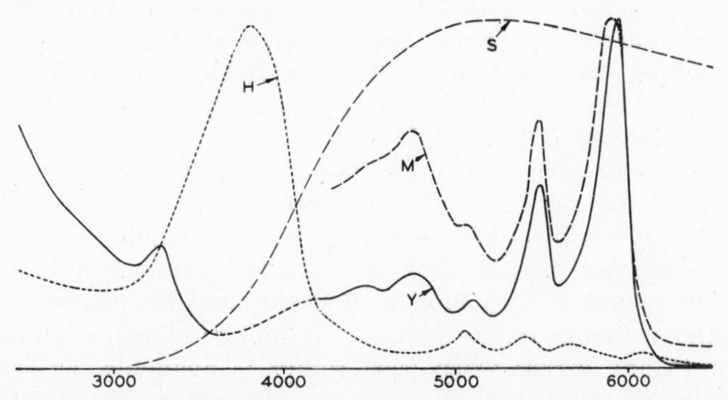

FIGURE 28. Abscissa: wave-lengths in Å. Ordinates: The ordinates for H, M, and Y are values of the absorption coefficient with the scales chosen to bring the maximum absorption to the same value.

H, absorption spectrum of hematoporphyrin. M, absorption spectrum of "Mykoporphyrin." (From Reinke, 1887.) Y, absorption spectrum of the "Y" component of *Hypericum* extracts. This is a composite curve; for wave-lengths greater than 4200 Å, the solvent was alcohol; for wave-lengths less than 3600 Å the solvent was acetone. (From Pace and Mackinney, unpublished.) S, spectral energy distribution of sunlight, ordinate units chosen to make the maximum intensity correspond to the maxima of the absorption spectra.

The Plant and its Distribution. The genus *Hypericum* is widely distributed; there are about 200 native species in the northern hemisphere, and a few more in the southern. The plants are herbaceous or low shrubs, usually deciduous, but sometimes evergreen. The leaves are dotted with pellucid or opaque glands which Siersch (1927) has shown to contain the hypericin. The flowers are generally yellow, but occasionally pink or purplish. Bailey (1915) lists 24 species which have been more or less extensively used as ornamentals. Many of these are delicate but the hardy *H. perforatum* which is responsible for much hypericism is among the group.

The Incidence of Hypericism. At least four species of *Hypericum* may produce photosensitivity. *H. perforatum* is the most common form, having spread to many parts of the world; *H. crispum* is responsible for the disease in the Mediterranean regions; and *H. ethiopicum* and *H. leucoptychodes* have been shown by Quin (1933b) to produce photosensitization. How many of the other species may produce hypericism is unknown. The hypericin-containing glands, which are characteristic of the genus, vary in number and distribution in different species, as pointed out by Siersch (1927), and since the ability to produce photosensitization should depend upon the quantity of hypericin in the plant it may run parallel with the number of these glands.

H. perforatum is the common European St. Johnswort. It is probably responsible for most hypericism, being the species which causes the disease in North America and Australia as well as in Europe. The extensive root system allows it to overcome other plants on uncultivated areas; its encroachment on the grazing lands of northern California is clearly described by Sampson and Parker (1930). *H. perforatum* seems to have arrived in the latter area only about thirty years ago, but today it covers thousands of acres. In the flowering season the great fields of yellow blooms offer a beautiful picture to the uninformed traveler, but a discouraging sight to the local sheep men who have seen their best pastures taken by the plant, which they call "Klamath weed." The principal complaint of these grazers is not the disease symptoms, which they seldom associate with sunlight, but the fact that the weed is driving out the valuable forage plants. Actually the animals do not eat the plant to any extent when good forage is available, but that it is eaten at times is clear from Sampson and Parker's report (1930). They state: "In northern California a proportion of the sheep grazed on areas heavily infested with St. Johnswort have each year, for a decade or more, been reported as sick animals. These reports have come particularly at shearing time in July, and again at the time of sorting and culling prior to breeding in September, when the animals are closely inspected. The sick animals typically exhibit blistering and a scabby condition about the nose, muzzle, eyes, and ears. The ears may swell to many times their natural size, portions sometimes actually sloughing off. In more severe cases, the animals lose their sight or their mouths become so sore that they die from starvation. . . . In still other instances they are seriously emaciated, with scabby skin about the legs, face, neck, withers, and back—even to sloughing of wool on affected areas. . . . Immediately after shearing, provided the day is sunny, a goodly proportion of a band will rub, bite, lick, and run as though crazed, the itching sensation apparently being so severe. In three or four days these reactions become less intense."

Although Marsh and Clawson (1930) were inclined to minimize the importance of the toxic properties on the basis of their more or less negative experiments, Sampson and Parker feel that hypericism is a real source of loss to the sheep rancher. They state: "The observations of the writers are in agreement with Marsh and Clawson to the extent that deaths among cattle and sheep are not common on the California

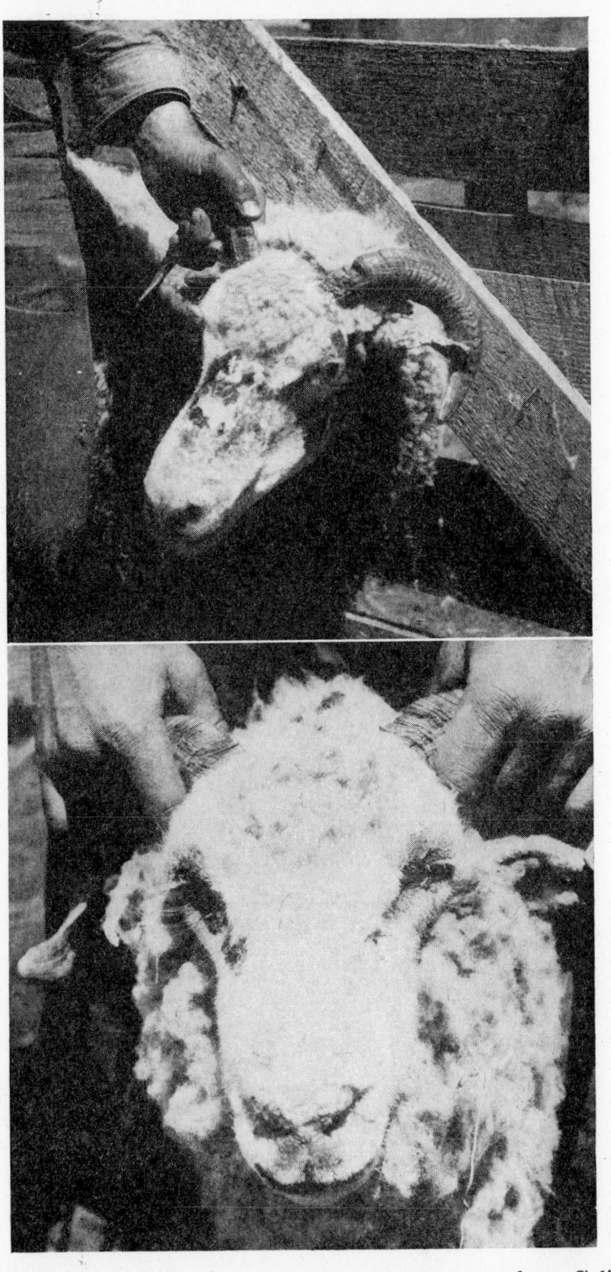

FIGURE 29. Lesions of face and eyes of sheep grazed on northern California pastures infested with *Hypericum perforatum*. [From A. W. Sampson and W. Parker, *University of California Agricultural Bulletin* No. 503 (1930).]

ranges. As already pointed out, however, several sheep in a band when grazed on infested areas may undergo slight to very serious loss of condition and rustling qualities; there may ensue severe dermatitis, resulting in a light clip of inferior wool on the parts of the animal affected. The deaths that do occur are in sick animals not being able to find food or eat it. On the basis of these facts, the writers cannot agree with the conclusion of Marsh and Clawson to the effect that St. Johnswort is not likely to be a source of much trouble in the grazing of livestock, particularly sheep, because of its toxic properties." Photographs of animals showing skin lesions are reproduced from Sampson and Parker's report (Figure 29).

The State of California recognized the importance of the problem to the extent that the legislature in 1929 made the sum of $3000 available for the study of methods of controlling the spread of the weed. Some progress has been made in the methods for the eradication of the plant since that time (Sampson and Parker, 1930; Raynor, 1937).

The plant is also a serious problem in Oregon, Washington, Idaho, Montana and Utah. It is known in these western states by various names, among which "Klamath weed" and "Tipton weed" are the most common. It is also widely distributed in the eastern states where it does not seem to be a serious problem, and it has undoubtedly spread to the west from these areas. It was mentioned as a pest in various eastern states in the early part of the last century; Marsh and Clawson (1930) present a number of interesting excerpts from such reports, which speak of lesions confined to the white areas of skin and to symptoms of sensory stimulation, leaving little doubt as to the nature of the condition. It seems probable that where cultivation has spread and grazing diminished, the incidence of hypericism has decreased until it is no longer a problem.

Similarly, hypericism does not seem to be a great problem in Europe where *H. perforatum* is native, although it is reported from time to time. When in Naples recently, I inquired about the plant but could obtain no information. Possibly it is no longer a problem as in the days of Cirillo or Manni di Lecce and Marinosci di Martina (see p. 128). I did not reach Apulia, however, and hence can give no direct testimony as to whether the Tarentine fields still support only black sheep. Nor could informants at Naples enlighten me on the matter.

In Australia the situation is comparable to that in the western United States. Summers stated in 1911 that *H. perforatum* was originally introduced into Australia as an ornamental plant at least twenty years before that date, and that it had just been declared a "noxious weed" by the Parliament. Dodd in 1929 (a) complains that nothing further had been done about the problem, calling to attention that the plant is a menace, not only because of its crowding out of good forage plants, but because of the results of its photosensitizing activity. The problem seems to be recognized at present, however, and attempts are being made to eradicate it.

The statement of Summers (1911) and an anonymous account in *Nature* (1933) indicate that *H. perforatum* was introduced into Australia as an ornamental plant. This species is often used to cover garden areas rapidly where other ornamental plants will not grow, potentially a very dangerous procedure. The Australian example should be called to the attention of horticulturists and garden amateurs.

H. Crispum, the species studied by Ray (1914) and Mélas-Johannidès (1928, 1930a, b) seems to be responsible for hypericism in Tunis and Greece, and is found also in Asia Minor. Ray states that the condition has been known to the Arabs of Tunis for a long time, and that they are in the habit of painting their white or piebald horses with tobacco or henna as a protection against the effects of sunlight. The etiology seems to be more clearly recognized there than by the sheep ranchers of the United States.

Quin (1933b) does not state whether hypericism is a problem in South Africa, although he has shown that two local species of *Hypericum* may produce photosensitization. Lange (1922) says that *H. maculatum* and *H. pulchrum,* as well as *H. perforatum,* are responsible for *hypericism* in Germany, but gives no further account.

Kinds of Animals Affected. There seems little doubt that any domestic animals which eat the *Hypericum* plants may be rendered sensitive to light, but some species may be more susceptible than others. White animals are, of course, more likely to be affected than dark-colored ones. Photosensitization of sheep, horses, cattle and various laboratory animals has been reported. The damage to sheep seems to be the most important from an economic point of view.

Pharmacological Properties Other than Photosensitization. Rogers (1914) reports that fluid extracts of *Hypericum* produced death of guinea pigs in about the same dosage as digitalis, and that there is a marked depressant effect on the heart of the dog. Presumably these effects were independent of the action of light. Marsh and Clawson (1930) found that cattle and sheep fed on *H. perforatum* developed high temperature, rapid pulse and respiration, and a tendency to diarrhea; these conditions probably were not secondary to the effects of light, since only very mild dermatitis is reported. Horsely (1934) reports that rats injected with hypericin and exposed to light showed intestinal hemorrhages, but these were probably secondary to the effects of light on the skin (p. 105).

Treatment and Control. The obvious method of preventing the disease is to keep the animals away from the plant or, if they should feed upon it, to protect them from light. Both these procedures may be difficult, and in many cases impossible. Grazing at night may be feasible in some circumstances, but not in most. In northern California the sheep tend to seek the shade of trees or bushes during the middle of the day, and this may tend to diminish the severity of the disease, but on less protected ranges where this is impossible the conditions may be more troublesome. Apparently the young plants are most readily eaten; and, although, according to Horsley (1934), they contain less hypericin than when older, they may be more often the cause of hypericism. This should be taken into consideration in grazing the animals.

Painting of animals, as practiced by the Arabs, may provide effective protection, but would be hard to carry out extensively. Black or dark-colored animals, being less susceptible, may be profitably introduced in some cases. This has been done in Australia with apparent success (see Daley, 1937), and was early practice in Italy according to Cirillo (p. 128). However, the most effective means of combating the disease is the same that may rescue the ranges for more profitable plants, namely, eradication of the weed. Interesting accounts of control methods in Australia have been published by Carn (1936), Reuss (1936) and Daley (1937). It is probable that different methods will have to be employed in different localities.

Since the wool of sheep offers protection against sunlight, they should be more sensitive after shearing. For this reason Daley (1937) recommends that shearing be done late in the summer so as to escape as much exposure to sunlight as possible. He also suggests that ear marking be delayed until after summer, since the surfaces wounded in this operation are particularly sensitive to light when the animals feed on *Hypericum*.

Chapter 14

Geeldikkop (Yellow Thick Head)

In the disease which is the subject of this chapter a photosensitizing substance is produced within the digestive tract of the animal from plant material taken in the diet. It thus differs from the condition just described (hypericism) in which the photosensitizing pigment is already present in the plant when it is eaten. The disease results from feeding on more than one kind of plant, and has been called by more than one name, although the etiology is apparently the same in all cases. First described in the Karoo veldt of South Africa it received there the name geeldikkop, which may be translated from the Dutch as yellow thick head.

According to Quin and Rimington (1935) geeldikkop is the oldest and best known small stock disease of the Karoo veldt of South Africa, where it causes heavy losses to the sheep-raising industry. Henning (1932) states that it was first observed by Hutcheon as early as 1886. It is generally supposed to result from eating certain plants of the genus *Tribulus,* and has been called "tribulosis." Its association with light has been recognized only very recently. Although Paine (1906) noted that affected sheep avoid direct sunlight, and suggested that they should not be allowed to graze in the daytime, he failed to recognize photosensitization, and believed the disease to be of a "malarial" nature. Theiler (1918) suggests an association of the disease with light, but Quin (1928) seems to have been the first definitely to claim the existence of such an etiological factor. As late as 1932 Henning, in a general discussion, barely mentions the possibility of photosensitization, but the admirable experiments of Quin and Rimington have demonstrated within the past few years that a part of the symptoms result from the action of light. While certain lacunae remain, the etiological picture as a whole may be regarded as clearly established.

The name *geeldikkop* gives the key to the major symptoms: intense icterus (jaundice) and edema of the skin of the head. It is the latter symptom which at once suggests photosensitivity. Quin and Rimington (1935) give the following general description of the course of the disease:

"The disease is definitely caused by excessive feeding on wilted 'dubbeltjie' plants,* which leads primarily to constipation in the large intestine and a stoppage or disturbance of movements in the rumen. At the same time the liver and gall-bladder are affected; there is a progres-

* *Tribulus sp.* (author's footnote).

sive accumulation of bile, which rapidly passes into the bloodstream, and which to a certain degree may be eliminated in the urine.

"It has been proved that the blood serum of many sheep may in this way turn yellow before any external symptoms of 'geeldikkop' can be detected. Some of these animals subsequently develop a strong sensitivity to sunlight, whereas amongst the rest the disease never progresses further than a slight jaundice in the blood serum.

". . . Sensitivity is localized to the unprotected parts of the skin, *e.g.*, the head and ears, and in mild cases results only in a superficial sloughing of the skin, closely resembling ordinary sunburn in man. In most cases, however, the symptoms are much more severe, the face and ears swelling tremendously, while the coronet of the hoof and the base of the horns assume a dark purple-red colour. Such animals appear very sick, running a high temperature and showing a certain measure of shock. They continually seek shade, and may easily be detected amongst a 'kraaled' flock where they can usually be found hiding themselves in corners and along the walls of the kraal. On further exposure to sunlight, the symptoms rapidly become aggravated, the animals appear half stunned and loath to move. The affected skin turns hard and dark brown. In many cases there is blindness and rupture of the eyeballs. Such animals cease feeding and drinking on account of the jaws becoming rigid through injury to the skin. The mucous membrane of the eyes and mouth assumes a deep yellow colour due to the extensive jaundice. The animals rapidly lose weight and present a pitiable spectacle; many ultimately die from starvation and thirst.

"It is especially the after effects of the disease and not the disease itself from which so many animals may succumb. Sheep which eventually recover may show a long period of convalescence. In this way large flocks frequently suffer a severe setback in condition accompanied by damage to the wool."

Marked icterus throughout the tissues seems to be the most constant post-mortem finding. Theiler (1918) describes changes in the liver, stating that the gall bladder is always distended and may be ruptured, but the bile duct is always open. Paine (1906) makes the statement that, while the bile duct is intact, it is impossible to force the bile into the duodenum, although when the duct is cut, the bile flows freely; this observation is somewhat difficult to explain.

The accompanying photographs (Figure 30) show the characteristic swelling of the head, and the attempt to avoid light. According to description, the symptoms of behavior and the skin lesions are similar to those of hypericism, the only important difference being the icterus in geeldikkop.

Outbreaks of the disease occur only during the months of December to March, which is compatible with the view that the disease is a photosensitization, since these are the summer months of the southern hemisphere. These outbreaks are sporadic, appearing as a rule in isolated paddocks among lambs under twelve months; but in some years they may be widespread and affect sheep of all ages; goats are also susceptible to the disease. The outbreaks are always sudden and involve a large portion of the flock. It is generally believed that they are related to climatic changes (Paine, 1906; Theiler, 1918; Quin and Rimington, 1935), involving a particular sequence of drought and rainfall. There seems to be no evidence that the disease is infectious.

FIGURE 30. Sheep with Geeldikkop. (From M. W. Henning's "Animal Diseases in South Africa," Central News Agency, Ltd., Johannesburg, 1932.)

The idea that geeldikkop is caused by eating certain species of *Tribulus* locally known as "Dubbeltje doorn" (devil's thorn), or "dubbeltjies" seems to have been long current among the sheep raisers. It was not accepted by Paine (1906) who cites certain unsuccessful attempts by Dixon to reproduce the symptoms of the disease by feeding these plants. Paine's own attempts to produce the disease by injecting the blood from affected animals, and by transferring the stomach contents from normal to diseased sheep, were likewise unsuccessful. Theiler (1918) performed a number of feeding experiments, many of which were unsuccessful, but in certain cases he found that feeding exclusively on *Tribulus terrestris* was followed by the symptoms of geeldikkop; and Quin was able to accomplish the same by feeding on *T. ovis* on one occasion, but not on others. The *Tribulus* species represent some of the best feed plants of the Karoo, where they are widely distributed and are eaten most of the time with no untoward effects. Thus there must be some particular precipitating factor which causes these common feed plants to become suddenly toxic.

The actual variability of the outbreaks is exemplified by a series of experiments of Quin (1929). In the region where an outbreak had just occurred he isolated lambs on three areas covered by *Tribulus ovis* in three different stages of development, young preflowering, flowering and late fruiting. True geeldikkop appeared in all areas; three cases appeared on the third day, and by the sixth day nine of the ten experimental animals had developed the symptoms of the disease. This indicates that the stage of development of the plant is of little importance, contrary to Theiler's conclusion that only the flowering stage was toxic. Heavy rains now occurred, and when another set of lambs was placed on the same plots four days later no geeldikkop appeared, although the animals were kept on the same pastures for eight days. Quin has made numerous other attempts to produce the disease by feeding *Tribulus sp.* shipped from regions in which outbreaks occurred, or grown on experimental plots, but without success.

Furthermore, conditions apparently identical with geeldikkop may follow feeding upon plants other than *Tribulus*. Quin (1933d) was able to produce all the symptoms of the disease by feeding two species of *Lippia, i.e., L. Rhemanni* and *L. pretoriensis*. He also points out (1930) that the disease known as dikoor has the same symptoms as geeldikkop (see Steyn, 1928), but is confined to the high Transvaal veldt where *Tribulus* is not to be found. More recently, Rimington and Quin (1937) observed the symptoms of geeldikkop among sheep grazing on two species of grasses, *Pannicum laevifolium* and *P. coloratum*. Apparently the disease may be produced by at least three genera of not closely related plants: *Tribulus sp. (Zygophyllaceae), Lippia sp. (Verbenaceae)* and *Pannicum sp. (Graminae)*, and hence the term *tribulosis* is too restricted to be applied to the disease.

Nevertheless, Quin and Rimington state that geeldikkop is "definitely caused by excessive feeding on *wilted* dubbeltje plants" (the italics are mine). They describe (1935) the peculiar climatic conditions of the Karoo which account for the toxicity of the Tribulus plants. Very local showers occur throughout the Karoo veldt during summer months (November to February) which, following periods of drought, cause rapid growth of small plants, particularly the "dubbeltjes." If the rainfall continues, these plants develop into excellent feed, but if short intervals of drought occur, the plants wilt and in this condition they are toxic. The plants are also parasitized by insect larvae which may assist the wilting process. The presence of these larvae has sometimes led to the local belief that they, rather than the plant, cause geeldikkop, but there seems no good evidence for this (see Theiler, 1918). The limitation of toxicity to wilted plants could explain the sporadic occurrence of the disease, and its apparent relation to climatic conditions which prevail in the Karoo veldt.

From the foregoing discussion it must appear that the exact precipitating cause of geeldikkop and dikoor, which from the similarity of their symptoms may be assumed to have a common etiology, is to a certain extent a mystery. While Rimington and Quin (1935), and Rimington, Quin and Roets (1937) have isolated a substance from *Lippia* which produces the symptoms of geeldikkop, they have not thus far isolated it from *Tribulus*. But whatever the precipitating cause, the course of subsequent events is quite clear, thanks to the experiments of Quin and Rimington. The various stages in the development of these experiments were published consecutively in South African veterinary journals; they will be considered in detail in the order of their appearance.

But for the constant accompaniment of icterus, the disease might be assumed to be due to photosensitization, similar to hypericism, by some substance occurring in certain plants under peculiar conditions of development. The first step was to make sure that the icterus was not secondary to the photosensitization. Quin first (1931, 1933a) studied the photosensitizing action of hematoporphyrin and photodynamic dyes, finding that they produced photosensitization similar to that of geeldikkop, but not the icterus which is characteristic of that disease. He also (1933b) produced photosensitization by administering ground dried plants of *Hypericum ethiopicum* and *H. leucoptychodes* to sheep by mouth, but found no signs of icterus in these animals, although the blood serum was carefully examined during eighteen days of feeding. This clearly established the distinction between geeldikkop and hypericism, showing that icterus is not a result of the photosensitizing process, but left open the possibility that icterus might be a cause of the sensitization since it was not a result.

In another study Quin (1933c, d) tried to produce photosensitization by administering numerous plants found growing in the region where attacks of geeldikkop occurred. Two plants, both species of *Lippia*, produced symptoms of photosensitization accompanied by icterus, comparable to but somewhat milder than that seen in geeldikkop (1933d). This further emphasized the correlation between icterus and photosensitivity in this condition.

The next step (Quin, 1933e) was a systematic experimental study of icterus produced by ligating the bile duct in young sheep and goats. Surprisingly, this operation produced not only icterus but photosensitivity, thus making it almost certain that the photosensitivity of geeldikkop results from the icterus characteristic of that disease, or from associated changes. The exact nature of the photosensitizing agent remained undetermined.

In 1934 Rimington and Quin reported the results of a series of studies to determine the nature of the photosensitizing substance. They found that a porphyrin was present in the bile of sheep, and that a porphyrin, apparently the same, was also detected in the blood after ligation of the bile duct. At this stage of the investigation an outbreak of geeldikkop made possible similar studies in animals suffering from the disease, the results of which were entirely comparable. The porphyrin was successfully isolated in crystalline form from bile obtained by means of a fistula, and identified as phylloerythrin. Injection of 0.1 gm. of the crystalline substance into a normal sheep resulted in symptoms of photosensitization *without* icterus. Phylloerythrin prepared from chlorophyll *in vitro* produced the same effect when injected.

Phylloerythrin was first described under that name by Marchlewski (1903) who found it in the feces of cows which had been fed a chlorophyll-containing diet. He found that no phylloerythrin was present in the feces when there was no chlorophyll in the diet, and hence assumed that the latter substance is produced during digestion as a breakdown product of chlorophyll. Later (1904, 1905), he showed the substance to be identical with the cholehematin of McMunn and the bilipurpurin of Loebisch and Fischler (1903) found in the bile of ruminants. Fischer and Hilmer (1925) identified phylloerythrin as a porphyrin, and later Fischer, Moldenbauer and Süs (1931), and Fischer and Riedmair (1931, 1932) determined its structural formula. Fischer (1937) discusses the relationship of chlorophyll to its derivatives, and gives the following formulas:

```
      CH3-C=======C-C2H5      CH3-C=======C-C2H5
          |       |       H       ||       ||
          |       |       |       ||       ||
          C       C=======C-------C        C
        /  \\    /                 \      / \
       /     N                        N       \
      /                               |         \
  H-C        H                        H          C-H
    \\       |                                  //
      \\     N                        N       //
        \\ /   \                     //  \   //
          C       C=======C-------C        C
          |       |       |       |        |
          |       |       |       |        |
      CH3-C=======C    H2-C       C=======C-CH3
                  |         \    /
                 CH2          C
                  |           ||
                 CH2          O
                  |
                 COOH
```

Phylloerythrin

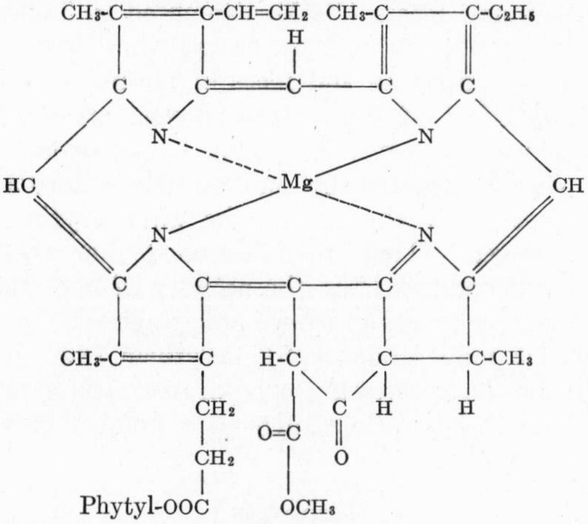

```
CH3-C=====C-CH=CH2  CH3-C--------C-C2H5
   |        |    H       ||       ||
   |        |    |       ||       ||
   C        C=====C------C        C
    \      /              \      /  \
      N.........            N        \
HC          .....Mg......                CH
  \\         /        .....               //
      N                     N
    /   \                 //  \
   C     C=====C------C        C
   |     |     |      |        |
CH3-C=====C   H-C      C--------C-CH3
         |     |  \   / |      |
        CH2    |   C   H       H
         |   O=C   ||
        CH2    |   O
         |     |
Phytyl-OOC    OCH3
```

Chlorophyll A

The porphyrins all have four pyrrole rings joined in the form of a ring by methin bridges. Most of them are photochemically active; they will be more fully discussed in Chapter 19.

Rothemund and Inman (1932) found phylloerythrin in the digestive tract of cows and sheep, particularly in the third and fourth stomachs, but not in the digestive tract of milk-fed calves, further demonstrating its derivation from chlorophyll. Rothemund, McNary and Inman (1934) found phylloerythrin in the walls of the stomach, indicating its absorption at this point in the digestive tract. Rimington and Quin (1934) found that animals with bile fistulas showed little or no phylloerythrin in the bile so long as they were kept on chlorophyll-free diet, but as soon as they were placed on green feed the phylloerythrin content mounted rapidly. All these findings serve to demonstrate that phylloerythrin is a normal product of digestion in the ruminant fed on a diet containing chlorophyll, and that it is taken through the walls of the digestive tract into the circulation and thence into the bile, where it is a normal constituent. The normal mode of excretion of the phylloerythrin absorbed from the digestive tract is through the bile, and so long as this is adequate no photosensitivity develops. If this excretion is interfered with, either by experimental occlusion of the bile duct, or by liver poisoning of the type found in geeldikkop, the concentration in the blood stream must be increased. Increase in the concentration of phylloerythrin in the blood stream leads to increased concentration in the skin, and photosensitivity results. Important evidence for this sequence of events is given by Rimington and Quin's (1934) observation that animals with the bile duct ligated do not display photosensitivity as long as they are maintained on a chlorophyll-free diet, but develop this condition as soon as put on green feed.

Quin, Rimington and Roets (1935) attempted to find the mechanism of phylloerythrin production. They found that this substance was always present in the intestine and feces of normal sheep. They eliminated the possible effect of saliva by exposing green plants to sheep saliva outside the body; no phylloerythrin was produced. The role of plant enzymes was eliminated by boiling the chlorophyll-containing plants before ingestion by the sheep; the phylloerythrin level in the feces was not changed. They then assumed that the formation of phylloerythrin was dependent upon the activity of bacteria, or infusoria, or both, in the digestive tract, and were able to show *in vitro* that either bacteria or infusoria may produce phylloerythrin from chlorophyll.

They carried out an interesting comparative study on the presence of phylloerythrin in the digestive tracts of a number of animals, which may be briefly summarized:

Herbivora

(1) Ruminants
- (a) Sheep. Phylloerythrin present throughout the intestine, particularly in posterior parts.
- (b) Cattle. Similar to sheep.

(2) Non-ruminants
- (a) Horses. Phylloerythrin absent in stomach and small intestines, but plentiful in caecum and large colon.
- (b) Rabbits. Phylloerythrin in stomach and intestine.
- (c) Guinea pigs. Phylloerythrin in stomach and intestine.

Omnivora

- (a) Pigs. Very small quantity of phylloerythrin in feces.
- (b) Rats. Appreciable quantities of phylloerythrin in stomach and small intestine.

Carnivora

- (a) Dogs. No trace of phylloerythrin in feces.

Birds

- (a) Fowl. Traces of phylloerythrin in large intestine.

The authors find a correlation in this series between the amount of phylloerythrin and the presence of bacteria and infusoria. The data are of interest as regards the probability of a condition similar to geeldikkop occurring in animals other than sheep and goats, as a result of biliary obstruction or liver dysfunction. Obviously such a condition is not possible in the carnivora and improbable in the omnivora. Among the herbivora the ruminants should be the most susceptible since the stomachs provide for greater phylloerythrin production because of the activity of microörganisms.

Rimington and Quin (1935), and Rimington, Quin and Roets (1937) finally succeeded in isolating the icterogenic principle of *Lippia rehmanni*. Three isomers of a substance having the empirical formula $C_{34}H_{52}O_6$ are found, to which the names Icterogenin A, B, and C have been given. The probable formula is:

$$C_{29}H_{43}O_2\begin{cases} -CO \\ -OH \\ -COOH \\ -CH\begin{matrix} \diagup CH_3 \\ \diagdown CH_3 \end{matrix} \end{cases}$$

Icterogenin produces severe constipation and icterus, and also photosensitization if the animals are fed a chlorophyll-containing diet. It stops the movements of isolated intestinal strips in a concentration of 1/200,000 and inhibits the perfused heart. It is a hemolysin *in vitro*. One gram or more given to a sheep by mouth will produce bilirubinemia within twenty-four hours. It has not as yet been isolated from *Tribulus*. Rimington and Quin (1933) isolated a toxic factor from the latter plant, but this was apparently NO_2^-, and did not produce geeldikkop.

Quin (1936) made a study of the bile excretion during the course of poisoning of sheep by *Lippia Rehmanni*. He found the excretion of bile to be greatly diminished after feeding this plant, the volume falling to almost nothing at the end of four days. At the same time, the bile which was normally brown became progressively paler, until at the end of about 24 to 30 hours it was water-clear. After this time the color gradually returned to normal. It seems thus that the liver, which shows little or no morphological changes, loses its ability to excrete the bile pigments. The volume of bile is also decreased, but to a lesser extent than the pigments, as shown by the loss of color. Parallel with these changes in bile excretion, the blood serum, normally water-clear, becomes yellow at the end of about 24 hours and brownish in about 48; it shows a strong direct Van den Bergh reaction. On the third day the animals become frankly icteric, and are sensitive to light. Liver damage produced by other poisons, chloroform, phosphorus, carbon tetrachloride and manganese chloride was accompanied by mild icterus at most, but never by photosensitivity.

Taken as a whole, the evidence indicates that geeldikkop and dikoor result from functional liver disturbances which prevent the excretion of bile pigments including phylloerythrin. Such changes result from eating *Lippia sp.*, wilted *Tribulus sp.*, certain grasses and probably other plants. The photosensitization results from the accumulation of the unexcreted phylloerythrin in the skin, to which it is carried by the blood stream; this happens only if the animal is receiving chlorophyll in the diet so that phylloerythrin is present in the blood stream. The source of the chlorophyll is of no importance, and it is possible to err in suspecting the plant supplying the chlorophyll as the toxic factor rather than the one contributing the icterogenic substance.

Similar photosensitivity might result from blockage of the bile duct, and recently Graham and Gordon (1937) have reported what appears to be a case of this kind. A sheep showed symptoms of photosensitization, and upon autopsy the bile duct was found to be occluded by a cyst. This being the only case of photosensitization observed in the herd, it is probably to be explained on the above basis. Similar occurrence might account for some of the isolated cases of photosensitivity which have been reported.

The etiological picture presented by the studies of Quin and Rimington is convincing enough, and adequately demonstrates that the first and second postulates are satisfied. The third postulate has not been demonstrated, *i.e.*, it has not been shown that animals suffering from geeldikkop are sensitive to the same wave-lengths as animals sensitized by injection of phylloerythrin. This is the only way of proving conclusively that phylloerythrin is the photosensitizing substance. The absorption spectrum of porphyrins, to which group phylloerythrin belongs, is discussed in Chapters 5 and 19; their principal absorption lies in the near ultraviolet, between 3000 and 4500 Å. Though the absorption spectrum of phylloerythrin has not been quantitatively determined in the ultraviolet, it is probable that its major absorption, and hence its greatest photosensitizing action, will lie in the same general region. Quin's observation (1930, 1933e) that the quartz-mercury arc failed to elicit symptoms of photosensitivity is in accord with the experiments on porphyrin sensitization discussed in Chapter 19.

Diseases Having Symptoms Similar to Geeldikkop. In a recent number of *The Pastoral Review* (1934)* an article appears by "a veterinary scientist" which discusses a number of diseases occurring in Australia that have symptoms resembling those of geeldikkop. Unfortunately no references are quoted in this article. The diseases listed are:

(1) Toxemic jaundice, an icteric condition of unknown cause in which there are no symptoms of photosensitivity.

(2) Toxemic jaundice in calves, a similar condition, but presumably of different origin.

(3) Yellow Bighead and photosensitivity, known as geeldikkop in Africa and as facial eczema in New Zealand. In Australia such outbreaks are usually associated with feeding on young growth of the panic grass *Pannicum effusum*.

(4) Swelled head in rams, develops as a swelling about the face, the skin becoming infiltrated with a straw-colored fluid. It is caused by *Bacillus oedematiens* infecting wounds of the head usually produced by fighting. This is also reported from South Africa (de Kock, 1927).

(5) Blackleg in rams, may result in swelling about the head, presenting an appearance similar to (4).

* Vol. 47, p. 665.

This account shows the possibility of confusion in diagnosis of photosensitivity of the geeldikkop type. Possibly this condition may be found much more widely distributed than is suspected, for the fact that a number of different plants have already been shown to produce the disease suggests that more will be found. Hopkirk (1936) says that "facial dermatitis" in New Zealand is of two types, one resulting from feeding on *Hypericum* and the other resembling geeldikkop. In the latter there is liver damage and icterus, but not in the former. The latter is associated with fast growth of pasture, followed by drought.

It is to be hoped that more experiments will be made in the future to determine whether animals are actually sensitive to light when this condition is suspected. Examination for icterus should be an important point in differential diagnosis, since it clearly separates the photosensitivity of geeldikkop from other types, *e.g.*, hypericism. Until more attempts are made to get at the etiology of these diseases, little can be hoped for in the way of treatment or prevention.

The plant chiefly implicated in outbreaks in South Africa, *Tribulus terristris,* is now widely distributed in the United States, where it seems to have been introduced about the beginning of the century in ships ballast. It is quite commonly known here as "puncture weed" because of the disastrous effect of its sharp-pointed dry fruit on automobile tires. In many regions of the western United States, where the plant is particularly prevalent, climatic conditions must be similar to those of the Karoo veldt, and it is probable that some of the outbreaks of disease resembling photosensitivity which are occasionally reported from this region owe their occurrence to this plant. *Tribulus terristris* is also found in the Mediterranean region of Eurasia and is reported from Australia, but there is no report from these regions of photosensitization associated with it.

Chapter 15

Fagopyrism (Buckwheat Poisoning) and Other Diseases Attributed to Photodynamic Action

Depending upon which of two nomenclatures is accepted, *Fagopyrum* is the name of the genus to which the buckwheats belong, or the species name of the most common buckwheat, which is called alternatively *Fagopyrum esculentum* or *Polygonum fagopyrum*. The term *fagopyrism* has long been used to designate the disease symptoms associated with feeding on buckwheat. This is the most widely known of the diseases produced by light in domestic animals, being the only one mentioned in many textbooks of veterinary science; but its etiology is less clearly established than that of either hypericism or geeldikkop.

The First Postulate. Bichlmaier (1912), Merian (1915) and Lutz (1930) quote numerous references to early observations that sensitivity to light follows the feeding of buckwheat. The symptoms, like those of the diseases already discussed, include signs of sensory stimulation and lesions confined to white areas of skin. The first controlled experiments seem to have been those which Wedding carried out in 1887. He states that lesions did not appear on sheep and cattle fed on buckwheat and kept in the dark. In one instance he protected one side of a cow from light by covering with tar, and found that the skin lesions appeared only on the uncovered side.

In 1908 Öhmke reported that he was able to sensitize mice, guinea pigs and rabbits to light by feeding them on buckwheat seeds and hulls, and Fischer gives a similar account in the following year (quoted from Lutz, 1930). Bichlmaier (1912) states, to the contrary, that his extensive studies on white guinea pigs, rabbits and sheep failed to demonstrate any evidence of fagopyrism and Hilz (1914) bears witness to these negative results. But Merian (1915) seems to have had no difficulty in producing symptoms of photosensitivity in guinea pigs and rabbits fed on buckwheat plants. Sheard, Caylor and Schlotthauer (1928) were able to sensitize some animals but not others; for instance, they could not produce symptoms of photosensitivity in rabbits and rats, contrary to some of the above results. On the other hand, guinea pigs, swine and goats were readily sensitized after a period of four or more days' feeding. Their results are summarized in Table 11.

Lutz (1930) performed a number of experiments on mice and guinea pigs, from which it appears that the whole buckwheat plant was much more effective in producing photosensitivity than was the buckwheat

Table 11. Results of Feeding Buckwheat.
(After Sheard, Caylor, and Schlotthauer, 1928)

Animal	Order of Suscepti-bility	Food	Number of days before symptoms *	Symptoms
Guinea pigs	1	Freshly cut buckwheat and dry seed	4	Shaking of head, excitation, squeaking, scratching of ears and intense agitation. At the end of 1 week, 20 min. sunlight produced symptoms similar to anaphylaxis.
Swine	2	Macerated green plant, milk and bread	10	Itching, weakness, extensive urticaria, followed by sloughing. First reactions after 2 hr. sunlight.
Goats	3	Freshly cut buckwheat and dry seed	14	Itching and weakness.
Rabbits	0	Freshly cut buckwheat and dry seed		No symptoms, although exposed every clear day for 30 days.
Rats	0	Freshly cut buckwheat and dry seed		None.
Dogs	0	Green macerated plant and chopped meat		None.

* Exposed to sunlight. (No symptoms in any case with quartz-mercury arc; mild symptoms with carbon arc.)

seed. He found that symptoms could be produced in the guinea pig after two days of feeding on buckwheat plants, and could persist for as long as 36 days after buckwheat was removed from the diet.

In spite of the degree of conflict, it seems certain that sensitivity to sunlight results from the feeding of buckwheat, at least in some species, so that the first postulate may be considered as fulfilled for fagopyrism.

The Second Postulate. Busck (1905) attempted, with the assistance of Professor Kofoed, to isolate a photodynamic substance from buckwheat, using whole flowering plants. By extracting with alcohol and precipitating the chlorophyll with lead acetate after boiling with lead oxide, he obtained a red fluorescent substance which he called "fluorophyll." Unfortunately Busck's researches came to an end at this time, and he was never able to test the photosensitizing properties of the substance.

Since the time of Busck's investigations, there have been a number of attempts to isolate the photosensitizing pigment from various parts of the buckwheat plant, the results of which have not shown marked agreement. They will be treated in their chronological order.

Öhmke (1908) extracted buckwheat seeds and hulls with alcohol, and obtained a fluorescent solution which produced photosensitization when evaporated and given by mouth to guinea pigs and rabbits; the seeds and hulls from which the material had been extracted were inactive.

J. Fischer (1909, cited by Bichlmaier, 1912; and Lutz, 1930) obtained extracts of buckwheat with alcohol, ether or chloroform, which were brownish or green in color and fluorescent. Upon evaporation, characteristic bodies were precipitated, which Fischer claimed to find also in the blood of animals fed on buckwheat or injected with the extracts. He did not believe the green material was chlorophyll. Fischer's hypotheses as to the mode of action of this material is not convincing, and will not be discussed.

Bichlmaier (1912) could not repeat the findings of Fischer, but observed the same effects after the injection of alcohol as after the injection of alcoholic extracts of buckwheat. In neither case did he observe photosensitization.

Fessler (1913) studied a green pigment which he extracted from buckwheat hulls, and came to the conclusion that it was a kind of chlorophyll. He believed it somewhat different from the chlorophyll ordinarily found in plants and concluded that it is the photosensitizer in fagopyrism. He offers no further proof to support this conclusion.

Lutz (1930) isolated a number of pigments from different parts of the buckwheat plant. These included chlorophyll, which he did not believe to be different from the chlorophyll of other plants, nor to be the photosensitizing substance in fagopyrism. Another substance was rutin, a quercetin glycoside; the experiments of Horsley, cited in Chapter 10, may be taken to show that quercetin is not a photosensitizer. A third substance was isolated from the blossoms of buckwheat, which Lutz thought to be an anthocyanin. Buckwheat seeds or plants extracted with alcohol did not cause photosensitivity when fed, but the alcoholic extract diluted to 30 per cent with water and injected subcutaneously produced such symptoms.

A summary of this material leads to the conclusion that the existence of a photosensitizing substance in the buckwheat plant which is responsible for fagopyrism has not been definitely established. It seems certain that most of the extracts which were found by different investigators to produce photosensitization contained chlorophyll, so the evidence derived therefrom is of little value, for reasons pointed out in Chapter 12.

The Third Postulate. Lacking a clear demonstration of the second postulate, the third cannot be established. However, certain evidence merits discussion. Sheard, Caylor and Schlotthauer (1928) found that the serum of swine which had been fed on buckwheat exhibited two absorption bands at about 5800 and 6000 Å respectively, which do not appear in normal serum. They found corresponding bands in alcoholic extracts of buckwheat, which they felt were not due to chlorophyll. Lutz (1930) found that such extracts show absorption bands at about the same position. This may be taken as evidence that a substance absorbing in these regions is present in the plant, and that it is able to get through the walls of the digestive tract into the blood stream. It remains to be determined whether this is the photosensitizing substance responsible for fagopyrism.

Sheard, Caylor and Schlotthauer made some attempt to study the spectral region of sensitivity for guinea pigs fed on buckwheat, but unfortunately their experiments are not sufficiently quantitative to be of great value in the solution of the problem. They used amber filters which cut off the violet and most of the blue, and blue filters which cut off the orange, red and a good deal of the green; transmission in the infrared was almost the same for both glasses, but transmission in the ultraviolet was not measured. Guinea pigs exposed to sunlight through the amber glass showed symptoms of photosensitization, while those exposed through the blue glass remained normal. They conclude, from a qualitative consideration of the distribution of the wave-lengths in sunlight and the transmission of the glass filters, that the spectral region of sensitization lies at wave-lengths longer than 5800 Å. This would agree well enough with the absorption bands which they demonstrated in the serum of pigs fed on buckwheat and in the extract of buckwheat plants; on the other hand, it would not agree with the absorption spectrum of chlorophyll, nor with that of hypericin. Matthews (1938a) used dyed gelatin filters whose transmission in the visible was determined, although the ultraviolet was neglected, and found photosensitization to be limited to a region near 5800 Å, thus agreeing with Sheard, Caylor and Schlotthauer. The latter investigators suggest that the photosensitizing substance may be phylloporphyrin or cholehematin (phylloerythrin); but this is highly improbable, since these are both porphyrins, which absorb most strongly in the near ultraviolet, and should be activated by wave-lengths passing through the blue glass used as a filter by these investigators. A discussion of the absorption spectra, and spectral region of photosensitization for the porphyrins appears in Chapter 19.

The finding of Merian (1915) and Lutz (1930) that sunlight is effective through window glass is in agreement with those of Mathews and Sheard, Caylor and Schlotthauer. The latter were unable to produce symptoms of fagopyrism with a quartz-mercury arc, which is compatible with their other findings, since the mercury arc emits relatively little of its energy in the suspected region (see Figure 7, p. 26). Lutz cites experiments which he concludes to be instances of fagopyrism induced by a quartz-mercury lamp, but from his description it seems much more probable that the symptoms were those of sunburn (see Chapter 17), and this renders some of his experiments uninterpretable. The experiments of Merian with colored filters are inconclusive.

Smetana (1939) could not demonstrate increased O_2 uptake by tissues, including skin, of guinea pigs which had been fed on buckwheat and were sensitive to light. He reasons that if a photodynamic substance were present, increase in O_2 uptake would occur.

It is obvious that the etiology of fagopyrism needs further experimental study, and the investigations which have been discussed offer numerous leads. At present, it must be admitted that the last two of the three postulates necessary to demonstrate a photodynamic etiology have not been established.

Other etiologies have been suggested for fagopyrism, and should be mentioned. Damann (1902, quoted from Lutz, 1930) suggests that the symptoms result from a fungus associated with the buckwheat, which is made more active by sunlight and thus enters the skin more readily, an explanation which hardly merits further discussion. Lutz (1930), who believes that the etiology is of a complex nature, examined the possibility that insufficiency in diet, *e.g.*, lack of vitamin B, might be a part of the picture, but obtained no evidence to support this. It is not probable that the etiology is the same as that of geeldikkop, since there is no mention of icterus in association with fagopyrism.

Geographical Distribution of Buckwheat and Fagopyrism. Merian (1915) and Heupke (1933) give historical discussions of buckwheat in Europe, where it was well known by the sixteenth century. It seems to have been brought from Asia, as the name "sarrasin" (saracen), by which it is commonly known in France, suggests. Buckwheat grows readily on poor soil, and it has been introduced into many parts of the world for this reason; it is common in parts of Asia, Europe and North America. Other common names are: French, sarrasin and blé noire; German, Buckweizen and Heidekorn. The whole plant, either green or dried, is used for forage, and the seeds are also fed to domestic animals. The seeds are ground into flour and used for human consumption, but it is probable that large quantities must be eaten before the symptoms of fagopyrism appear, so that those addicted to buckwheat cakes need not be greatly concerned (p. 280).

From the experiments of Bruce (1917) and Bull (1927), it appears that wild plants closely related to buckwheat do not produce photosensitization although they have been frequently suspected. Six species of buckwheat are cultivated, of which *Fagopyrum esculentum* is the most common. *Fagopyrum tartaricum,* a more cold-resistant form, is grown principally in Northern Asia.

Symptoms and Occurrence. The symptoms of fagopyrism are so like those described for hypericism that they do not need further discussion. Symptoms which may not be directly associated with photosensitivity have been described by various writers. Bichlmaier describes certain hyperemia, swelling of the intestinal canal, and minor changes in other organs. Merian also reports gastrointestinal hemorrhage which recalls the symptoms found by Horsley (1934) in rats sensitized with hypericin.

Fagopyrism is generally considered as a disease of swine, sheep, cattle and horses, but Reuter (1920) lists it as a disease of chickens. It is probable that all animals are subject to the disease if they eat enough buckwheat, but the susceptibility varies greatly as indicated by the experiments of Sheard *et al.* (see Table 11).

It is probable that fagopyrism occurs more often in animals fed on buckwheat than is suspected. Pichon and Baissas (1935) describe mild symptoms in cattle which may have somewhat serious results. They observed that cows feeding on buckwheat often showed signs of more or less severe skin irritation, manifested by rubbing and scratching, when no outward signs of skin damage were seen; the udders and teats might be so irritated that the milk duct was occluded. Careful observation might reveal such conditions where not suspected, or where they had been attributed to other causes. The long period of time (days or weeks) during which animals may remain sensitive after eating the plant is shown by experiments of Lutz (1930). This might lead to puzzling situations where animals which, because of the protection of cloudy weather, had not shown symptoms of fagopyrism while eating buckwheat might develop such symptoms upon the advent of sunny weather after being removed from such pasture.

Lechuguilla Poisoning (Goat Fever, Lechuguilla Fever)

Mathews (1937, 1938a, b) finds part of the symptoms which follow feeding on *Agave lechuguilla*, found in the arid regions of New Mexico, Mexico and Texas, to result from photosensitization. Stock feed on this agave only when other forage is not available. They develop icterus, liver and kidney lesions, and edema of the face and ears. Mathews separates two toxic principles, one responsible for the liver and kidney lesions, and the other for the photosensitivity.

The First Postulate. His experiments to demonstrate the photosensitivity in sheep, goats and white rats fed on this agave are somewhat variable; they could not be obtained during the winter months.

The Second Postulate. Mathews found that water or alcohol extracts of *A. lechuguilla* produced photosensitivity when given by mouth; he did not inject them, nor purify and determine their chemical nature.

The Third Postulate. Window glass, or a filter composed of acridine yellow, did not diminish the sensitivity to sunlight, whereas a number of other filters did. The spectral limits in the visible are given for the filters, but no transmission values are given, and unfortunately the ultraviolet has been disregarded so that it is difficult to draw conclusions from these experiments. Mathews found that animals fed on buckwheat were sensitive to sunlight passing through the same filters.

"Clover Disease"

Early descriptions of various symptoms which appear to have been the results of sensitivity to light, and which were associated with feeding on lucerne and vetch, have been reported in Chapter 12 (p. 127). Merian (1915) lists a few accounts of similar outbreaks in the same region, *i.e.*, northern Germany, and numerous other reports from various parts of the world, Canada, United States, Australia, as well as Europe, have appeared in the literature of veterinary science since that time; a number of these are cited by Lutz (1930). The symptoms are generally associated with feeding on leguminous plants which are ordinarily assumed to be important forage plants. Lucerne or alfalfa *(Medicago sativa)*, burr clover or trefoil *(Medicago denticulata)*, some true clovers *(Trifolium sp.)* and vetch *(Vicia sp.)* have all been suspected. Numerous textbooks of veterinary science list "clover sickness," or "Kleekrankheit," as a specific disease of domestic animals. So little experimental evidence has been produced that it is impossible to know the etiology. There seems little doubt that some cases at least involve sensitivity to light, but it is possible that more than one etiology is involved.

Merian (1915) assumed that the condition reported from north Germany by Erdt, Schrebe, Steiner, Burmeister and others (see Chapter 12) was really fagopyrism, although the presence of buckwheat is not mentioned in these articles. The probability of the presence of *Hypericum* among the leguminous plants seems much greater, but Merian was apparently unaware of the existence of hypericism and thought fagopyrism to be the only form of photosensitivity. In certain instances where photosensitivity has been described, *Hypericum* was found to be present among leguminous plants and the symptoms were undoubtedly due to the former. This may be the explanation of other instances of the disease (see Paugoué, 1861). However, Dodd (1916) reports the condition among sheep in South Australia which were feeding entirely on trefoil, where no *Hypericum* was present. Other reports from this region (Byrne, 1937a, b) where the presence and effects of *Hypericum* are recognized, leave little doubt that this is not the cause in all cases of photosensitivity associated with feeding on leguminous plants.

The accounts of Schrebe (1843), Steiner (1843) and Burmeister (1844) report the condition as more or less epidemic, associated with heavy infestation of the plants with aphis. They describe the plants as covered with honey dew, an excretion of the aphis, and "mildew." Saprophytic fungi are generally found growing upon the honey dew, and these were probably described as "mildew." According to Dodd (1916) the condition which he describes was generally known among the sheep raisers of Australia as "aphis disease" because of the association of these insects with the symptoms, although Dodd could not confirm this. The possibility that the insects, or more probably the fungi, play a part in the disease cannot be discarded without definite evidence, but it seems more likely that their presence was associated with a particular condition of the plants in which they are toxic. This recalls the fact that the presence of certain insect larvae on the *Tribulus* plants is associated with outbreaks of geeldikkop, in which disease the relationship to climatic conditions and a particular state of the plant has been emphasized (see Chapter 14). This all points toward the possibility that at least some of the outbreaks of photosensitivity following feeding on leguminous plants are of the same nature as geeldikkop. The apparent epidemic nature of the outbreaks supports this view, since they may be precipitated by climatic conditions favorable to the development of toxicity by the plant. In fact, the early writers, Schrebe, Steiner and Burmeister, mention that drought conditions existed during the years in which the outbreaks occurred. On the other hand, there are reports of the occurrence of such outbreaks when the growth of the plants was particularly luxuriant. Fröhner (1927) lists icterus as a symptom, but although I have looked with some care for the mention of icterus in these reports, I have found nothing definite. Some of the cases reported may have been geeldikkop, but it cannot be assumed that all are instances of this disease.

A third possibility is that, like *Hypericum,* the leguminous plants contain a specific type of sensitizer. Such a substance might be present in greater quantity at some times than at others, thus accounting for the sudden outbreaks of the disease, or, on the other hand, the condition might be present at all times, but not noticed except when the animals are suddenly exposed to sunlight. Dodd (1916) claims to have produced photosensitivity in guinea pigs by feeding on trefoil, but these results were not confirmed by Bull and Macindoe (1926). Byrne (1937a) found that swine fed on *Medicago sativa* and *M. denticulata* became sensitive to light, but recovered when changed to other pasture; he states (1937b) that the only way to combat the disease among pigs in Australia is to change to feed other than the leguminous plants.

Certainly the problem merits a thorough study since the plants suspected are so commonly used for forage. The necessity of fundamental studies to supplement observations in the field cannot be too much emphasized.

Other Conditions Reported

It is not improbable that other plants are capable of producing photosensitivity either of the hypericism or the geeldikkop type. Mention can be made of only a few other conditions which have been reported but upon which there seem to be no experimental data available.

The following quotation is from Charles Darwin's "Origin of Species:" "From facts collected by Heusinger, it appears that white sheep and pigs are injured by certain plants, whilst dark-colored individuals escape: Professor Wymann has recently communicated to me a good illustration of this fact; on asking some farmers in Virginia how it was that all their pigs were black, they informed him that the pigs ate the paint-root (Lachnanthes), which coloured their bones pink, and which caused the hoofs of all but the black varieties to drop off; and one of the 'crackers' (*i.e.,* Virginia squatters) added, 'we select the black members of a litter for raising, as they alone have a good chance of living.'" This has frequently been cited as an example of photosensitivity but all the references which I have found lead back to this original statement. I have made inquiry as to the occurrence of such a condition from the late Dr. A. B. Clawson of the United States Department of Agriculture, who specialized in the toxic plants of this country. He told me that he had searched for further evidence of this condition but had never found it. I would be very thankful if any reader could inform me in this regard.

Schindelka (1903) lists the skin eruption which follows feeding of potato peelings or raw potatoes, as a type of photosensitivity, but he gives no further information and I have been unable to find any original reports of such a condition.

There seems adequate evidence that photosensitivity may follow the exclusive feeding of maize to laboratory animals, a topic which will be discussed in Chapter 24 in connection with the etiology of pellagra. I have been able to find no references to indicate the importance of such a condition among domestic animals.

Lubberink (quoted from Lutz, 1930) reports a disease from the Dutch East Indies known there as "blaische ziekte," which may be geeldikkop. I have not had opportunity to read the original article, but Lutz mentions icterus among the other symptoms.

Howarth (1931) describes an outbreak of photosensitivity in white-faced sheep pastured on sudan grass. The symptoms were the typical sensory manifestations, accompanied by swelling and lesions of the face, head and ears. Blackfaced sheep on the same pasture were not so affected. No experiments were attempted.

Clawson and Huffman (1937) have described a disease in the state of Utah, "Bighead in sheep," which they suggest to be photosensitization caused by feeding on *Tetradymia glabrata*. No experiments are quoted to prove that the sheep are truly sensitive to light, and there is no report of the presence of icterus.

Dr. Robert Jay of the United States Department of Agriculture informs me * that animals which have fed upon bracken fern become sensitive to sunlight.

The possibility that some of the cases which have been reported as abnormal sensitivity to light are merely instances of sunburn resulting from excessive exposure has been suggested in Chapter 12 (see also Chapter 17).

Photosensitivity after Administration of Drugs, etc. Photosensitivity may follow the administration of certain substances for therapeutic or other reasons. The topic will be discussed at some length in Chapter 23, and only those references will be given here which apply particularly to domestic animals.

Bourbon and Cholett (1927) and Cholett (1933) report that photosensitivity follows the injection of male fern, and that precautions should be taken to protect the animals from light when this therapeutic agent is used.

* Personal communication.

Schantz (1915) and Hauptmann (1922) discuss the results of widespread feeding of barley treated with eosin in compliance with a regulation of the German government. This is also indirectly mentioned by other writers, but the exact details seem somewhat uncertain. It appears that a ruling was passed in 1908 which permitted the customs authorities to mark with eosin that barley which was intended for the feeding of domestic animals, to distinguish it from barley to be used for brewing purposes. Researches were carried out which indicated that the eosin-treated barley had no bad effects, and the order went into effect in September, 1909. According to Schantz, complaints began to appear immediately after. These reported detrimental effects on health and on the quality of meat of slaughtered animals, particularly swine, resulting from the feeding of the eosin-treated barley; even sudden death of such animals was reported. As a result, further feeding experiments were undertaken, which purported to show no detrimental effects from feeding the eosin-treated barley. According to Schantz, these experiments were not conclusive because they were carried out during the winter and ended at the beginning of summer, so that the animals may not have been exposed to sufficient sunlight to permit the appearance of photosensitivity. At any rate, the enforcement of the ruling seems to have been relaxed.

Schantz' article, which appeared during the World War, seems to have been written as a warning against the proposed marking with eosin of a large quantity of barley intended for the feeding of animals, and points out the danger of photosensitization. The article is accompanied by a rebuttal by E. Rost, who flatly denies the importance of Schantz' claims. From Hauptmann's article (1922) and from reading between the lines, or indeed from reading the lines themselves, one gathers that there was a good deal of personal feeling involved which war hysteria did not improve. One gathers from Hauptmann that the proposal for marking barley with eosin was not put into effect.

Whatever the merits of the particular case, it seems certain that if eosin is fed, it will cross the intestinal wall into the blood-stream and be carried to the skin (see the observation of Prime p. 268). If the animals are kept in the dark or in weak light it should be possible to feed the substance in large quantities without any ill effects, but if they are exposed to sunlight symptoms of photosensitivity might appear even though the quantity of eosin fed was relatively small. The details of the experiments upon which the opinion that eosin is not toxic was based are not given, so that it is impossible to estimate their value nor that of Schantz' warning. Schantz was certainly right, however, in pointing out the possible danger, particularly when so many complaints had appeared regarding untoward effects resulting from feeding of the eosin-treated barley.

Part IV

Diseases Produced by Light in Man

Chapter 16

Abnormal Sensitivity to Light in Man

General. The existing confusion of ideas concerning diseases caused by light can be understood and appreciated only when the historical aspects of the subject are considered. Much interest and confusion centers around one disease which is often considered as a "type" example although its etiology and relationship to light are far from certain. Detailed historical account of this disease, Hydroa vacciniforme, must be postponed to Chapter 19 where its various aspects will be treated. In this chapter only a brief historical discussion of classifications and nomenclature of diseases allegedly caused by light can be presented. This will be followed by a tentative arrangement based upon experimental evidence.

Historical. Although Willan described sensitivity to light under the name Eczema solare in 1798 (see Bateman, 1818, and Rasch, 1926), and Rayer was acquainted with such conditions in 1826, it was not until the latter part of the last century that they began to attract attention. In 1860 Bazin described, in a general work on dermatology, three types of bullous lesions which he grouped under the name Hydroa. One of them, Hydroa vacciniforme, was represented by the case of a young boy in whom the attacks appeared only after he had been taken for walks in the open air. Bazin stated that light was a precipitating factor, although he made no attempt to reproduce the lesions by exposure of the patient to light. Since that time a considerable number of similar cases have been reported, and the term Hydroa has come to be associated with sensitivity to light, its use in the broader sense of Bazin having almost disappeared.

In 1878 Hutchinson described a condition which manifested itself as itching papules distributed chiefly on the exposed parts of the body, but in some instances extended to parts covered normally by clothing. This he called summer prurigo. He attributed the eruption to the heating effects of the sun's rays, but like Bazin made no attempt to reproduce the lesions by exposing the skin to sunlight under controlled conditions. In 1889 he described a case of bullous eruption on the exposed parts of the body, which he considered similar to Xeroderma pigmentosum, a disease attributed to sensitivity to sunlight by Unna (1894). Hutchinson, who was apparently unacquainted with the work of Bazin, applied the name summer eruption to his case, but Brocq (1894) claimed that the condition was the Hydroa vacciniforme of Bazin.

Confusion soon arose regarding the two conditions described by

Hutchinson, and the term summer eruption came to be applied to all sorts of skin conditions in which sensitivity to light was suspected. From time to time physicians called attention to this confusion, pointing out that the macroscopic appearance of the lesions would not allow them all to be placed in the same category, without, however, making any important attempt to separate them on the basis of their possible etiologies. Many new names were introduced, which were based on the superficial characteristics of the various individual cases, such as Eczema solare, Hydroa aestivale, and others.

Most early attempts to classify these diseases were based on the macroscopic and microscopic appearance of the lesions; usually the only evidence that light was an etiological factor was the limitation of the eruption to parts of the body not protected by clothing. Sometimes the patient might give a history of exposure to sunlight just before an attack, but seldom was there an attempt to reproduce the lesions by exposure. In a few instances effort was made to determine the wave-lengths to which the skin was abnormally sensitive, but these were usually confused by the belief that only the so-called "chemical" or "actinic" rays (p. 29) could be effective. It had been shown in the latter half of the past century that sunburn of normal skin is induced by ultraviolet radiation (Charcot, 1858; Widmark, 1889; Hammer, 1891; and others), having been previously believed to result from heat, and hence there was a tendency to interpret all abnormal effects as due to "chemical" rays. The exact delimitation of the radiations which produce erythema in man, which will be discussed in the following chapter, was not accomplished until much later.

Interest was aroused by the discovery of porphyrin in the urines of two Hydroa patients studied by McCall Anderson in 1898, and it was soon found in other cases. In 1908 Hausmann demonstrated that hematoporphyrin is a photosensitizer for photodynamic action, and this was immediately related by Ehrmann (1909) to the association of porphyrins with Hydroa. Following this there was a tendency to attribute *all* abnormal sensitivity to light to the photosensitizing action of these compounds. In fact, in current literature, description of cases of photosensitivity are often accompanied by the categorical statement that porphyrin is the sensitizer, even when no attempt has been made to demonstrate abnormal quantities of porphyrin, and no real evidence is advanced to support the contention.

More recently, attempts have been made to explain photosensitivity as an allergic response (Duke, 1925b, 1935; and Jausion and Pagès, 1933). As will be pointed out in succeeding chapters (*e.g.*, Chapter 18) it is unnecessary to invoke the concept of allergy to explain abnormal sensitivity to light. In some cases an allergic response may form a part of the total picture, but any generalization from this seems only to confuse the issue.

Existing Classifications of Diseases Influenced by Light. There have been numerous attempts to classify the pathological photosensitivities of man. The first of importance was that made by Möller in 1900, who separated two classes of abnormal sensitivity to light under which he placed subgroups. Möller exposed some of his patients to light under various conditions. He concluded that in all these diseases the lesions were brought about by light from the ultraviolet region of the spectrum, or what he designated as "chemical rays." Since he recognized no difference in the exciting radiation his classification, which follows, is based only on the macroscopic appearance of the lesions:

(1) Conditions in which the skin appears normal except for its sensitivity to light.
  (a) Dermatitis papulo-vesiculosa, or Eczema solare (under this grouping Möller includes Hutchinson's summer prurigo, although he is doubtful that this is definitely a photosensitivity).
  (b) Hydroa aestivale.
    First type: Hydroa aestivale vacciniforme
    Second type: Hydroa aestivale vesiculo-bullosum
    Third type: Atypical cases.
  (c) Xeroderma pigmentosum.
(2) Conditions in which the sensitivity of the skin to light depends upon a previously existing pathological condition.
  (a) The action of light on pellagra.
  (b) The action of light in variola.

A later classification is given by Günther (1919), who was particularly interested in the relationship of porphyrins to light sensitivity.

(1) Hematoporphyria.
(2) Hydroa aestivale.
(3) Xeroderma pigmentosum.
(4) Other light sensitivities associated with skin symptoms.
(5) Light sensitivity of the eyes.
(6) Light sensitivity as a symptom accompanying other diseases.

Here no particular attempt is made to separate fundamental etiological factors. It is not quite clear whether Günther wishes to consider poryphrins as the casual factor in all these diseases or not; he does not find porphyrins associated with all diseases, but an occasional remark makes it appear that there is some confusion of concepts.

Hausmann and Haxthausen (1929) present a more extensive classification which may be summarized as follows:

I. The action of light on pathologically light-sensitive skin.
  (A) Hydroa aestivale.
    (1) Hydroa aestivale with congenital porphyrinuria.
    (2) Hydroa aestivale with chronic porphyrinuria.

(3) Hydroa aestivale without porphyrinuria.
(B) Chronic polymorphic light eruption.
(C) Pellagra.
(D) So-called pellagroid eruption due to light.
(E) Unclassified eruptions due to light, accompanying metabolic and other internal disturbance.
(F) Sensitization of the skin to light by medicaments.
(1) Exogenous
(2) Endogenous
(G) Xeroderma pigmentosum.
(H) Late Xeroderma pigmentosum, Seemanshaut other photogenic forms of skin cancer.
(I) Degeneration of elastic and collagen tissues of the skin.
(J) Photoelectric dermatitis.

II. Skin diseases exacerbated or precipitated by light.
(A) Action of light on acute exanthemic infections.
(1) Variola
(2) Other acute exanthemata
(B) Intrinsic skin diseases.
(1) Acne vulgaris
(2) Eczema
(3) Multiform exudative erythema
(4) Lupus erythematosis
(5) Pityriasis streptogenes faciei
(6) Psoriasis
(7) Rosacea
(8) Toxic exanthemata
(9) Urticaria
(10) Other skin diseases.

One sees here a considerable extension of the number of skin affections in which light is assumed to be a more or less important factor. The separation into two groups, those in which the skin is assumed to be pathologically sensitive to light, and those in which light is an excaberating factor only, is interesting and recalls the earlier classification of Möller.

The classification of Jausion and Pagès (1933) is based on quite different conceptions of the etiology of these diseases; he admits that it is only a schematization, for in his opinion the etiologies may be mixed and may present very complex pictures. Some aspects of other classifications will be discussed elsewhere in the following chapters. Those presented suffice to show how extensively light has been considered an etiological factor.

Unfortunately there has been relatively little experimental effort to establish the real importance of light in these various diseases, and by far the greater part of the evidence is based upon clinical observation only. Without wishing to detract from the importance of clinical observation, which must be the starting point for all experimental medicine, I must point out that further evidence is necessary before the etiology of a given disease can be definitely established. The classification of these diseases from purely clinical observation is an important step, but one which can be considered only tentative until experiment has established their true etiology. In the development of the study of diseases produced by microörganisms, the first step was clinical classification, but with the development of the germ theory of diseases and accompanying experimentation, it was possible to establish an etiological factor for many diseases. Sometimes this resulted in a complete re-classification on an etiological basis rather than on the basis of clinical resemblance. In the same way, it is to be expected that an experimental study of light-sensitive diseases will result in their re-classification, so that none of the existing classifications can be more than tentative.

The present study is an experimental one, and will deal only with experimental material. This is quite meager, and in reviewing it one finds that many alleged light sensitive conditions have not been experimentally demonstrated. This should not necessarily mean that they are unworthy of consideration, but rather that they require more careful examination before they may be accepted or discarded. In reviewing the literature of the subject, it has seemed to the author that the scantiness of experimental studies in proportion to clinical observation is an important justification for the preparation of such a work as the present. The task of winnowing the material has been an arduous one, but it is hoped that the service may be of value to those placed in a position to carry on further investigation, and that it may help to point the direction in which progress may be made.

A Simple Tentative Classification. A much simpler classification of diseases caused by light in man may be constructed upon a theoretical and experimental basis. This classification, like those which have preceded it, must be regarded as tentative only, to be modified at any time the experimental evidence requires.

As has been pointed out repeatedly in the preceding chapters, the first step in any photobiological process is the absorption of a quantum of radiation by a particular substance in the living system. Likewise, in all diseases produced by light, the first step must be the absorption of a quantum by some substance in the skin. Since all substances absorb radiation of certain wave-lengths (certain sized quanta) only, a clew to the nature of the light absorbing substance should be given by the wave-lengths which produce the lesions. The relationship between action and absorption spectra has been discussed at length in Chapters 3 and 5. Thus if the wave-lengths producing the lesions can be determined, it should be possible to make some sort of classification having true etiological significance.

In all attempts to separate these diseases on the basis of wave-length, it must be remembered that normal skin reacts to a definitely limited spectral region in the ultraviolet having a long wave-length limit at about 3300 Å. Wave-lengths shorter than this produce the sunburn which occurs in normal individuals, whereas longer wave-lengths have but little effect upon normal skin (see Chapter 17). It seems obvious that derangement of the sunburn mechanism might result in abnormal response to wave-lengths shorter than 3300 Å, and since this mechanism is rather complicated and involves more than one tissue, lesions having different morphologies might be expected. It is probable that the majority of abnormal photosensitivities are of this type.

On the other hand, conditions occur in which the lesions are produced by wave-lengths outside the sunburn spectrum. In these, the presence of a photoactive substance not normally found in the skin, or normally inactive, must be admitted. Such conditions are more or less comparable to those of hypericism and geeldikkop in domestic animals. It is useful to group all such conditions together in a general category.

The following general classification may be made:

(A) Skin lesions produced only by radiation which induces the sunburn of normal skin (wave-lengths shorter than 3300 Å) (see particularly Chapters 20 and 21).

(B) Skin lesions due to abnormal photosensitizing substances in the skin (best exemplified by the conditions described in Chapter 18).

While it is possible that a given condition belonging to type B might have the same wave-length limits as type A, this is not highly probable. A convenient fact is that ordinary window glass cuts out wave-lengths shorter than about 3200 Å, but transmits the longer wave-lengths (see Figure 36, p. 194); thus it may be used in the experimental separation of A from B. If it is found that the lesions can be reproduced by wave-lengths which pass through window glass, the case may be placed in category B; but if the lesions are produced only by radiation which does not pass through window glass, the case almost certainly belongs to category A. In the following chapters, further subdivision of the two types will be considered.

It may be pointed out that such classification is of more than "academic" interest, for progress in the treatment of these diseases cannot be hoped for until more is known of their etiologies. Too often the assumption is made that all the diseases caused by light may be thrown into a common etiological group and a single method of treatment adopted. Again, the efficacy of ameliorative treatment (specific treatment seems thus far lacking) may depend upon a knowledge of the wave-lengths to which the patient is sensitive; for example, it is possible in some cases to protect the patient by shielding with window glass, but only in conditions belonging to category A.

A glance at the older classifications given above will indicate one difficulty in the analysis of experimental studies. The somewhat confused nomenclature * and the difficulty of making accurate written descriptions of dermatological conditions, which always leave a certain degree of uncertainty as to whether a given number of cases are really comparable, make analysis of the published literature difficult. It is believed by many dermatologists that the same etiological factor may produce quite different clinical pictures, particularly in the group of conditions described as eczema, and it is also thought that the same disease may present quite different lesions in different phases of its development (see, *e.g.*, Scholtz, 1936). With specific regard to diseases in which light is an etiological factor, Haxthausen (Hausmann and Haxthausen, 1929) states that in chronic polymorphic light eruptions the morphological character of the lesions may vary in the same individual from one attack to another.

In many case reports the only evidence presented to show that the lesions described are evoked by light is their limitation to those parts of the body not covered by clothing. This is not always a reliable criterion, because uncovered parts of the body are exposed not only to light, but to such other environmental factors as wind, air-borne substances, and trauma. Allergy to pollens is reported to result sometimes in lesions at the region of contact, which is limited to the uncovered portions of the body as a rule, and hence might be confused with abnormal sensitivity to light.

The need of making experiments to determine whether the lesions may be reproduced by light, before diagnosing abnormal photosensitivity, must be obvious. Unfortunately such experiments offer difficulties, and there is no standardized method at present. The design and results of such experiments will be treated in subsequent chapters, but the mechanism of sunburn will first be discussed.

* For the nomenclature of dermatological lesions see, *e.g.*, Ormsby (1927).

Chapter 17

The Response of Normal Skin to Light: the Sunburn Spectrum

A certain limited part of the solar spectrum, consisting of wave-lengths shorter than about 3300 Å, produces a specific series of responses in normal skin. The first is erythema (reddening), which, with moderate doses of radiation, appears only after a few hours; this may persist for some days. It is ordinarily followed by pigmentation (tanning) which begins to appear before the erythema has completely subsided, and may persist for months. Both erythema and pigmentation are limited to the area exposed to the radiation. The amount of erythema and the latent period before its appearance vary according to the intensity and duration of the irradiation, as does the amount of pigmentation. With large doses of radiation there may be some edema of the skin, and there is usually a certain amount of desquamation (scaling off) which carries some of the pigment with it. Excessive irradiation may result in blistering and desquamation, and may be attended by severe secondary effects. In this book, the term *sunburn* will be used to describe this total picture, including all degrees from the mildest erythema to severe blistering.

Wave-lengths longer than those which produce sunburn, *i.e.*, ultraviolet longer than 3300 Å, visible and infrared, seem to have little specific effect on the skin of normal individuals (see p. 186 however). Radiation of any wave-length may produce a heat erythema or burn, and even pigmentation if intense enough, but this cannot be regarded as a specific reaction (see p. 40).

The antirachitic spectrum lies in the same general region as the sunburn spectrum (see Bunker and Harris, 1937), but this process has no relationship to sunburn. There is no good evidence that the antirachitic mechanism is of importance in abnormal sensitivity to light (but see p. 255), whereas the sunburn mechanism is undoubtedly involved in many instances of such abnormality.

Before discussing abnormal photosensitivity in man, which will occupy the remainder of this book, it will be necessary to know something of the mechanism of sunburn; and the present chapter is therefore devoted to this subject. It is probable that erythema and the principal phase of pigmentation result from the same initial process, but they will be treated separately for convenience of discussion.*

* Certain systemic responses may follow irradiation with the sunburn spectrum (*e.g.*, see Laurens, 1933; Mayerson, 1935; Ellinger, 1935), but these are for the most part outside the province of this book.

Erythema

The Action Spectrum. The action spectrum for the production of erythema in normal human skin has been determined by several groups of investigators whose data, which are gathered together in Figure 31, are in rather good agreement. The methods for determining the erythemic response were somewhat different. Hausser and Vahle (1922) and Haus-

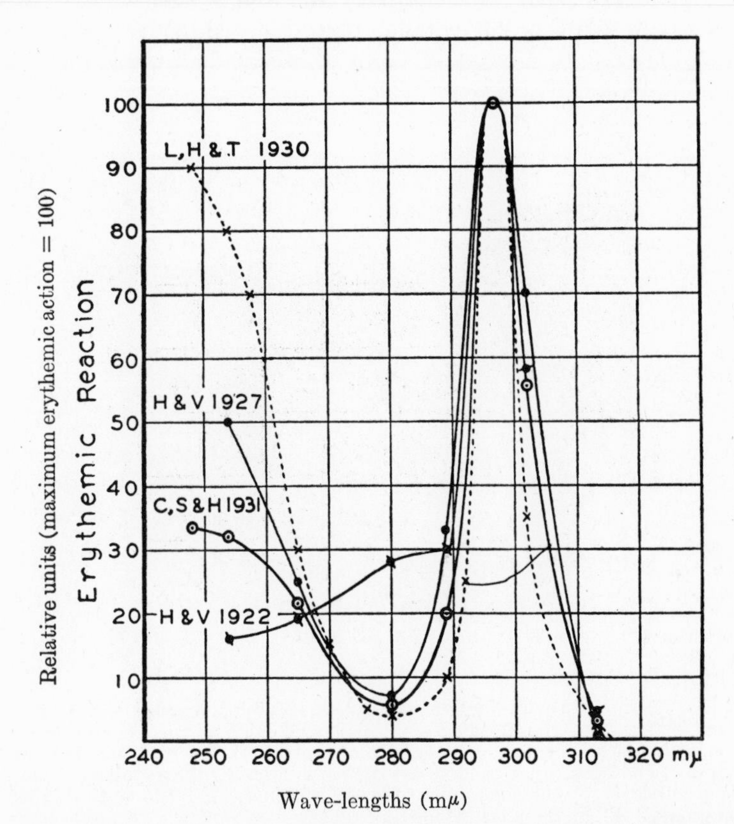

FIGURE 31. Action spectrum for erythema production. L, H, & T, 1930; Luckiesh, Holladay and Taylor (1930). H & V, 1927: Hausser (1928). C, S & H, 1931: Coblenz, Stair and Hogue (1931). H. & V, 1922: Hausser and Vahle (1922). [From W. W. Coblenz, R. Stair, and J. M. Hogue, *Bureau of Standards Journal of Research,* 8, p. 541 (1932).]

ser (1928) used the degree of erythema developed, while Lukiesh, Holladay and Taylor (1930) and Coblenz and co-workers (1931, 1932) used the appearance of minimal perceptible erythema. The latter seems the more accurate method. Coblenz and Stair (1934) have revised all these data, and their set of values has been adopted in this book, for example in Figures 8 D, 36 and 46. The action spectrum shows a maximum at about 2500 Å, a minimum at 2800 Å, and a very sharp maximum at about

2970 Å; it falls sharply from the latter maximum, nearly reaching zero at 3200 Å, but tailing off gradually from this point toward higher wave-lengths. The selection of a long wave-length limit is arbitrary, but there can be little effect beyond 3300 Å.

Coblenz uses the term "erythemic radiation" to describe those ultra-violet wave-lengths which produce erythema in normal human skin, but since x-rays, radiation from radium, and other agents, also produce erythema, the name is not without objection. Other terms have been proposed but none is universally accepted, and I shall use the terms *sunburn radiation* and *sunburn spectrum.*

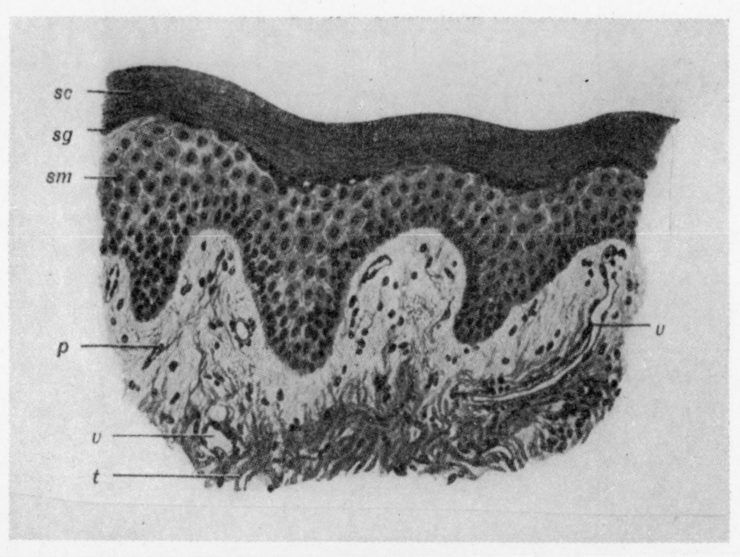

FIGURE 32. Section through human skin from shoulder region. sc, stratum corneum; sg, stratum granulosum; sm, prickle cell layer of the mucosum; p, papillary layer of the corium; v, vessels in the papillary layer; t, reticular layer of corium. The dark layer of close-set nuclei just above the papillary layer is the basal cell layer of the mucosum. (From Maximov and Bloom's "Textbook of Histology," W. B. Saunders Company.)

Transmission and Absorption of the Sunburn Radiation by Human Skin. Figure 32 represents a section through a piece of human shoulder skin cut perpendicular to the surface. Several layers may be distinguished. The corneum or horny layer (stratum corneum) forms the outside surface and is made up of non-living material derived from the layers lying beneath. Next, proceeding inward, is the granular layer (stratum granulosum), represented in Figure 32 by a few scattered masses of cells. Beneath this is the prickle cell layer (stratum spinosum), and below this the basal cell layer which appears in Figure 32 as a wavy line of cells with darker nuclei. These layers make up the epidermis; the last two are often called the stratum germinativum (also mucosum or Malpighian layer) because they represent the region of the epidermis in which new cells are being formed by division, to degenerate subsequently and form the corneum. Below the epidermis is the corium, or dermis. The wavy line which forms the sharp boundary between epidermis and corium at the lower border of the basal cell layer in Figure 32 represents sections through tiny teat-like projections of the corium, the dermal papillae, which contain the most superficial of the blood vessels; sections through these vessels may be seen in the figure at *v*. This layer is called the papillary layer of the corium and overlies the reticular layer. Beneath the corium is subcutaneous tissue.

The measurement of absorption of radiation by skin is attended with difficulties. A great portion of the incident light is scattered in the skin; and if one attempts to measure the amount passing through by placing an excised layer between source and spectrograph, as is the usual practice for measuring absorption by homogenous systems, the radiation entering the slit of the spectograph is only that which leaves the skin at right angles to the surface, and that which has been scattered in other directions is not measured. This results in absorption values which are too high. To estimate the true absorption of the epidermis, Lucas (1931) "cleared" the tissue with glacial acetic acid or glycerol, presenting experimental evidence to show that these clearing agents did not permanently alter the absorption by the epidermis. He used scattered as well as parallel incident radiation in his measurements. He concludes that the true absorption spectrum of epidermis is approached by those measurements in which epidermis cleared in glacial acetic acid, and scattered incident radiation were used.

Some of Lucas' absorption spectra for epidermis are re-drawn in Figure 33. They show a maximum at about 2700—2800 Å, with a minimum at about 2500—2600 Å, an increasing absorption toward the shorter wave-lengths, and extremely little absorption by wave-lengths longer than 3300 Å. As has been pointed out in Chapter 3, such an absorption spectrum is characteristic of many proteins and the absorption spectrum of serum albumin has been introduced in Figure 33 to show this resemblance. Protein absorption in this region of the spectrum is probably due to benzenoid amino-acid structures, and the absorption spectra of tyrosine and trytophane have been introduced in Figure 33 to show their general resemblance to those of proteins. It is quite probable that the principal absorption of sunburn radiation by the epidermis is due to proteins containing benzenoid amino-acid structures in their configuration, although other substances may contribute significantly.

Measurements by Hausser (1928) and by Bachem (1929, 1930) show the same type of absorption for the epidermis. Bachem's measurements show some differences in the absorption by different layers, but the absorption spectra of all the epidermal layers are similar, and resemble those of proteins. It is difficult to make comparisons between the measure-

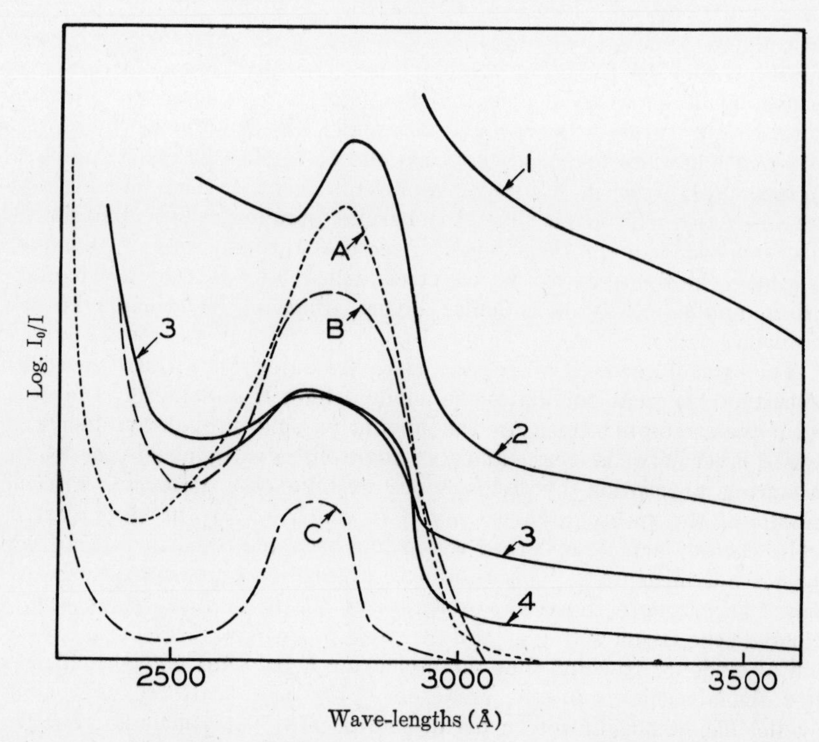

FIGURE 33. Absorption spectrum of human epidermis. Ordinate units chosen to bring the data into comparison. (See 6 to 11, Chapter 2.) The units for curves 1, 2, 3, 4 are the same.

Human epidermis 0.08 mm. thick. 1, In water, incident light parallel. 2, Cleared in glycerin, incident light parallel. 3, Cleared in acetic acid, incident light parallel. 4, Cleared in acetic acid, incident light scattered (from Lucas, 1931).

A, Absorption spectrum of Tryptophane (from Smith, 1928). B, Absorption spectrum of Serum albumin (from Hicks and Holden, 1934). C, Absorption spectrum of Tyrosine (from Smith, 1928).

ments made by various investigators because different methods have been used. The thickness of the various layers is not the same for different regions of the body surface, and the individual layers are not of uniform thickness, *e.g.*, the thickness of the stratum germinativum is much less over the papillae than between them, as is shown in Figure 32. Thus it is impossible to make more than rough estimates of the depth of penetration of the sunburn radiation. It is certain, however, that very little radiation of wave-lengths shorter than 2900 Å reaches the Malpighian layer, and such radiation does not penetrate to the corium. Wave-lengths shorter than 2900 Å produce erythema, which is the macroscopic manifestation of dilation of the minute blood vessels of the corium, but since they do not penetrate to these vessels this cannot be a direct effect. The longer sunburn wave-lengths (*e.g.*, 3000—3300 Å), on the other hand, may penetrate to the corium, but even these are absorbed before they reach the subcutaneous tissue (see Figure 41, p. 202).

Site of the Primary Changes. The photochemical changes which lead to sunburn must be most intense in the epidermis, since the sunburn radiation penetrates very little below this layer. Keller (1924a) and Guillaume (1927) agree that the principal histological change is a degeneration of the prickle cells, although this does not appear until some time after the irradiation, indeed long after erythema has appeared. The basal cell layer is not appreciably affected even at the tips of the papillae, which may be closer to the surface than some of the prickle cells (see Figure 32). Thus it seems probable that the principal site of the initial photochemical changes is in the prickle cell layer, although there is a certain degree of accumulation of blood cells in the papillary layer and even more superficially, which is probably secondary. Both the Malpighian layer and the corneum subsequently become thickened.

The Latent Period. The long latent period which ordinarily intervenes between exposure to light and appearance of erythema indicates a certain degree of separation between the initial photochemical process and the final result. This separation is further attested by the effects of temperature on the erythemic process, which have been studied by Clark (1936). She finds that the threshold time, *i.e.*, the period of irradiation required to produce a just perceptible erythema, is very little influenced by temperature; but that, on the other hand, the latent period for the appearance of erythema is markedly affected by temperature, having a temperature coefficient (Q_{10}) of 2.3 (see p. 17). The simplest explanation is that the erythemic mechanism is composed of two rather distinct parts. The first is a photochemical reaction which determines the amount of radiation necessary to produce the erythemal response, which like most photochemical reactions has a low temperature coefficient. This reaction takes place principally in the prickle cell layer, but sets off other reactions which result in the production of substances which cause the dilation of the minute vessels in the papillary layer. It is these secondary reactions which occupy the latent period, and which are dominated by a thermal process so that the overall temperature coefficient is relatively high.

The Absorbing Substance. In Chapter 5 it was shown that close resemblance between action spectrum and absorption spectrum of the light absorber is to be expected only if a relatively small fraction of the incident light is absorbed. In the case of sunburn, close resemblance could hardly be expected since the epidermis absorbs such a large portion of the incident radiation.

It is therefore somewhat difficult to arrive at the absorption spectrum of the absorbing substance for the photochemical processes which initiate sunburn from a consideration of the action spectrum for erythema production. If the photochemical processes take place only in the prickle cell layer of the epidermis, the corneum and the granulosum must form a light filter which absorbs some wave-lengths to a greater extent than others. Thus the radiation which reaches the prickle cell layer is quite different in spectral distribution from that incident at the surface of the skin. The prickle cell layer itself is relatively thick, and should modify the action spectrum of a photochemical reaction taking place within it (see p. 59).

Mitchell has recently (1938) attempted to account for the shape of the action spectrum by assuming that the corneum acts as a filter and that the proteins of the Malpighian layer are the light absorbers. He estimates the latter as equivalent in absorbing power to 0.005 cm. of a 6 per cent solution of a "typical protein" (ovalbumin is chosen as such). He arrives at the conclusion that proteins are the light absorbers for the erythemic mechanism, and that the displacement of the maximum of the action spectrum (about 2970 Å) from the maximum absorption of protein (about 2800 Å) can be adequately accounted for. Mitchell's analysis must be read to appreciate his argument fully, but examination of Figure 33 shows that the strong absorption of the corneum at short wave-lengths with a maximum at about 2800 Å must shift the maximum of the action spectrum to longer wave-lengths.

Since absorption by the epidermis seems due chiefly to proteins, it is highly probable that these are the absorbing substances for the reaction which results in erythema. On the other hand, the evidence may not be incompatible with the view that nucleic acids are the absorbing substances, although this seems less probable (see Figure 10, p. 36 for a comparison of protein and nucleic acid absorption spectra). In either case, there is little to suggest that the fundamental mechanism underlying erythema production by ultraviolet radiation is greatly different from that underlying the destruction of unicellular organisms by such radiation, and as a reasonable working hypothesis it may be assumed that erythema is a sequel to injury to the prickle cells produced by wave-lengths shorter than 3300 Å.

Theories of the Mechanism. It is generally agreed that the initial photochemical reaction brings about the elaboration in the epidermis of a substance which migrates to the region of the minute vessels in the papillary layer and brings about their dilation. Since the sunburn wave-lengths do not penetrate to the papillary layer to any appreciable extent, there can be little or no direct action on these vessels, and the long latent period between exposure to radiation and appearance of erythema also indicates an indirect effect. The nature and mode of elaboration of this dilator substance has been the subject of a number of hypotheses.

Lewis (1927) has presented considerable evidence that erythema always results from the local elaboration of a histamine-like * substance.

* The term histamine-like will be frequently used in the following pages as referring to the pharmacological action of histamine but not necessarily its chemical constitution.

This hypothetical substance he calls "H" substance to indicate its similarity to histamine, and at the same time to avoid too definite a statement as to its chemical nature. If histamine is pricked into the skin, erythema immediately appears in a small area surrounding the prick; this is followed shortly by edema of the same area, and a "flare" of erythema which spreads outward from the edematous region (Lewis' "triple response"). The same sequence of events is observed in spontaneous urticaria ("hives"), in dermatographia (abnormal response to mechanical stimulation), and in certain other responses of the skin; and Lewis believes that such phenomena all result from the action of a histamine-like "H" substance. Further discussion of such responses will appear in the following chapter.

Lewis explains the erythema of sunburn in the same way, assuming that the dilation of the minute vessels results from the production of "H" substance in the skin. Krogh (1929), however, believes it is necessary to assume more than one dilator substance to explain all the different types of erythema, and in the case of sunburn thinks it cannot be the same as Lewis' "H" substance. The character of the erythema of sunburn certainly differs from that of urticaria, in that edema is not a common sequel but only follows severe dosage, and in that the latent period is much longer than that which elapses between the introduction of histamine into the skin and the appearance of erythema, or between mechanical stimulus and response in dermatographism. The long latent period of sunburn might represent either the time required for the elaboration of the dilator substance or for its penetration to the vascular layer of the skin; and the non-appearance of edema might be due to the fact that, because of its slow elaboration, the dilator substance never reaches a very high concentration before it is removed. Thus it seems unwise either to accept or reject the "H" substance hypothesis for the erythema of sunburn without further evidence. That histamine-like dilator substances are present in greater quantity in irradiated than in normal skin has been shown by Ellinger (1928) and Kawaguchi (1930).

Ellinger (1929, 1930, 1932a), following a suggestion of Trendelenburg, proposed that the erythema of sunburn results from the direct production of histamine from histidine by a photochemical reaction:

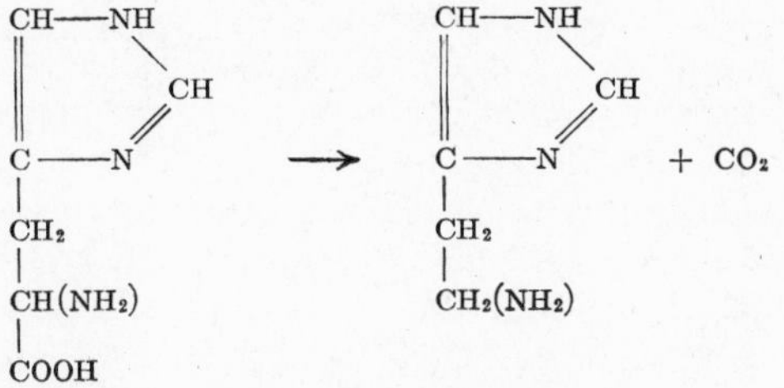

```
CH——NH                    CH——NH
||    \                   ||    \
||     CH                 ||     CH
||    //                  ||    //
C———N          ——>        C———N          + CO2
|                         |
CH2                       CH2
|                         |
CH(NH2)                   CH2(NH2)
|
COOH
```

Histidine *Histamine*

In investigating this hypothesis, Ellinger (1929) studied the absorption spectrum of a histidine preparation in the region of the erythemic spectrum, and found (1930) that exposure to a mercury arc resulted in the formation of a substance having the biological properties of histamine. This seemed good evidence that the light-absorbing molecule which initiates the erythemic response is histidine, and that the mechanism is a simple photochemical production of histamine from this substance in the skin. However, Bourdillon, Gaddum and Jenkins (1930) were unable to confirm Ellinger's measurements of the absorption spectrum of histidine, finding no appreciable absorption of wave-lengths longer than 2550 Å. Ellinger then repeated his measurements on purified histidine and confirmed the findings of these investigators. However, he was able (1932a) to separate from his original histidine preparation an iron-containing substance whose absorption spectrum corresponded quite well to that of the erythemic spectrum; and he suggested that this impurity acted as a photosensitizer in the production of histamine.

The hypothesis that histamine was formed by the direct action of light upon histidine was also entertained by Frankenberger (1933) whose ideas will be discussed in subsequent pages in connection with pigment formation. Szendrö (1931) identified the vaso-dilator substance produced when histidine is irradiated with ultraviolet light as imidazolacetaldehyde instead of histamine. However, the reaction which he proposes requires O_2, and Frankenberger (1933) and Holz (1934) find the dilator substance to be formed in the absence of O_2. Moreover erythema production, like other biological effects of ultraviolet radiation, proceeds in the absence of O_2 (see p. 116).

Besides the wave-length difficulty, another objection to the hypothesis of direct formation of histamine from histidine presents itself. It would be necessary to schematize the mechanism as follows:

$$\text{histidine} + h\nu \longrightarrow \text{histidine}'$$
$$\text{histidine}' \longrightarrow \text{histamine} + CO_2$$

But the long latent period which intervenes between the action of radiation and the appearance of erythema would demand that the histidine molecule retain its activation for periods as long as several hours before breaking up into histamine and CO_2. Regarded from a photochemical point of view, this is an extremely remote possibility. The reaction proposed is a direct one with no intermediate steps which might be delayed in one manner or another. This difficulty might be overcome if it could be assumed that the histamine is immediately formed by the action of light on histidine in the prickle cell layer, but that a long period of time is needed for the diffusion of the histamine to the papillary layer where it can effect dilation of the minute blood vessels. There are certain objections to this argument. The formation of histamine would have to occur in very short periods of time, since only a few minutes, or even seconds, are required to produce erythema if the intensity of the sunburn radiation is very high, yet both Ellinger and Frankenberger (1933) found it necessary to irradiate histidine solution for long periods with intense quartz-mercury-arc radiation before detectable quantities of histamine could be obtained. The distance for diffusion is very short, so that the rate of diffusion must be very slow to provide the long latent period which characterizes the response under ordinary conditions. The temperature coefficient found by Clark (1936) for the latent period could hardly be characteristic of a diffusion process, which should have a temperature coefficient (Q_{10}) near unity. Because of the latter objection it is difficult to account for the erythema of sunburn by any hypothesis which demands the formation of a dilator substance by a direct photochemical process.

Mitchell (1938) postulates that the "H" substance is a high molecular weight breakdown product of protein, which diffuses very slowly. This explanation is more plausible than that of direct histamine formation, but leaves the high temperature coefficient of the latent period unexplained.

An alternative to such hypotheses may be formulated upon the well known fact that injured cells are known to release histamine-like dilator substances. If it is assumed that the action of light is to injure cells in the epidermis, dilation of the blood vessels should result due to the release of such dilator substance by the injured cells. The latent period should then represent the time required for the liberation of the dilator substance by the injured cells, which process might well have a temperature coefficient such as that found by Clark for the latent period. Such an hypothesis will also account for the development of pigmentation, which is a common response of epidermal cells to injury regardless of the injurious agent. The introduction of histamine into the skin results in the triple response, but does not result in pigmentation, whereas injury to the epidermal cells may be followed by a similar triple response and then by pigmentation (see p. 105).

Clark proposed that sunburn results from the coagulation and denaturation of the skin proteins by light. She carefully studied the temperature coefficients for such processes (1935a) and compared them with those for the threshold and latent period of sunburn erythema (1936); but finding no agreement, she abandoned the hypothesis. However, the rate of coagulation may be less important than the total amount of protein altered, which should depend upon the number of light quanta absorbed. The amount of protein altered should determine the extent of cell injury, and hence the threshold of erythema, which Clark found to have a low temperature coefficient. The latent period, on the other hand, would be dominated by the rate of production of dilator substance by the cells, according to the foregoing hypothesis, and might be expected to have a temperature coefficient of the order of that found by Clark ($Q_{10} = 2.3$). Thus there seems no irrefutable objection to the hypothesis that the photochemical basis for sunburn is coagulation or denaturation of the protein of the prickle cells of the epidermis.

Hausser's (1928) attempt to explain the shape of the erythemic spectrum should be mentioned. He attributed the minimum of activity at 2800 Å to the minimum transmittance of the epidermis in this region of the spectrum, and explained the long wave-length limit at 3300 Å by the assumption that the minimum size of quantum which will supply sufficient energy for the process corresponds to this wave-length. His explanation assumes that the initial mechanism in the erythema production is similar to a photoelectric effect. As has been pointed out in Chapter 3, the general long wave-length limit for the destructive effects of radiation at about 3300 Å seems better explained on the basis of specific absorption spectra, and a hypothesis such as that proposed by Hausser requires further evidence.

Pigmentation

Pigmentation is usually observed much later than erythema; as a rule the reddish color changes almost imperceptibly to brown. Sometimes the erythema may be imperceptible, slight pigmentation occurring after a few days with little or no earlier indication of sunburn. Since the degree and rate of pigmentation varies with the dosage of radiation, considerable differences in the histological picture may be expected. Keller (1924a) finds that the first event in pigmentation is not the formation of new pigment *in situ,* but the migration of pigment granules already formed in the basal cells to more superficial regions, including the corneum. It is found in the bits of desquamated corneum which are cast off following a mild degree of sunburn. The actual production of new pigment by the cells of the basal layer does not commence until a considerable time after the irradiation.

It seems certain that the primary changes which lead to pigmentation are the same as those which cause erythema, and that the former is a normal sequel to the latter. The action spectrum for the underlying photochemical changes is probably identical for the two processes, although there may be apparent deviations in the observed action spectra. Recently a mechanism which results in darkening of the skin pigment has been described, which seems to be of secondary importance, but which has probably led to confusion in earlier attempts to compare the action spectra for erythema and pigmentation (see p. 186).

Theories of Pigment Formation. The pigment which follows exposure to sunburn radiation consists of granules of melanin, a common type just *et al.*, 1928) and Verne (1930). Bloch (1916) believed that melanin is formed by the oxidation and polymerization of phenolic compounds through the action of an enzyme; this has been reviewed by Dejust (Dejust *et al.*, 1928) and Verne (1930). Bloch (1916) believed that melanin is formed only from the amino-acid dihydroxyphenylalanine, commonly called "dopa," by a specific enzyme "dopaoxidase." It is more probable, however, that other phenolic compounds, particularly tyrosin, may react to form melanin, and that tyrosinase or perhaps oxidizing enzymes may promote this reaction (see Verne, 1930).

Frankenberger (1933) suggested that pigment is formed directly by the action of ultraviolet radiation on tyrosin in the skin. He also suggested, like Ellinger, that histamine was formed by the action of ultraviolet radiation on histidine. Frankenberger proposed that the sunburn spectrum was made up of two parts, erythema production corresponding to the absorption spectrum of histidine, and pigmentation to the absorption spectrum of alkaline tyrosin. He found that histamine or a histamine-like substance could be formed from histidine and that a brown substance was formed by long irradiation of tyrosin. The former reaction took place in an atmosphere of nitrogen, but the latter only in the presence of molecular oxygen. He presents an analysis based on a comparison of the absorption spectra of these two substances with the action spectrum of sunburn.

It may be pointed out that an analysis which indicates similarity between the action spectrum of sunburn and the absorption spectrum of tyrosin must also show similarity between the absorption spectrum of protein and this action spectrum, since tyrosin and protein absorption spectra are very similar (see Figure 33). Furthermore, the same argument may be brought against Frankenberger's hypothesis for the formation of pigment by the direct action of ultraviolet radiation on tyrosin which has already been advanced against Ellinger's proposal that histamine is formed from histidine in like manner. Both hypotheses demand that a molecule, either tyrosin or histidine, is activated by a quantum of radiation, and remains in some sort of activated state during the long latent period which elapses before pigmentation or erythema occurs. This is incompatible with photochemical theory.

This difficulty is avoided by the hypothesis introduced by Arnow (1937), who follows the contention of Bloch that pigment is formed only by the direct action of "dopaoxidase" on "dopa." He suggests that ultraviolet radiation causes the oxidation of tyrosin to "dopa," which reaction he has shown to take place *in vitro*. Once formed, the "dopa" can be acted upon by the "dopaoxidase" in the skin to form melanin; it might be this process which occupies the latent period.

One important difficulty confronts both Frankenberger's and Arnow's hypotheses. Both require the presence of molecular oxygen for the formation of pigment, and pigment can be formed when the oxygen tension in the skin is considerably reduced during the time of irradiation (see page 116).

Like erythema, pigmentation may be readily explained on the basis of cell injury. Pigmentation may develop following injury to the cells of the epidermis by other noxious factors, such as heat, chemical or mechanical injury, and is generally preceded by erythema. This hypothesis is very general and suggests nothing as to the intimate mechanism of pigmentation. One may assume that the migration of the pigment from the undamaged basal cells into the injured cells of the more superficial epidermis is due to some "tropic" action of the injured cells. The subsequent formation of new pigment in the basal cells themselves may be explained by the action of some substance elaborated by the injured cells. This explanation is noncommittal, and should serve to demonstrate our ignorance rather than to hide it; it appears to be more useful as a working hypothesis than the others which have been mentioned.

Darkening of Preformed Pigment. It has been recently shown that a blackish coloration of the skin is brought about by wave-lengths longer than those of the sunburn spectrum (Hausser, 1938; Henscke and Schulze, 1939a). The action spectrum for this process extends from 3000 to about 4400 Å, having a broad maximum at about 3400 Å. It may appear within a few minutes and passes its maximum within an hour after the radiation, thus differing markedly in its time relationships from the pigmentation caused by the sunburn spectrum. About one thousandfold greater dosage of radiation is necessary to elicit this response than is required for sunburn erythema. It is most pronounced in previously tanned skin.

Miescher and Minder (1939) offer evidence that this process is identical with the darkening of dead pigmented skin brought about by heat (Meirowsky, 1909) or by ultraviolet radiation (Lignac, 1923). All three phenomena occur only in the presence of O_2 (the sunburn mechanism is independent of O_2), and Meischer explains them as due to the oxidation of pigment already present in a colorless, reduced state. It is thus a reversible process quite independent of the formation of new pigment.

This finding probably explains many earlier observations that tanning is more intense when a source rich in wave-lengths longer than the sunburn spectrum is used. Thus, tanning by sunlight or carbon arc is much greater as compared to erythema than that produced by a mercury arc (see Figure 36, p. 194). Henschke and Schulze (1939b) have shown that this darkening effect represents a relatively great amount of the observed tanning produced by sunlight because of the great proportion of longer wave-lengths as compared, for example, to the mercury arc.

The recent interesting observation by Hamilton and Hubert (1938), that the degree of coloration of a previously sunburned area may be increased or decreased by hormonal action, may be an example of the same type of effect, or may represent new pigment formation in altered skin.

Quantitative Aspects

Development of Protection against Sunburn. It is common experience that skin which has been exposed to sunlight becomes less susceptible to subsequent irradiation. The action of pigment as a light filter seems so obvious that it is commonly assumed that this decrease in susceptibility is due to the formation of pigment. However, studies of the position of the pigment in the skin and of the locus of action and penetration of the skin by sunburn radiation make it improbable that pigment can offer a great deal of protection. Moreover, skins which show no trace of pigmentation may lose their sensitivity to light after frequent exposures, as shown by With (1920) and Meyer (1924) for non-pigmenting vitiligo areas. However, it seems certain that pigment plays some part in the protection of skin, although perhaps a minor one. A discussion of this subject is given by Laurens (1935).

Miescher (1932) has pointed out that the pigment of negro skin offers a considerable protection against sunburn radiation since it is distributed in the more superficial layers, in contrast to the pigment of the white races, which is found principally in the basal cells and can offer little protection. The absorption spectrum for melanin is almost uniform in the sunburn spectrum (see Miescher, 1932), so that its presence in the superficial layers should not markedly alter the shape of the action spectrum, although decreasing the general sensitivity. This is in accord with Hausser's (1928) finding that the action spectrum for a negro was the same as for his white subjects, although the threshold was much higher.

Hausmann and Spiegle-Adolf (1927) have shown that irradiated proteins lose their transparency in the region of the sunburn spectrum, and have suggested that coagulation of proteins in the epidermis protects in this way. That there are numerous physical changes in the skin after irradiation is certain, as the studies of Keller (1931) show.

Certainly the hypothesis of Guillaume (1927) offers the explanation for a large part of the protection following irradiation. This writer suggests that thickening of the corneum, as a result of damage to the underlying layer of prickle cells, decreases the intensity of the sunburn wavelengths which reach the latter cells, and hence the sensitivity of the skin to sunlight. Measurements of transmission of sunburn radiation by the skin show that a considerable degree of protection would be afforded by relatively slight thickening (see Figures 41, 42, p. 202). Miescher (1930), who has followed up Guillaume's hypothesis, shows that the thickest parts of the skin are the least sensitive to sunburn, and that thickening of the skin occurs after irradiation. It is probable that this mechanism provides one of the principal means of protection against sunburn.

Crew and Whittle (1938) have recently shown that human sweat, due to its absorption of wave-lengths shorter than 3300 Å, provides some protection against sunburn.

Normal Variations. It is common observation that certain individuals are more susceptible to sunburn than others, so that it is difficult to determine the norm. Ellinger (1932b, 1934, 1935) has made an extensive study of the threshold dosage for erythema production, which provides very interesting data. The method which he employed was to expose small areas of the skin to doses of mercury-arc radiation of the same intensity but different duration, and to examine the exposed areas for erythema twenty-four hours later. By using equal increments of the irradiation period, the number of areas which show erythema may be taken as a measure of the sensitivity of the skin to light. The data from this type of measurement, when considered statistically, show changes with season of the year (sensitivity being least during the summer), with age of the individual, and with other factors. Ellinger finds that blonds are more sensitive as a rule than brunettes, and light or reddish blonds more sensitive than darker blonds. There are also statistical variations among women with relationship to the menses and pregnancy. Ellinger (1932c) also finds that individuals who exhibit certain characteristics—"vegetative stigmatics"—are much more sensitive than the average. The individuals having these characteristics are assumed to have mild hyperthyroidism, and Ellinger believes that sensitivity of the skin to sunburn radiation is related to the activity of the thyroid gland.

Qualitative differences in individuals are also observed. Schall and Alius (1926) find the curve of development and disappearance of erythema to vary widely. Differences in the production of pigment in different individuals are common knowledge; some develop a beautiful mahogany brown after repeated sunbathing, while others, particularly reddish blonds, find great difficulty in developing pigment. Some individuals deposit the pigment in isolated spots to form freckles.

Under such circumstances it is difficult to state where normal sunburn ends and hypersensitivity begins. It is necessary to recognize a wide latitude between the extremes of what is considered normal.

Light Sources and Sunburn. A good deal of confusion has resulted from failure to recognize the character of the spectral distribution of energy from different sources used to study the effect of radiation on normal skin. The spectral distribution of energy from sources commonly used—sunlight, carbon arc, and mercury arc—are shown in Figures 6 and 7 (pp. 24 and 26).

That part of sunlight which produces sunburn, *i.e.*, wave-lengths shorter than 3300 Å, is a very small and variable fraction of the total (see Figure 50, p. 241). The short wave-length limit varies with time of day, season of year, locality, etc. At no time of the year is it appreciably shorter than about 2900 Å, and frequently is not longer than 3300 Å, even at midday; in the latter case the sunlight has no sunburn-producing power. Smoke is very effective in absorbing the short wave-lengths of sunlight, and thus in preventing sunburn, but on the other hand water vapor allows these wave-lengths to pass, so that it is often possible to be badly sunburned on a cloudy day. Clouds over the vicinity of a large city may contain a considerable amount of smoke, and should be much more effective in filtering out the sunburn radiation than clouds over a rural area, or particularly over the sea. This probably accounts for the fact that sunburn frequently occurs at the sea-shore on overcast or foggy days, particularly when an onshore wind removes any traces of smoke.

Water, snow, and ice are good reflectors for the sunburn radiation, and sunburn is very common close to the sea or other bodies of water, or to snow or ice, *e.g.*, above glaciers. The terms *snowburn* or *glacierburn* are often applied to the severe sunburn that may occur after exposure on snowfields or glaciers.

All these variable factors in the character of the incident sunlight, together with the variations of the individual threshold, give sunburn a capricious behavior which frequently mystifies the scientific as well as the popular mind.

Chapter 18

Abnormal Response to Blue and Violet Light (Urticaria Solare)

The most clear-cut instance of photosensitivity outside the normal sunburn spectrum is a very rare condition which has sometimes been called urticaria solare. Since this term has been applied to other conditions which obviously have different etiology, I prefer, until the etiology is more clearly established, to refer to it merely as a triple response produced by blue and violet light. Actually there are only a few cases of this type reported in the literature; those of Ward (1905), Duke (1923), Frei (1925), Blum, Allington and West (1935), and two of Vallery-Radot *et al.* (1926, 1928) seem all to belong to one type. The analysis at the end of this chapter will demonstrate the necessity of distinguishing this group from other conditions which have been classed with them as urticaria solare.

The case which I have studied in collaboration with Dr. Herman Allington and Mr. Robert J. West seems to be typical of this condition, and will serve as the example for the following discussion. Exposure of the skin to sunlight for brief periods of time is followed within a few minutes by the appearance of erythema, limited strictly to the exposed area. This erythema is replaced in a short time by edema, likewise limited to the area exposed to sunlight. As the edema develops, an irregular flare of erythema spreads outward into the surrounding unexposed skin. These events are accompanied by intense itching. The appearance of the response at its height is shown by the photographs reproduced as Figures 34 and 35. It subsequently recedes and at the end of a few hours has disappeared, leaving no trace. This sequence of events might be taken as a typical example of Lewis' (1927) triple response. Other examples are: "hives," which appear spontaneously; dermatographia, which appears after stroking of the skin in certain individuals; and the response which follows the pricking of histamine into normal skin. The visible signs are manifestations of dilation and increased permeability of the walls of the minute vessels in the papillary layer of the skin, which Lewis attributes to the release of a histamine-like "H" substance by cells in the vicinity in which the response is observed. The term *urticaria* is usually applied to this type of response, and the raised area of edema is often called a wheal. While triple response should probably be synonymous with true urticaria, which is transitory and leaves no trace after a short time, the latter term is often applied to skin conditions which are quite

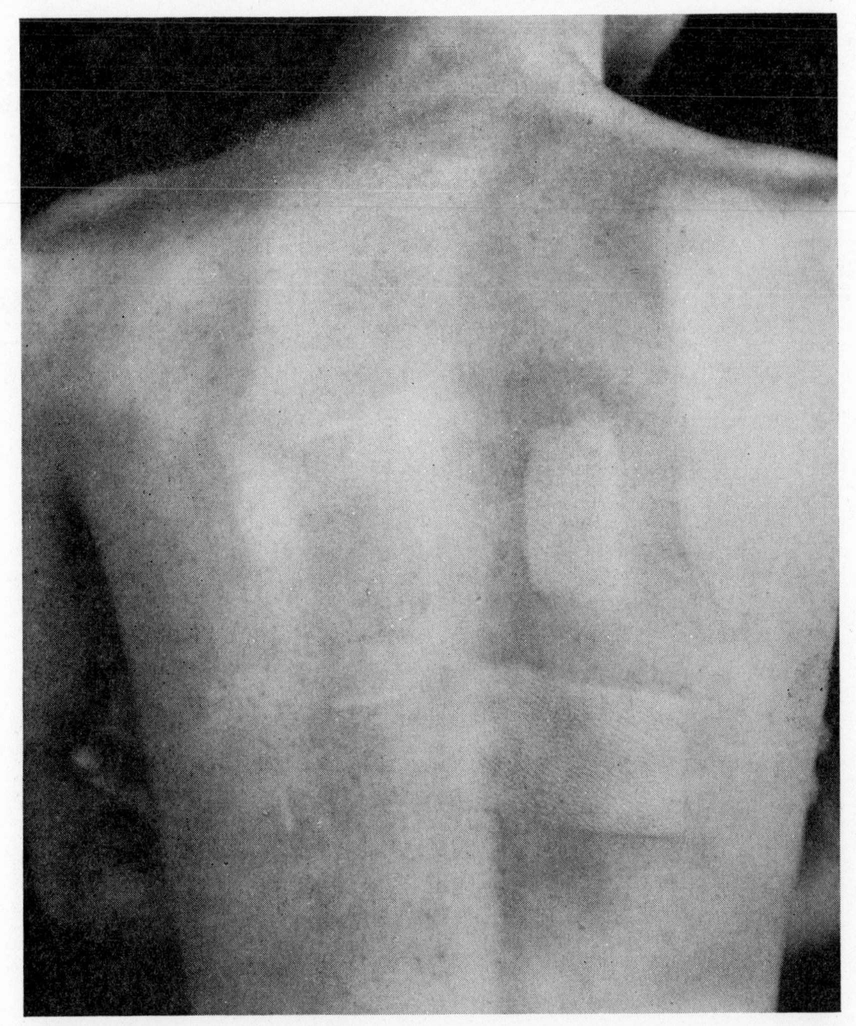

FIGURE 34. Edematous patches with surrounding erythema (triple response) resulting from exposure to sunlight. Different exposures were applied to different areas, resulting in differences in the intensity of the triple response.

unlike it, and far more persistent. Thus it seems preferable to employ the term *triple response* to describe the type of reaction under discussion in this chapter.

The amount of sunlight necessary to produce the response is very small. Exposure of the abdomen for one minute is sufficient to produce it in this individual, but fortunately the hands and face are much less sensitive, although sufficiently so to be a source of severe annoyance. The response shown in Figure 35 resulted from a three-minute exposure.

A region of the back was exposed through a square hole in a large cardboard which protected the surrounding skin; the outline of the hole is clearly marked out by the square area of edema, showing that the initial response is limited to the area exposed to light.

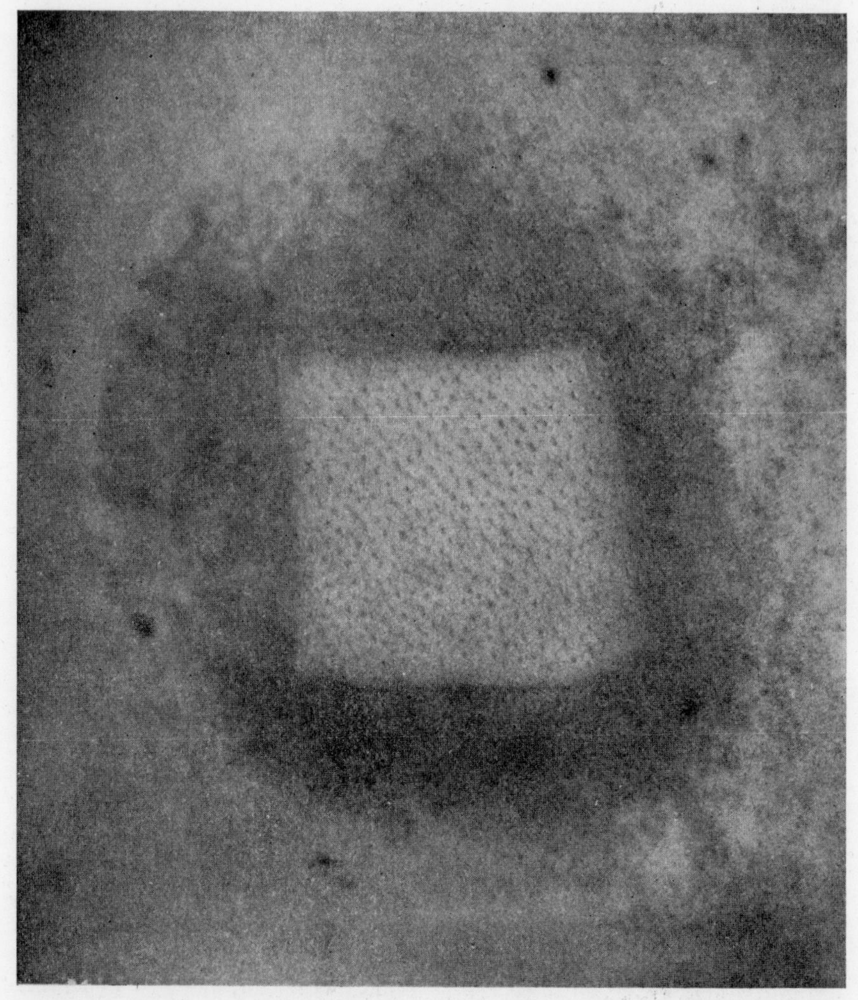

FIGURE 35. "Triple response" resulting from exposure to direct sunlight for 3 minutes. The square area of edema is that exposed to light. Note the flare of erythema spreading outward. [From H. F. Blum, H. Allington and R. J. West, *Journal of Clinical Investigation,* 14, 435 (1935).]

It may be difficult to believe that such an intense response can result from so brief an exposure. At my first observation of this individual, I was skeptical of his statement that two or three minutes' exposure to sunlight would produce a response. Feeling sure that nothing would occur in less than half an hour, I suggested that areas of the back be exposed for ten minutes, thinking to play safe and that longer exposures would have to be made subsequently. The response was so severe at the end of ten minutes that I never cared to repeat such a long exposure. Fortunately there was no untoward after-effect, for it is characteristic that all traces of the response disappear in the course of a few hours, the time varying with the intensity of the response, as is likewise true of hives, the histamine prick, and all other examples of the triple response. On another occasion, the individual went to sleep in the sun thinking himself protected by coverings which slipped off and left one arm exposed for about one hour. The arm became so swollen that the movement of the elbow joint was restricted. The swelling did not subside until after forty-eight hours, but no lasting effects on the skin were noted, nor any systemic symptoms.

In all cases studied, the health has been otherwise good, and the symptoms of sensitivity to light have appeared suddenly without a known precipitating cause. In the present case, the first sensitivity to light was noted just after the individual had suffered a bee sting which produced severe edema of the face. Certain experiments were performed pursuant of the guess that this might have precipitated the condition; but these were without result, and it seems improbable that there was any connection between the bee sting and the onset of photosensitivity, particularly since no such event is reported by other patients suffering from this condition. In all cases described, the sensitivity to sunlight has continued without abatement from the time of its sudden appearance; in one of Vallery-Radot's cases the condition had persisted for twenty-eight years. There is no correlation with sex or with age of onset.

The outward manifestations of this triple response differ in two important ways from sunburn. The former appears "immediately," *i.e.*, it follows the exposure to sunlight by an interval of only a few minutes at most, whereas sunburn appears only after an hour or more, unless the dosage is excessive. The triple response disappears within a relatively short time, usually a few hours, leaving no visible trace of any change in the skin, whereas sunburn is followed by pigmentation which may last for weeks or months.

These differences alone indicate that this response is quite independent of normal sunburn; but the most important point of distinction between the two processes is the difference in their action spectra, shown in Figure 36. The determination of the action spectrum for the triple response will be described later; the Figure shows that it lies in the violet and blue portions of the visible spectrum, and is separated from the sunburn wave-lengths by nearly 1000 Å. Thus it should be possible to elicit the two responses separately from the same individual by restricting the incident wave-lengths. This was accomplished by Blum, Allington and West (1935), who produced only normal erythema and tanning with mercury-

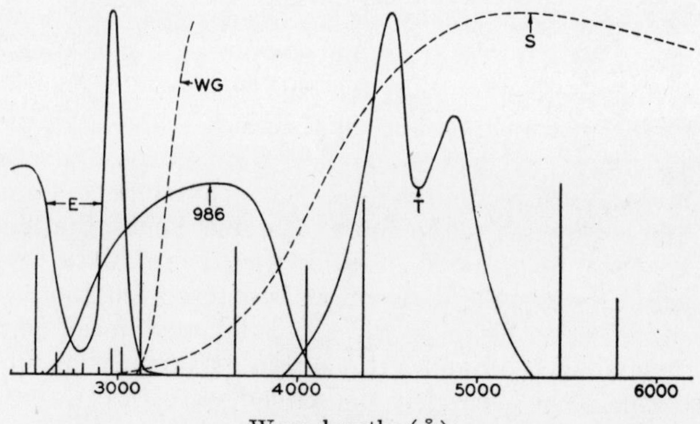

FIGURE 36. Ordinates are arbitrary units differing for each curve, chosen to bring the data into relationship for comparison. The vertical lines represent the position of the principal mercury lines and their relative intensities (see Figure 7, p. 26). T, action spectrum of the abnormal triple response (see p. 205). E, action spectrum of sunburn. 986, transmission of Corning filter 986. Mercury-arc radiation through this filter produced the erythema and pigmentation of sunburn without eliciting the triple response. WG, lower wave-length limit of transmission by window glass. S, spectral distribution of sunlight.

arc radiation passing through a filter cutting out virtually all the triple response wave-lengths (see Figure 36) in the same individual who developed a triple response when exposed to sunlight through window glass or other filters which remove the sunburn radiation.

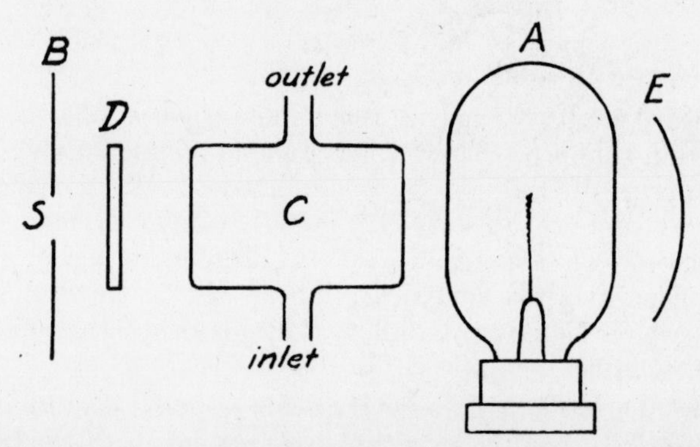

FIGURE 37. Diagram of apparatus for studying the abnormal response to blue and violet light. A, 500-watt, projection type Mazda lamp. B, Opaque screen with circular opening, S. C, Water filter. D, Glass color filter. E, Concave mirror. [From H. F. Blum and R. J. West, *Journal of Clinical Investigation,* **16,** 261 (1937).]

There can be no doubt that the triple response to light shown by this individual and a few others (see Table 16, p. 209) is entirely distinct from sunburn. It is safe to say that this response occurs in skins which are otherwise apparently normal, because of the presence of a photoactive substance which absorbs in the blue and violet regions of the spectrum. This substance is probably not present in normal skins, although it may be present but not active under normal circumstances.

Threshold Time as an Index of the Response. Blum and West (1937) made a series of experimental studies, the apparatus for which is diagrammed in Figure 37. A 5000-watt tungsten filament lamp (Mazda, projection type), *A*, is placed at a given distance from an opaque screen, *B*, which has a circular opening about two centimeters in diameter, *S*, through which the skin may be irradiated. The screen *B* is rigidly fixed so that the skin may be pressed against it without disarranging the apparatus, thus making it possible to establish and reproduce the distance between lamp and skin. A "Pyrex" glass chamber, *C*, ten centimeters in length, through which water is circulated, is introduced between the lamp and the skin to cut off infrared radiation. Glass filters may be introduced at *D*, to further restrict the wave-lengths when this is desired. *E* is a concave mirror which was used in some of the experiments. The distances between the various elements were altered according to the requirements of given experiments, and in some cases one or more of the elements was removed.

The period of irradiation required to produce a just perceptible erythema, the threshold time, was determined under various conditions, and its reciprocal taken as a measure of sensitivity of the skin. It was measured as follows: The skin was placed against the screen, *B*, and irradiated through the opening, *S*, for a measured number of seconds. The skin was then moved away from the screen, and observed for erythema of the area which had been exposed to light. At room temperature in the region of the threshold, the erythema usually required several minutes to appear, and it was found that if none was observed within fifteen minutes it might be safely assumed that the time of exposure had been less than the threshold time. The procedure was repeated on other, neighboring areas of skin using different periods of irradiation, and the shortest period which produced erythema was taken as the threshold time.

Lamps of the type used burn at higher color temperatures than most incandescent lamps, and as a result emit more blue and violet; the emission of one of these lamps is shown in Figure 38 for several voltages. The curves were calculated by means of Wien's equation (see p. 23), from color temperatures obtained by photometric comparison with a standard; they give an idea of the importance of maintaining the lamp voltage within reasonable limits when exact measurements are to be made.

If the lamp is placed too close to the skin a heat erythema is produced which appears immediately but, unless severe, fades rapidly. Since at threshold values there is always a delay of several minutes before the appearance of the erythema resulting from light, there is little danger of confusing the two responses. However, the increase of temperature effects the threshold time, and for this reason the water filter shown in Figure

37, or in some experiments a glass filter (Corning 395, extra-light shade Aklo), was used to remove the infrared which constitutes the major part of the energy emitted by the lamp.

The Reciprocity Law. Table 12 indicates close adherence to the reciprocity law (see p. 15), justifying the use of the threshold time as a measure of the sensitivity of the skin. For these experiments, the water filter *C*, which focuses the light rays to some extent, was replaced by the

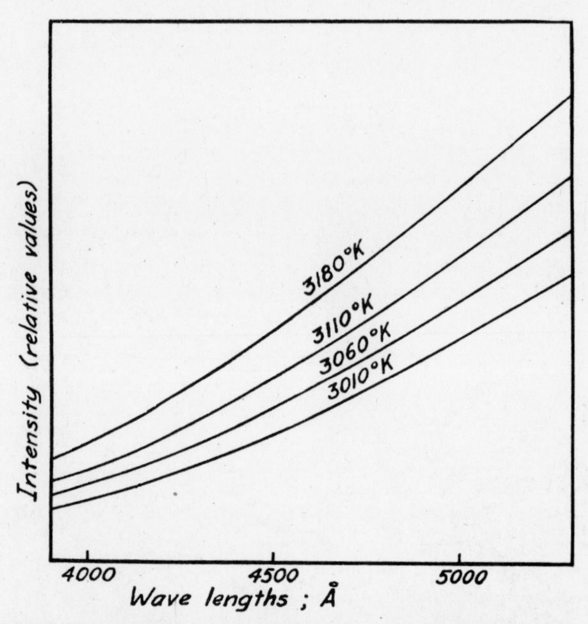

FIGURE 38. Spectral distribution of emission for a black-body at different temperatures. The temperatures are the color temperatures (see p. 24) of a given Mazda lamp when operated at the following voltages:

Volts	Color temperature (°K)
105	3010
110	3060
115	3110
120	3180

[From H. F. Blum and R. J. West, *Journal of Clinical Investigation,* 16, 261 (1937).]

Aklo filter, and the intensity was adjusted by changing the distance of the lamp from the skin. The intensities were checked by means of a photocell. Experiment 8-31 was the most carefully conducted, and probably offers the best estimate of agreement with the reciprocity law; the other experiments, made on other days, also show good agreement and indicate no deviation.

Incident Energy Required to Produce the Triple Response. From the data of Table 12 a rough estimate may be made of the amount of light

Table 12.

(Data from Blum and West, 1937)

Experiment	d *	t †	k ‡
8-29	29.8	95	.103
	21.9	65	.136
	15.5	30	.125
	10.0	12	.120
8-31	10.0	14	.140
	21.6	65	.139
	50.3	300	.119
	15.3	30	.128
	33.2	130	.118
9- 5	10.0	14	.140
	50.0	315	.126
9-19	15.0	31	.138
	30.0	100	.111
	10.0	13	.130
	50.0	300	.120
	20.0	55	.137

* d = distance from lamp in cm.
† t = time in seconds.
‡ $k = \frac{1}{d^2} \times t \propto I \times t$, where I = intensity (see p. 27).

energy which must fall on the skin to elicit the triple response. A black body at 3200° K emits 6.12 microwatts per sq. cm. per foot candle* in the spectral region between 4000 and 7600 Å, according to Lukiesh (1930 p. 165). From the curve for the emission of a black body at that temperature, which is approximately the color temperature of the lamp used in these experiments, it may be estimated that one-sixth to one-tenth of the emission between 4000 and 7600 Å lies between 4000 and 5000 Å. Hence, emission in the latter region, that which elicits the triple response, is about one microwatt per sq. cm. per foot candle. The lamp used in establishing the reciprocity law emits about 1300 horizontal foot candles, and from the data given in Table 12 the threshold time at one foot distance is about one minute. Thus approximately 1300 microwatts of radiant energy of wave-lengths 4000 to 5000 Å must fall on one sq. cm. of the skin of this individual over a period of one minute to produce minimal erythema.

This gives a value of 78,000 microwatt-seconds per sq. cm., which may be compared with the estimate of 4300 microwatt-seconds per sq. cm. which Lukiesh, Holladay and Taylor (1930) give for the amount of radiant energy of wave-length 2967 Å required to produce minimal erythema of sunburn. Rough as the first estimate is, it is obvious that the amount of incident energy required to produce the triple response is of an order of magnitude greater than that required to produce sun-

* This unit relates total radiant energy output (measured in microwatts per sq. cm.) to luminous radiant energy output (*i.e.*, energy perceived by the human eye, which is measured in foot candles).

burn. This explains the observed fact that mercury-arc radiation may elicit the erythema of sunburn without producing an appreciable triple response, even though the 4037 Å and 4358 Å lines which lie within the action spectrum of the triple response are relatively intense (see Figure 36).

Effect of Temperature. To examine the effects of temperature on the urticarial response, the screen, *B,* (Figure 37) was replaced by a thin black paper stop with a hole 2 cm. in diameter fastened against the

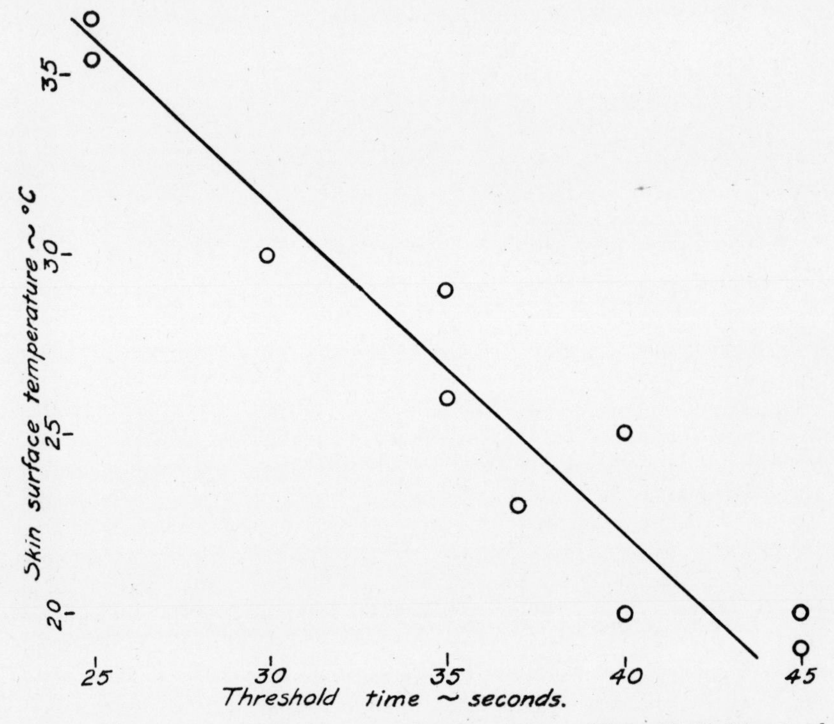

FIGURE 39. Effect of temperature on threshold time. [From H. F. Blum and R. J. West, *Journal of Clinical Investigation,* 16, 261 (1937).]

water filter, *C*. The skin of the abdomen was held tightly against this black paper, so that the area of skin exposed to light would take on approximately the temperature of the water in the filter, which could be altered at will. By thermocouple measurements, described in detail in the original paper (Blum and West, 1937), a direct relationship was found between the temperature of the skin and the temperature of the water at the outlet of the filter, provided the skin was allowed to remain in contact with the filter for not less than five minutes. Figure 39 shows the effect of temperature on the threshold time; for these measurements the skin was held in contact with the filter for a preliminary period of five minutes to establish the temperature, plus the period of irradiation, then moved away from the filter and observed for the appearance of erythema. This was repeated for several different irradiation periods in order to estimate the threshold time at a given temperature. The Q_{10} lies between 1.3 and 1.4, a reasonable value for a photochemical reaction. These data show that a variation of 3 °C would only produce a shift of about 10 per cent in the threshold time in the region of room temperature.

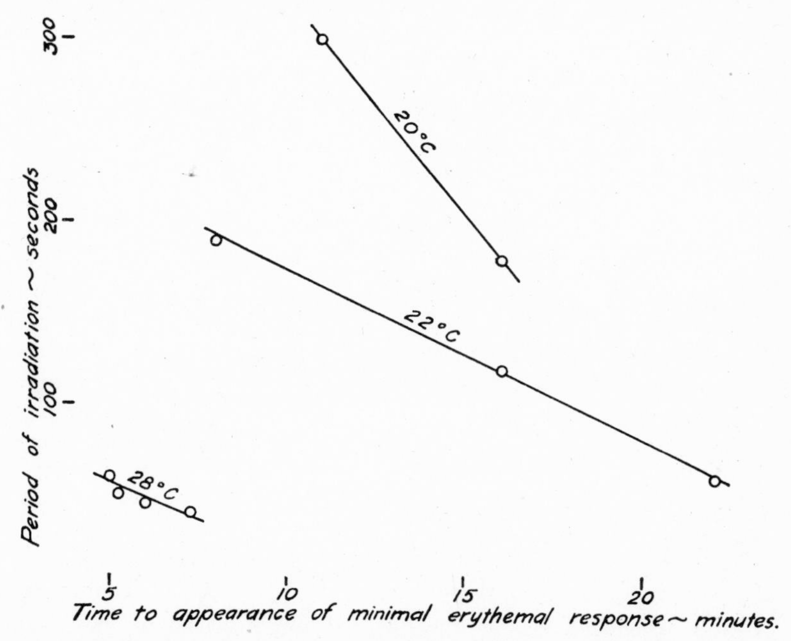

FIGURE 40. Effect of temperature on latent period. [From H. F. Blum and R. J. West, *Journal of Clinical Investigation,* 16, 261 (1937).]

While temperature has relatively little effect on the threshold time, it has a great effect on the latent period which precedes the appearance of erythema. At a given temperature the latent period is shortened as the period of irradiation is increased. To study the effect of temperature on the latent period, it was necessary to measure the latent period for more than one duration of irradiation at several temperatures. This rather lengthy process was accomplished as follows: The skin was held against the filter for a preliminary five-minute period, and then irradiated for a period which was known, from the data of Figure 39, to be greater than the threshold time. The skin was allowed to remain in contact with the filter for a further, measured period of time; and was then moved away and observed for erythema. If erythema was present the experiment was repeated using the same period of irradiation, but a shorter total period of contact with the filter before observation for erythema; if erythema was not present, the total period was increased. By such a series of trial approximations, the least time for appearance of erythema was estimated, and was taken as the latent period for erythema development at the given temperature and period of irradiation. Further trials were made to measure the latent period for other irradiation periods, and for other temperatures.

The results of a series of such measurements, presented in Figure 40, show that the rate of development of erythema after irradiation is markedly affected by temperature. The variation is so great that it was not feasible to obtain comparable data over a wide range of temperatures and hence no attempt was made to calculate the temperature characteristics. When the temperature of the skin was raised to about 40 °C, no erythema appeared at all if the skin was kept in contact with the filter for several minutes, although it appeared when the period of contact was short. This was undoubtedly due to increased blood flow stimulated by this high temperature, which carried away the erythema-producing substances so rapidly that erythema disappeared in a very short time. These results seem comparable to those which Lewis (1927) describes for the triple response after histamine pricks in normal individuals, and after stroking in some urticarial subjects. He found that either low (12 to 15 °C) or high (45 to 47 °C) temperatures inhibit these responses. It is probable that in his experiments, and in those described above, low temperature inhibits by slowing the reaction to the "H" substance, whereas high temperature inhibits by increasing the blood flow in the skin and thus hastening the removal of this substance.

The following schematization helps to explain the effects of temperature on the urticarial response produced by light. The reactions are written according to the scheme introduced in Chapter 2 (p.19).

$$D + h\nu \longrightarrow D_r \quad (1)$$
$$D_r + \text{cells} \longrightarrow \text{"H"} \quad (2)$$
$$\text{"H"} + \text{blood vessels} \longrightarrow \text{triple response} \quad (3)$$

In the first reaction, D is a molecule of a light-absorbing substance in the skin whose identity is not important for the present discussion, and D_r is the same molecule which has been rendered reactive as a result of absorbing a quantum of radiation, $h\nu$. This reaction is photochemical and independent of temperature; it dominates the threshold time, and thus accounts for the slight effect of temperature on that feature of the response.

The second step (2) may represent one or a number of reactions in sequence, which result in the formation or release of "H" substance by cells in the skin. This must be a rapid step, and is probably dominated by step (1). "H" is a dilator substance which causes a physiological response similar to that produced by histamine, but need not be related chemically to that substance.

The third step (3) is the production of the triple response by action of the "H" substance on the minute blood vessels of the skin. This step dominates the latent period. It is markedly affected by temperature, showing its lack of direct dependence on the photochemical reaction represented by (1). Since the effect of temperature on the latent period seems to be the same as on the latent period of the triple response following the introduction of histamine into the skin, it is probable that the latent period is determined by the action of "H" substance on the blood vessels, rather than by the formation of the "H" substance.

Topographical Distribution of Photosensitivity. Threshold times for various regions of the body, measured for constant light intensity, are shown in Table 13. The photosensitivity of different parts of the body varies widely, the abdomen and lumbar region of the back being over twenty times as sensitive as the face and hands. Although it was impossible to produce a response on the latter areas with radiation from a 500-watt tungsten lamp, exposure to sunlight produced a severe triple response.

Table 13. Sensitivity on Various Regions of the Body on December 3, 1935.
(From Blum and West, 1937)

Region	Threshold time
Ventral surface of abdomen	60 seconds
Lumbar region of back	60 seconds
Back over scapula	70 seconds
Medial surface of thigh near knee	165 seconds
Outer surface of forearm (15 cm. above wrist)	150 seconds
Inner surface of forearm (15 cm. above wrist)	105 seconds
Inner surface of forearm at wrist	180 seconds
Palm of hand	180 seconds (itching only, no observable erythema)
Dorsum of hand	Longer than 20 minutes
Cheek	Longer than 20 minutes

Regular measurements of the sensitivity of the face and hands were not made, but there seems to be no doubt that their sensitivity to light decreased as a result of repeated exposure which occurred after the onset of the disease. Both Duke (1923) and Vallery-Radot (1926) found that successive exposure of a given area to sunlight resulted in decrease of the sensitivity of that area. Several explanations for this are possible. One is that the epidermis thickens as a result of repeated production of the triple response. As was said in the last chapter, thickening of the corneum is an important factor in the decrease of sensitivity to sunburn which results from repeated exposure to the sunburn spectrum. The transmission of the skin for the wave-lengths which produce the triple response (4000 to 5000 Å) is much greater than for the sunburn wave-lengths (about 3000 Å) as is shown in Figure 41, and a much

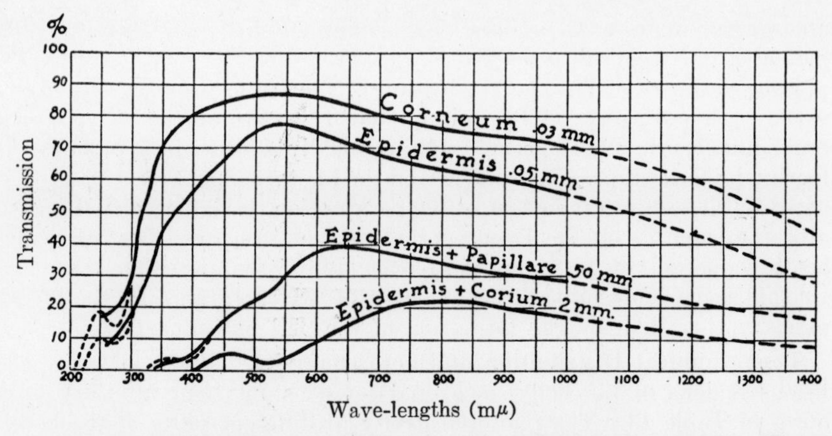

FIGURE 41. Transmission of radiation by the various layers of skin. [From A. Bachem and C. I. Reed, *American Journal of Physiology*, **97**, 86 (1931).]

greater thickening would be required to provide the same degree of protection for the former than for the latter. Figure 42 shows curves which relate thickening of the epidermis and penetration of radiation, calculated from the data of Bachem and Reed (1931); these give an

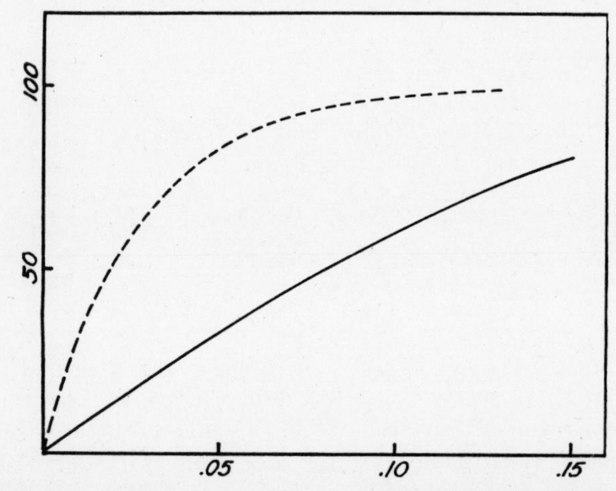

FIGURE 42. Effect of thickening of epidermis on penetration of radiation. Abscissa: thickening, in millimeters. Ordinate: percentage reduction in radiation reaching the photosensitive layer. Broken line, 3000 Å malphigian layer (sunburn). Solid line, 4500 Å papillary layer (triple response). [From H. F. Blum and R. J. West, *Journal of Clinical Investigation*, **16**, 261 (1937).]

idea of the relative effectiveness of such thickening. Sunburn occurs in the region of the Malpighian layer and the curve shows that an epidermal thickening of .03 mm. would cut out about 70 per cent of wave-length 3000 Å reaching this depth; this should afford a considerable degree of protection. The rapid appearance of the triple response indicates that the "H" substance is produced close to the minute blood vessels in the papillary layer. About 30 per cent of the incident radiation of wave-lengths 4500 Å reaches this layer, where about fifty per cent is absorbed. A thickening of the epidermis of 0.15 mm. would be required to decrease the light of this wave-length which reaches the papillary layer by 70 per cent. Thus thickening of the skin would offer much greater protection against sunburn than against this triple response, though in the latter case it should also be a contributing factor to the decrease of sensitivity to light which develops from repeated exposure.

Pigmentation may afford better protection against the triple response than against sunburn, since it is better placed to protect the papillary layer than the prickle cells (see pp. 184, 186). Blum, Allington and West (1935) produced a strong pigmentation by successive exposures to mercury-arc radiation from which the longer wave-lengths were removed by means of a Corning 986 filter (see Figure 36). The sensitivity to blue and violet light was somewhat decreased by this treatment, but not to a sufficient extent to provide important protection.

It is possible that repeated exposure to light renders the skin less sensitive to the products of the photochemical reaction, *e.g.*, "H" substance in the foregoing scheme. As a test of this, histimine pricks were made at a number of points on the subject's body, and the resulting

Table 14.

(From Blum, Allington and West, 1935)

Source	Filter	Wave-lengths (Å) passed by filter	Duration of exposure (min.)	Result
Sun	246	Longer than 5800	31	No reaction
Sun	351	Longer than 5300	30	No reaction
Sun	586	3300-3900	30	No reaction
Sun	338a	Longer than 4900	12	Erythema (slight)
Sun	338a	Longer than 4900	31	Erythema (very slight)
Sun	338b	Longer than 4800	7	Erythema (slight)
Sun	986	2500-4100	12	Erythema
Sun	986	2500-4100	31	Erythema
Sun	Window glass	Longer than 3200	1.5	Erythema
Sun	970	Longer than 2500	9	Edema
Sun	None	All	9	Edema
Sun	Window glass	Longer than 3200	7	Edema
Sun	352	Longer than 5000	30	Edema (very slight)
Hg arc* 15 cm.	None	All	15	Erythema (slight)
Hg arc 15 cm.	Window glass	Longer than 3200	15	Erythema (slight)
Hg arc 15 cm.	986	2500-4100	15	No reaction

* Therapeutic-type mercury arc.

responses compared with those of a series of normal individuals. The responses of the patient were found to be within normal limits for all the regions tested, which included the hands and face. In normal individuals the latter regions are somewhat less susceptible to histamine than most other areas. Reactions to adrenalin pricks were within normal limits. These tests do not indicate any abnormal vascular responses in the skin, nor provide any evidence that the skin of the exposed areas has developed increased tolerance to histamine.

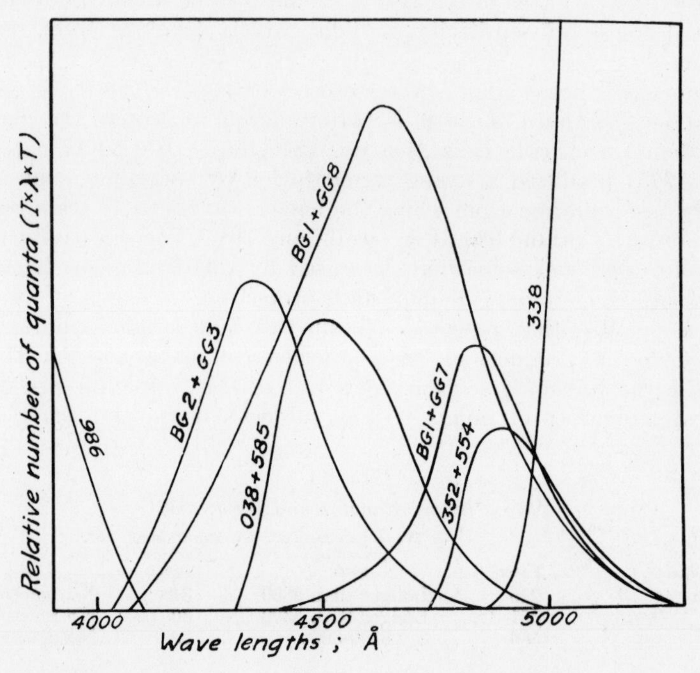

FIGURE 43. Light passing through filters used to isolate spectral regions. [From H. F. Blum and R. J. West, *Journal of Clinical Investigation,* **16**, 261 (1937).]

It has been suggested that repeated production of the triple response by exposure to light exhausts some part of the mechanism, resulting in reduced sensitivity (*e.g.*, see Duke 1923). This is possible but difficult to demonstrate, because of the existence of the above-mentioned mechanisms which may accomplish the same result. Each of the factors mentioned may play its role in the reduction of the photosensitivity of the exposed parts.

Variation of Photosensitivity. Although there is some evidence of local decrease in sensitivity, sensitivity of the body as a whole remained at about the same level during the three and a half years during which the individual was under observation. Frequent tests on the skin of the abdomen indicated minor fluctuations, but no major change in the amount of light required to produce a response. Numerous alterations of diet and other forms of treatment, were hopefully tried from time to time, but without noticeable effect on the photosensitivity of the skin.

The Action Spectrum. The results of preliminary tests by Blum, Allington and West (1935), using sunlight as a source, are summarized in Table 14. A slight response was obtained with wave-lengths shorter than 4100 Å, but none with wave-lengths shorter than 3900 Å, even after long exposure. A slight response was elicited by wave-lengths longer than 5000 Å, but none by wave-lengths longer than 5300 Å. This sets the short and long wave-length limits at about 4000 Å and 5200 Å respectively.

Table 15. Sensitivity to Restricted Spectral Regions.
(From Blum and West, 1937)

Filter	λ Maximum	A	Experiment	t Seconds	$A \times t$	$\frac{1}{A \times t}$
BG2 + GG3	4350 Å	.44	1	150	66	.015
			2	150	66	.015
			3	195	86	.012
						Ave.=.014
038 + 585	4500 Å	.38	1	90	42	.025
			2	90	42	.025
			3	85	39	.026
						Ave.=.025
BG1 + GG8	4650 Å	1.00	1	90	90	.011
			2	70	70	.014
			3	70	70	.014
						Ave.=.013
BG1 + GG7	4850 Å	.32	1	165	53	.019
			2	195	62	.016
			3	165	53	.019
						Ave.=.018
352 + 554	4900 Å	.19	1	330	64	.016
			3	300	58	.017
						Ave.=.016
986*				300 (no response)		
338†				100 (heat response)		

* Longest wave-length passed, 4100 Å.
† Shortest wave-length passed, 4900 Å.

Blum and West (1937) undertook to determine the action spectrum more exactly. They employed the same apparatus described in Figure 37, using various combinations of color filters to isolate rather narrow spectral bands. The transmission of these filters, corrected for the spectral distribution of the source, which was the lamp whose characteristics are described in Figure 38, operated at 120 volts, are shown in Figure 43. The intensities are expressed as relative number of quanta, so that the area, A, under any given curve is a measure of the number of quanta reaching the skin when the corresponding filter is used (see p. 60). No correction is made for absorption by the water in the light filter since this is uniform over the wave-length range studied. Table 15 shows the data obtained in a series of experiments in which the threshold time, t, was determined when each of the filter combinations was used. Since the reciprocity law holds, $1/At$ must be an index of the sensitivity of the skin to the radiation passing the particular filter employed.

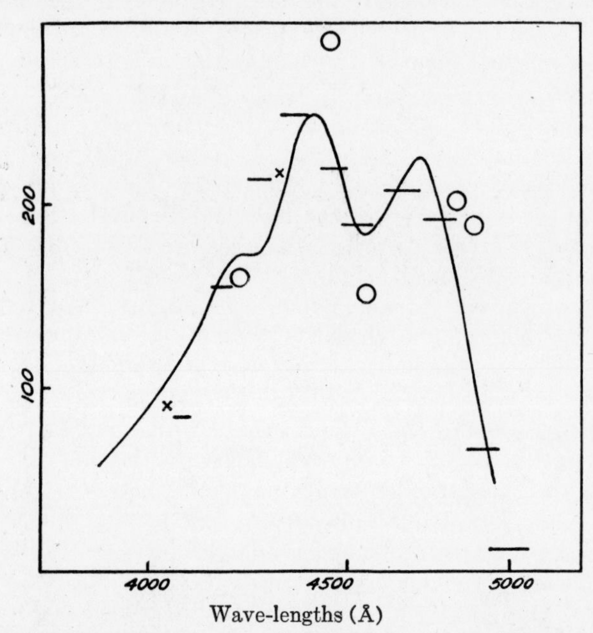

FIGURE 44. Action spectrum of the triple response. Ordinate: arbitrary values, chosen to permit comparison of the different sets of data. Circles, relative sensitivity of the "triple response." Horizontal lines and crosses, relative sensitivity of phototropic bending of the oat seedling. Curve, absorption of α-carotene in alcohol. [From H. F. Blum and R. J. West, *Journal of Clinical Investigation,* **16**, 261 (1937).]

The values, which are plotted as open circles in Figure 44, are subject to considerable error in both parameters. The error along the ordinate should be about the same as that for the other measurements which have been discussed, but the error along the abscissa is difficult to estimate. The points are plotted at the wave-length of maximum intensity transmitted by the filter combination, but might obviously lie on either side of this. However, careful inspection of the data leads to the conclusion that the action spectrum displays two maxima separated by a minimum. Bachem's and Reed's data show some variation in the transparency of the epidermis in this spectral region (see Figure 41) but not of a type which would modify the general characteristics of the action spectrum.

Nature of the Photosensitizing Substance. Such an action spectrum recalls the absorption spectra of the carotenoid pigments which were mentioned in Chapter 3, where their resemblance to the action spectrum for phototropic bending was discussed. Johnston's (1934) measurements for the action spectrum of phototropic bending of the oat coleoptile, corrected to relative number of quanta, are plotted in Figure 44, together with the absorption spectrum of a typical carotenoid, α-carotene. There is a certain agreement in the outside spectral limits and in the appearance of two strong maxima separated by a minimum in both action spectra and in the absorption spectrum.

It is perhaps unwise to stress these resemblances too much, but they suggest the tentative hypothesis that the photosensitizing substance responsible for the triple response produced by blue and violet light is a carotenoid of the same type as carotene or xanthophyll. The evidence for this is as good, in a sense, as that which is accepted to demonstrate that the photoactive substance in phototropism of the plant is of the same nature. No other types of compound which are known to be common in living systems display the characteristics of the carotenoids, either in respect to their spectral limits or the possession of two strong maxima, an important point in both cases.

Against the acceptance of this hypothesis is the fact that carotene may be taken into the human body in quantity sufficient to give the skin a yellow color, without the occurrence of photosensitivity. Moreover, the writer has made several attempts to sensitize human skin by the injection of carotene after the manner employed for sensitization with rose bengal or hematoporphyrin (see p. 104), but without success. Thus, if a carotenoid is the photosensitizer of the triple response, either it must be different from carotene, or it is for some reason distributed at a different place in the skin from that at which such compounds are normally deposited. If the photosensitizer is introduced in the diet, the sensitivity to light might be expected to show considerable fluctuations as the diet changed; but those which have been observed are not of great magnitude, and a deliberate reduction of the carotenoid content of the diet did not result in reduction of the photosensitivity. Blum and West (1937) suggested the possibility that a carotenoid might be introduced into the skin by a parasite, since the subject was infected with a fungus which was thought to be *Malasezia furfur*, but experiments designed to examine this possibility were without success. Dr. Vallery-Radot has informed me that he saw no evidence of such infection in the patients which he studied, and it is not known that other individuals suffering from this malady have had fungus infections.

Independence of O_2. Very little more is known about the nature of the mechanism of this photo-response. An important fact is that it is not dependent upon the presence of molecular oxygen, and hence does not belong to the photodynamic type of reaction. This was determined by Blum, Allington and West (1935) by the following experiment. The circulation to the skin of one forearm was occluded by means of a sphygmomanometer cuff on the upper arm. After about ten minutes' occlusion, areas of both forearms were exposed to sunlight, the circulation remaining cut off in the one arm during the period of exposure. Triple responses developed on both arms and comparison failed to show any difference between the arm with circulation intact and that with circulation occluded. Occlusion of the circulation in this manner reduces the oxygen tension in the skin to a sufficiently low level to prevent a photodynamic response (see p. 116).

Relationship to "Allergy." It is quite common to attribute all reactions of the triple response type to "allergy." The definition of the term is somewhat vague, being often used in a sense synonymous with idiosyncrasy. Certainly the ability to produce a triple response is not an idiosyncrasy, since all human skins respond to histamine pricks in this way. If, as suggested by Lewis, the triple response results from the elaboration of a histamine-like "H" substance, any idiosyncrasy must reside in the manner of production of the "H" substance, which may be altogether different in the case of "hives" resulting from a food idiosyncrasy or a bacterial allergy from the case of the triple response produced by blue and violet light. The scheme presented on page 200 would seem to have very little relationship to any conceivable mechanism for food idiosyncrasy.

Duke (1925a, b) has applied the term "physical allergy" to those conditions in which a triple response or other skin disturbance is precipitated by heat, cold or light. This term seems to be hardly justifiable for there can be little in common between the mechanisms of skin responses to heat or cold and those to light, the difference being analogous to that between thermal and photochemical reactions (see p. 17). Abnormal sensitivities to light may have more than one etiological background, and need be no more closely related than diseases resulting from the action of different bacteria.

Duke and others stress the importance of heredity in the diagnosis of allergy since a tendency toward this condition is supposed to run in families; but in the individual examined by the writer and his collaborators there was no evidence of allergy in the individual or in the family. Altogether, the introduction of the "allergy" concept into the subject of diseases caused by light, as has been done by Duke and particularly by Jausion and Pagès (1933), seems to add little to the clarity of the subject.

Other Conditions Called Urticaria Solare. A number of instances of triple response resulting from light have been reported which probably do not belong to the type which has just been discussed. All those cases in which there is experimental evidence as to the wave-lengths which elicit the response are treated in Table 16. The measurements of the spectral limits are not of equal value in all these cases; most of the wave-lengths have been estimated by the writer from the general characteristics of the sources and filters used since the investigators usually give no spectral measurements. Careful examination of the table will show that in the first seven cases the response was evoked by light including 4000 to 5000 Å. Cases 1, 2, and 4 are almost certainly of the same type. There is somewhat greater latitude in the wave-length limits which have been estimated for cases 3, 5 and 6, but all include the region 4000 to 5000Å, and probably belong to the same group. Case 7 is less certain, but could be of the same type.

Table 16. Limits of Action Spectrum for "Urticaria Solare."

Number	Investigators	Outside wave-length limits of action spectrum (Å)*	Approximate time between irradiation and response (min.)
1	Blum, Allington and West (1935) Blum and West (1937)	3900-5200	0-10
2	Vallery-Radot *et al.* (1926)	4000-5000	1-10
3	Vallery-Radot *et al.* (1928)	4000-7000	1
4	Duke (1923)	4000-5000	2½
5	Ward (1905)	3200-6500	Apparently only a few
6	Frei (1925)	3200-6000	Apparently a few seconds
7	Cummins (1926)	Longer than 3200	Brief
8	Beinhauer (1925)	Shorter than 3200	3-20
9	Schmidt La Baume (1929)	Shorter than 3200	20
10	Weiss (1932)	Shorter than 3200	240

* In most cases my estimate; based on the investigator's description of the sources and filters used.

In the first seven cases, the response is produced by light passing through window glass, or at least some glass which should have the same short wave-length transmission limit, *i.e.*, should cut out those wave-lengths which produce sunburn (see Figure 36). Case 8, in contrast, was not sensitive to those wave-lengths which pass through window glass; *i.e.*, the abnormal triple response is produced in this case by the sunburn wave-lengths, in all probability represents an abnormality of the sunburn mechanism, and must be placed in a separate group from the first seven. In cases 9 and 10 the triple response was produced by moderate doses of mercury-arc radiation. This should place them in the same group as case 8, *i.e.*, as abnormalities of the sunburn mechanism.

The length of the latent period serves as another distinction between the first seven cases and the last three. In the first six, and probably the seventh, the latent period is a matter of a very few minutes, whereas in cases 9 and 10 it is a matter of hours, as for normal sunburn. In case 8 it is only a matter of minutes, but it is probable that the dosage was excessive in this case, and it must be remembered that the sunburn response in normal skin may have a very short latent period if the dosage of radiation is high. It seems probable then that the last three cases represent some abnormality related to the sunburn mechanism, and are distinctly separate from the first seven.

There can be no doubt that more than one etiology is included under the term *Urticaria solare*. Other cases which have been so diagnosed are those of Ochs (1910), Duke (1924), Sellei (1930), and Kruspe (1931), but there is no means of estimating the spectral sensitivity in these cases.

Dr. Harold A. Abramson has described a case to me which appears to belong to neither of the above groups, and it is entirely possible that other types of "Urticaria solare" exist. Since the etiology may be very different for cases which exhibit much the same outward signs, it is important to delimit the action spectrum in all such cases, since this should be the most important diagnostic characteristic. The length of the latent period may also be an important diagnostic sign, since it may distinguish between the two groups represented in Table 16.

Chapter 19

The Photosensitizing Action of Porphyrins in Man

Anderson was the first to suggest the association of porphyrins with abnormal sensitivity to light. In 1898 he described two brothers suffering from recurrent bullous lesions of the uncovered parts of the body, which he diagnosed as Hydroa aestivale. The urine of both took on a burgundy-red color at times, and one of them stated that he could avoid the attacks of skin eruptions by remaining indoors during those periods when the urine appeared dark. Anderson, following this lead, submitted the urine to a chemist who reported the presence of a pigment with spectral absorption bands coinciding with those of hematoporphyrin, a substance known to be derivable from hemoglobin or hemin. It has since been shown that hematoporphyrin does not appear in either normal or pathological urines, but other porphyrins whose absorption spectra resemble that of hematoporphyin may be present. According to Günther (1911a) red-colored urine had been previously observed to accompany Hydroa, but the presence of porphyrins was not recognized before Anderson's report.

Porphyrin was soon demonstrated in the urine of a number of cases of this type, and the summaries of Günther (1911a, 1919, 1922) show a certain degree of correlation between porphyrinuria and Hydroa.

In 1908 Hausmann found that hematoporphyrin is a photosensitizer of the photodynamic type, and Ehrmann (1909) connected this with the occurrence of porphyrinuria in Hydroa, suggesting that the lesions of this disease might result from the sensitization of the skin to light by porphyrins. In 1913 Meyer-Betz sensitized himself to light by injecting hematoporphyrin into his own vein—his famous "Selbstversuch." He remained sensitive to sunlight for two months after the injection; edema and pigmentation resulted from exposure to light, but not hydroaform lesions. This experiment was quite generally assumed to settle the question of the etiology of Hydroa. However, attempts to reproduce the lesions of Hydroa by exposure of patients to light have yielded uncertain results, which leaves the role of light in this disease open to question. These experiments will be subjected to analysis and their significance discussed in subsequent pages.

The interest in the relationship of porphyrins to human disease has increased progressively since the time of these first discoveries. Porphyrins have been shown to be present in very small quantities in normal urines, and in abnormally large amount in a number of pathological conditions. The chemical structure of these substances has been firmly established, principally due to the work of Fischer and his co-workers, and the method of their elaboration in the body has received considerable attention. Before continuing, a brief discussion of the chemistry of the porphyrins must be introduced.

Chemistry of the Porphyrins

Reviews by Fischer (1930), Rimington (1936), Carrié (1936), Fischer (1937), and Steele (1937), make an extensive discussion of this subject superfluous; therefore only a brief account is given here. The essential structure of the porphyrins is a nucleus of four substituted pyrrole rings joined by methine bridges. Etioporphyrin, the formula of which is shown below, is not found in nature, but may be taken as the starting point for the formation of all the porphyrins. In this compound four methyl and four ethyl groups are substituted on the carbons of the pyrrole rings not attached to methine bridges.

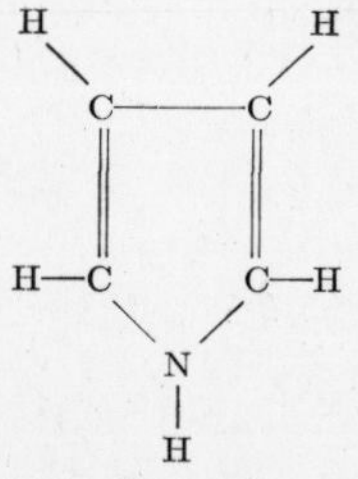

```
      H           H
       \         /
        C-------C
        ‖       ‖
        ‖       ‖
      H—C       C—H
         \     /
            N
            |
            H
```

Pyrrole Ring

```
                 CH3-C=======C-C2H5
                     |       |
          H-C-------C       C=====C-H
            ‖        \\   /        |
            ‖          N           |
  C2H5-C-----C                     C=====C-CH3
       ‖       \                 /       |
       ‖        N-H        H-N           |
       ‖       /                 \       |
   CH3-C-----C                     C=====C-C2H5
             ‖         N           |
             ‖       //  \         |
          H-C-------C       C=====C-H
                     |       |
                C2H5-C=======C-CH3
```

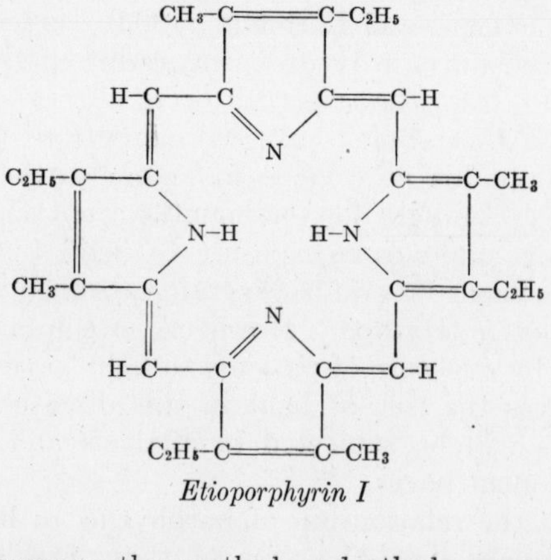

Etioporphyrin I

It is possible to arrange the methyl and ethyl groups of etioporphyrin in four different ways to obtain four isomers. For convenience the four pyrrole rings may be represented by brackets, and the formula of the four isomers written in the following manner:

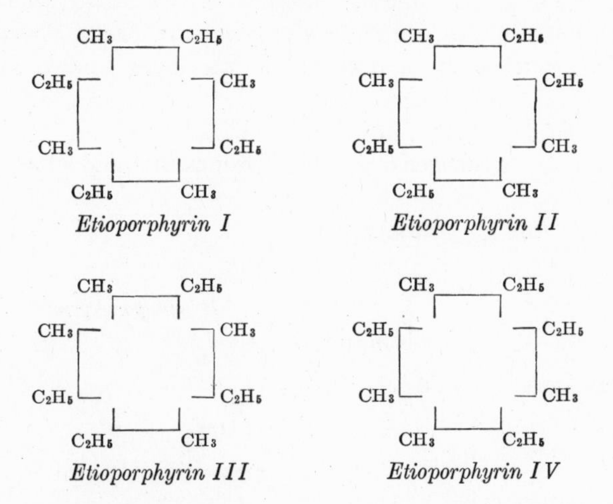

Etioporphyrin I *Etioporphyrin II* *Etioporphyrin III* *Etioporphyrin IV*

Numerous other porphyrins may be derived from etioporphyrin by the substitution of other radicals for the methyl and ethyl groups, and many have been synthesized in the laboratory, but up to the present only isomers of the I and III series have been found in nature.

Only such porphyrins as may enter into the present discussion will be described here. Protoporphyrin has two vinyl groups ($CH=CH_2$) and two proprionic acid groups (CH_2-CH_2-COOH), arranged as follows in protoporphyrin III, the only isomer of this formula found in nature*:

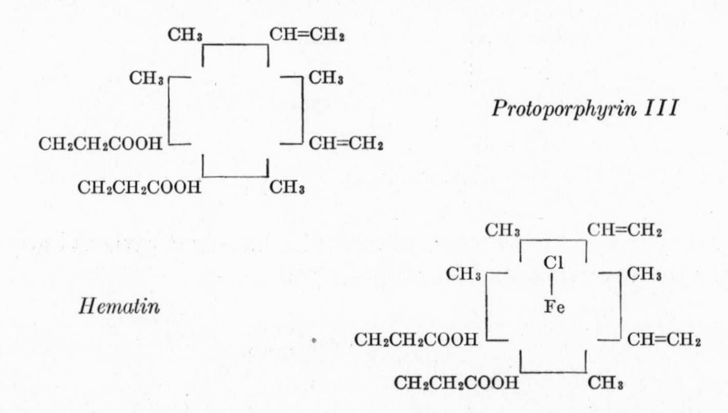

Protoporphyrin III

Hematin

The formulas above show the close relationship between protoporphyrin III and hematin, the prosthetic group of hemoglobin. Proto-

* The nomenclature employed here is that used by Carrié (1936), Karrer (1938) and others. The structures and nomenclature of these compounds are subject to revision at the present time (see, *e.g.*, Fischer and Orth, 1937).

porphyrin has also been called oöporphyrin, Kämmerer-porphyrin, Snapper-porphyrin, Hämatoporphyroidin, and Hämaterinsäure.

Hematoporphyrin, the first porphyrin to be discovered, has two $CHOHCH_3$ groups in place of the vinyl groups of protoporphyrin. This compound does not occur in nature. It was once known as iron-free hematin.

Deuteroporphyrin, also known as copratoporphyrin, has two ethyl groups replaced by hydrogens, and two proprionic acid groups.

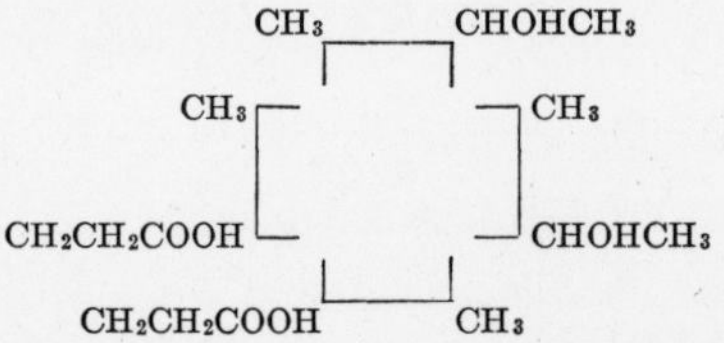

CH_3 — $CHOHCH_3$
CH_3 — CH_3
CH_2CH_2COOH — $CHOHCH_3$
CH_2CH_2COOH — CH_3

Hematoporphyrin

CH_3 — H
CH_3 — CH_3
CH_2CH_2COOH — H
CH_2CH_2COOH — CH_3

Deuteroporphyrin III

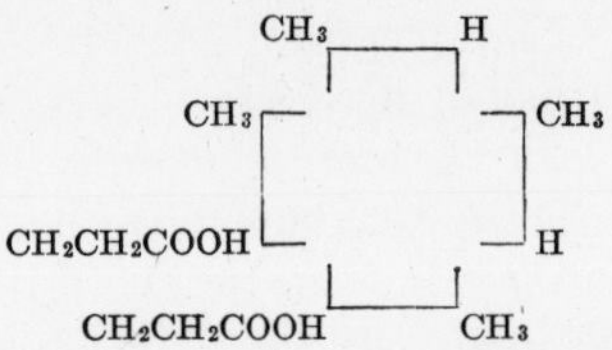

Coproporphyrin, also called Kotporphyrin, Enteroporphyrin or Stercoporphyrin, has four proprionic acid groups replacing the ethyl groups of etioporphyrin.

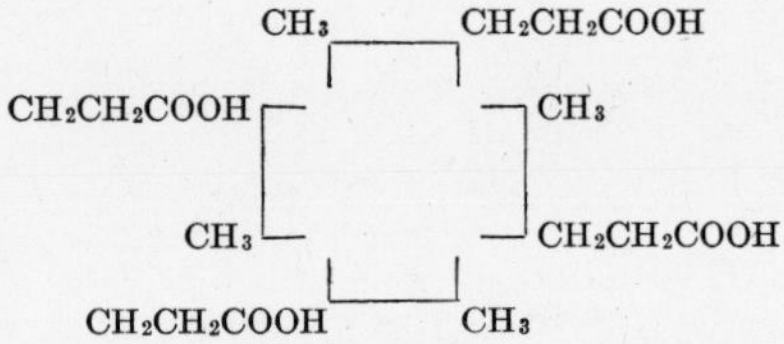

CH_3 — CH_2CH_2COOH
CH_2CH_2COOH — CH_3
CH_3 — CH_2CH_2COOH
CH_2CH_2COOH — CH_3

Coproporphyrin

Uroporphyrin, also known as urinporphyrin, has eight carboxyl groups arranged in the following way in uroporphyrin:

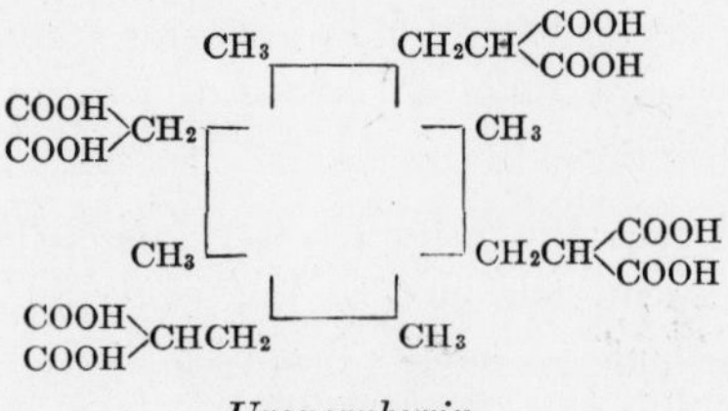

CH_3 — $CH_2CH(COOH)_2$
$(COOH)_2CH_2$ — CH_3
CH_3 — $CH_2CH(COOH)_2$
$(COOH)_2CHCH_2$ — CH_3

Uroporphyrin

The various porphyrins are distinguished principally by their solubilities, their absorption spectra, and their melting points (usually obtained for the methyl ester), which have been carefully determined upon synthesized material, so that their identification can be made with relative ease. They are found widely distributed in nature; protoporphyrin is found in the shells of brown hen's eggs, coproporphyrin in the shells of certain molluscs, uroporphyrin in the bones of some rodents (see Turner, 1937), and the feathers of some birds. They are found in the blood and bones of the human fetus, and in small quantities in other normal human tissues; in some pathological conditions their quantity may be greatly increased.

The Photosensitizing Action of Porphyrins

Hausmann contributed much to our knowledge of the photosensitizing action of the porphyrins, to which subject he devoted a long series of investigations. He showed in 1908 that hematoporphyrin may act as a photosensitizer for red blood cells and paramecia (Hausmann and Kolmer, 1908), and in 1910 that white mice may be rendered sensitive to light by this compound. The general similarity between the behavior of porphyrins and other photodynamic substances has been shown in earlier chapters.

From the standpoint of photosensitization in man, it is very important to know the wave-lengths to which the porphyrins sensitize living systems, but the evidence on this point has been confused. It was pointed out in Chapter 5 that the action spectrum for sensitization by a photodynamic dye may closely approximate the absorption spectrum of that dye, and that some degree of agreement is to be expected unless limiting factors interfere. Some of these factors were discussed in that chapter; and to these may be added, in the present instance, absorption by the epidermis, the importance of which as regards sunburn has been pointed out in Chapter 17.

The absorption spectrum of hematoporphyrin and its measurement were discussed in Chapter 5, where it was shown that this compound has its principal absorption in the near ultraviolet at about 3900 Å. Few quantitative measurements of the absorption spectrum of porphyrins which cover both the ultraviolet and visible have been made, but all agree in placing the maximum absorption in this general region. In Figure 45 a number of absorption spectra of porphyrins have been plotted in a way which illustrates this agreement. The measurements of Hausmann and Krumpel (1927) have not been corrected for the characteristics of the photographic plate, but indicate good agreement in the ultraviolet for a number of porphyrins, and the same is true for the spectra shown by Treibs (1932). The major absorption for all these compounds lies in the region 3600-4000 Å, and hematoporphyrin may be taken as an example of the group in this respect.

Blum and Pace (1937) carried out experiments to determine whether the action spectrum for human skin sensitized by hematoporphyrin agrees with the absorption spectrum of this substance, and found reasonable agreement although the experiments were admittedly rough. They used sunlight, carbon arc, and mercury arc as sources, with filters to delimit spectral regions. For the first two the amount of energy passing the

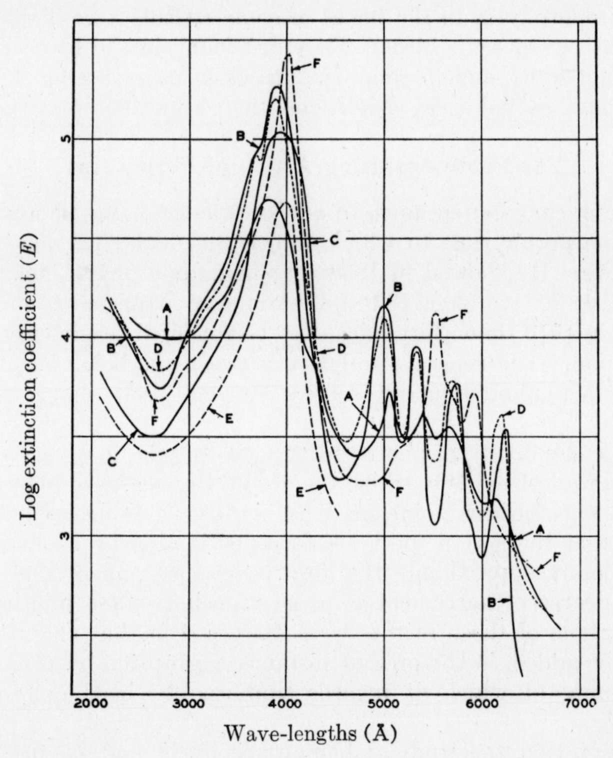

FIGURE 45. Absorption spectra of porphyrins.* The logarithm of the extinction coefficient, *E,* is plotted as the ordinate to show the general agreement in magnitude of absorption in the different spectral regions.

A, Hematoporphyrin hydrochloride in water (Blum and Pace, 1937). B, Hematoporphyrin dimethylether-dimethylester, in methyl alcohol (Clar and Haurowitz, 1933). C, Hematoporphyrin dimethylether in ethyl alcohol (Friedli, 1924). D, Mesoporphyrin (diammonium salt) in ethyl alcohol (Clar and Haurowitz, 1933). E, Mesoporphyrin hydrochloride in ethyl alcohol (Friedli, 1924). F, Mesoporphyrin in alcohol + 2% concentrated HCl (Clar and Haurowitz, 1933).

* See also Fischer and Orth (1937, p. 588).

filters was estimated in the manner described in Chapter 5.* The distribution of energy in the sources, and the estimation of absorption of light from these sources by hematoporphyrin with and without filters is

* In the present instance no correction for relative number of quanta was made because of the low degree of accuracy of the measurements.

shown in Figures 46, 47 and 48. The relative values for sun's energy absorbed per unit time appear as A in Table 17.

Local areas of the skin of the inner surface of the forearm were sensitized by injection of 0.01 per cent hematoporphyrin hydrochloride (photodyn) intradermally. Exposure of such sensitized areas to an appropriate light source for a few minutes resulted in appearance of an itching wheal with surrounding flare. The time to appearance of the wheal, t, was taken as an inverse measure of the intensity of the response, and these values are listed in Tables 17 and 18.

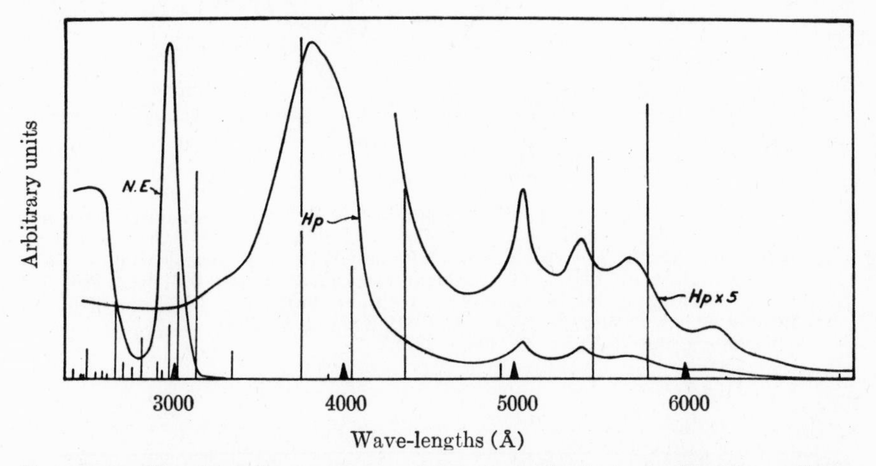

FIGURE 46. Hp, absorption spectrum of hematoporphyrin hydrochloride; ordinates are values of E; Hp × 5, ordinate values increased to show the bands of alkaline hematoporphyrin in the visible region more clearly. N.E., sunburn spectrum. Vertical lines show position and relative values of the lines of a mercury arc. [From H. F. Blum and N. Pace, *British Journal of Dermatology and Syphilis,* 49, 465 (1937).]

As pointed out in Chapter 5, the product of time and intensity should be the same in all cases, if the reciprocity law holds and there is complete agreement between absorption and action spectrum. There is a considerable disagreement between these values when sunlight is used as the source, (At, Table 17); but this is no greater than between determinations made at different times, and the absorption and action spectra may be assumed to be in essential agreement.* It should be remembered that the estimation of the distribution of sunlight is very rough (see p. 25). Comparison of Tables 17 and 18 shows that carbon-arc radiation is more effective than sunlight, which is readily explained by the fact that the principal emission of the carbon arc is in the same spectral region as the maximum absorption by hematoporphyrin (see Figure 48).

*No attempt has been made to correct for absorption by the epidermis, which would tend to shift the action spectrum toward longer wave-lengths, as may be seen by reference to Figure 48 where the spectral transmission of that layer is shown.

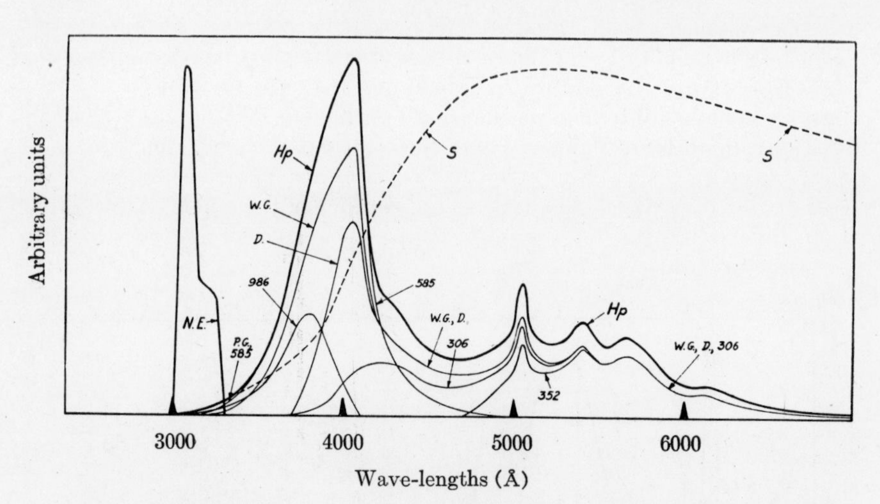

FIGURE 47. S, distribution of "average" sunlight. N.E., sunburn spectrum corrected for sunlight; Hp, absorption of sunlight by a thin layer of hematoporphyrin. W.G., absorption by hematoporphyrin of sunlight passing through window glass; P.G., through plate glass; D., through "document" glass; 585, 986, 306, 352, through Corning filters corresponding to these numbers. (See pp. 60-61 for methods of calculating.) [From H. F. Blum and N. Pace, *British Journal of Dermatology and Syphilis,* **49,** 465 (1937).]

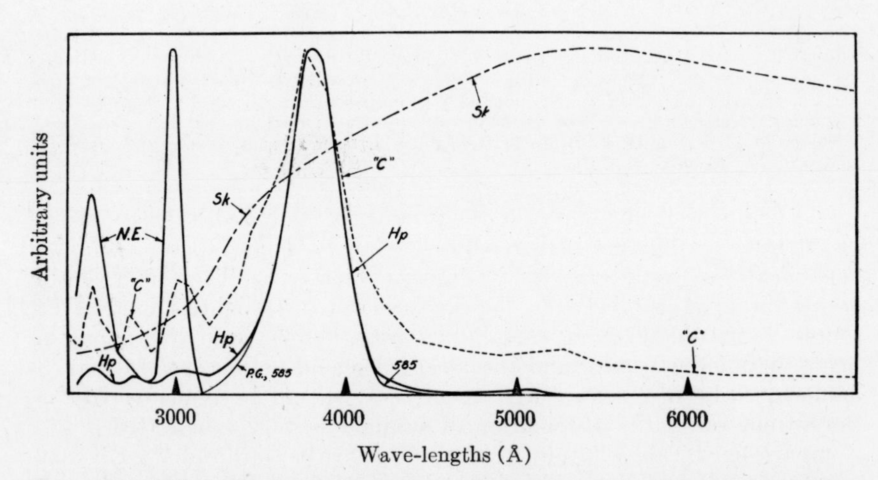

FIGURE 48. "C," spectral emission curve of "therapeutic C" carbon arc. N.E., normal erythema curve corrected for this source. Hp, absorption of this radiation by hematoporphyrin. P.G. and 585, absorption by hematoporphyrin of this radiation passing through plate glass and Corning 585 respectively. Sk, spectral transmission of the epidermis. [From H. F. Blum and N. Pace, *British Journal of Dermatology and Syphilis,* **49,** 465 (1937).]

The mercury arc * caused no photodynamic effect, exposure for periods sufficient to produce severe erythema of the area surrounding the point of injection failing to produce a wheal on the sensitized area. This is explained upon reference to Figure 46, which shows that relatively few mercury lines are appreciably absorbed by hematoporphyrin.

Table 17. Sensitization to Sunlight by Hematoporphyrin.
(From Blum and Pace, 1937)

Date	Subject	Time of day p.m.	Filter	λ	*t* min.	*A*	Response	*At*
21.i.36	P	3:00	None	All	3	1.00	++	3.0
21.i.36	B	3:00	None	All	3	1.00	++	3.0
22.ix.36	B	2:00	None	All	6	1.00	(+)	6.0
10.x.36	P	1:27	None	All	5	1.00	++	5.0
10.x.36	W	2:45	None	All	5	1.00	++	5.0
10.x.36	W	3:45	None	All	6	1.00	++	6.0
21.i.36	P	3:00	W.G.	3150→	4	0.75	++	3.0
21.i.36	B	3:00	W.G.	3150→	4	0.75	++	3.0
10.x.36	P	1:48	P.G.	3100→	7	0.89	++	6.2
10.x.36	W	3:00	P.G.	3100→	9	0.89	++	8.0
21.x.36	P	3:00	986	2600-4100	4	0.14	(+)	0.6
21.x.36	B	3:00	986	2600-4100	10	0.14	+	1.4
22.ix.36	B	2:00	986	2600-4100	10	0.14	(+)	1.4
10.x.36	P	2:21	986	2600-4100	18	0.14	+	2.6
10.x.36	W	2:00	986	2600-4100	22	0.14	+	3.1
22.ix.36	B	2:00	306	3800→	5	0.36	(+)	1.8
22.ix.36	B	2:00	585	3100-4600	10	0.57	(+)	5.7
10.x.36	P	1:34	585	3100-4600	12	0.57	++	6.9
10.x.36	W	3:52	585	3100-4600	6	0.57	++	3.4
10.x.36	P	1:59	D	3700→	7	0.54	++	3.7
10.x.36	W	2:45	D	3700→	5.5	0.54	++	2.9
10.x.36	P	2:08	352	4700→	11	0.21	++	2.4

λ = wave-lengths in Å passed by filter. *t* = period of irradiation. *A* = relative amount of sun's energy absorbed per unit time. ++ = immediate strong whealing response. + = immediate slight whealing response. (+) = delayed whealing response. W.G. = window glass. P.G. = plate glass. 986, 306, 585, 352 = Corning glasses characterized by these numbers. D = document glass. (Compare with figure 47.)

From this evidence it seems certain that photosensitization by the porphyrins is produced principally by wave-lengths lying between 3000 and 4500 Å, and to a much less extent by visible radiation extending to about wave-length 6500 Å. An examination of earlier data shows them to be in agreement with these findings, although often interpreted to the contrary. Early studies by Hausmann (1910) showed that hematoporphyrin sensitizes to visible wave-lengths shorter than the red; this is what should be expected from the absorption spectrum, and since these measurements did not include the ultraviolet they do not deny a greater sensitization in the latter region. At the same time, Hausmann found that mercury-arc radiation produced about the same effects in hematoporphyrin-sensitized and in normal mice; it is probable that the sunburn response was elicited in both cases, and that little or no energy was

* This was a "therapeutic type" arc, and it may be assumed that the arcs used in all the experiments discussed in this chapter were of this type.

Table 18. Sensitization to Carbon Arc Radiation by Hematoporphyrin.
(From Blum and Pace, 1937)

Date	Subject	Filter	λ	t min.	Response
8.xii.36	P	None	All	1	(+)
8.xii.36	B	None	All	1	(+)
10.x.36	B	P.G.	3100→	3	+
8.xii.36	P	P.G.	3100→	3	++
8.xii.36	P	P.G.	3100→	2	++
8.xii.36	B	P.G.	3100→	2	++
8.xii.36	B	P.G.	3100→	1	(+)
10.x.36	B	585	3100-4600	3	+
10.x.36	B	585	3100-4600	5	(+)
8.xii.36	P	585	3100-4600	3	++
8.xii.36	P	585	3100-4600	2	++
8.xii.36	B	585	3100-4600	2	++
8.xii.36	B	585	3100-4600	1	(+)
10.x.36	B	306	3800→	5	None

λ = wave-lengths in Å passed by filters. *t* = period of irradiation. ++ = immediate strong whealing response. + = immediate slight whealing response. (+) = delayed whealing. (Compare with figure 48.)

absorbed by the hematoporphyrin. It has been claimed that porphyrins sensitize laboratory animals to mercury-arc radiation (*e.g.*, see Perutz, 1910), but it is probable that the sunburn radiation has not been eliminated in such instances, or that temperature, a factor to which rodents are particularly susceptible (see p. 106), has not been controlled.

Later, Hausmann was led by experiments of Möller (1900) and Freund (1912), which appeared to show that the lesions of Hydroa were produced by wave-lengths in the ultraviolet, to search for sensitization in that region. Hausmann and Sonne (1927), using a quartz monochromator, dispersed the mercury-arc spectrum upon agar plates containing red blood cells, and found hemolysis in the regions of certain mercury lines when the plates were examined after a suitable period of time. Plates which were not sensitized showed hemolysis only at wave-lengths shorter

Table 19. Photodynamic Effectiveness of Porphyrins in Killing Albino Mice with Direct Sunlight.*
(After Dankemeyer, 1930)

Porphyrin injected (mg.)	Time to killing (minutes)		
	Uroporphyrin	Deuteroporphyrin (Kopratoporphyrin)	Coproporphyrin
1.0	40- 45	60- 65	70- 75
0.5	45- 50	80- 85	95-100
0.25	100-110	120-130	150-165
0.125	150-160	170-175	225-240
0.062	210-245	about 18 hours	

* Slight sensitivity to a tungsten filament lamp was observed, but none to the mercury arc.

than 3300 Å; the action spectrum for hemolysis in this spectral region appears in Figure 8A (p. 30). Plates sensitized with hematoporphyrin showed sensitization in the near ultraviolet and visible. Further studies using this method (Hausmann and Kuen, 1933) have demonstrated the photosensitizing action of other porphyrins, but since these measurements were not quantitative it is impossible to compare the action and absorption spectra. Lassen (1927) studied the sensitization of paramecia by hematoporphyrin to ultraviolet radiation and found the greatest sensitizing action in the region of 3670 Å, with little sensitization in the region of 3000 Å; this is in general agreement with the absorption spectrum of hematoporphyrin.

It is probable that the ability of the different porphyrins to sensitize human skin depends upon the extent to which they are taken up and held by that organ. They vary widely in their photosensitizing action on albino mice as shown by the experiments of Dankemeyer (1930) summarized in Table 19.

The Hydroa Riddle

More or less categorical statements to the effect that the condition known as Hydroa vacciniforme (or Hydroa aestivale) is due to the photosensitizing action of porphyrins are often found in reviews and textbooks. However, if one examines the original papers describing studies of this disease, one finds many expressions of doubt, including some suggestions that light plays no part whatsoever in its etiology. If one attempts an analysis of the experiments which have been made on patients one is even more uncertain; and the reader who has labored through any considerable number of papers on this subject will appreciate the above caption.

Nomenclature. The word *Hydroa* was originally employed to describe vesiculous lesions in general (see Lafitte, 1907). Bazin (1860) used the term more specifically to describe three types of bullous lesions, all of which he believed to be associated with arthritis, and to have been previously diagnosed as syphilis. One of these types, which he designated as Hydroa vacciniforme, was represented by the case of a young boy in whom the attacks appeared after he had been taken for walks in the open air (see Bazin, 1868). The lesions appeared on the exposed parts of the body at first, but later on parts which were covered by clothing. Bazin states that light is the precipitating factor in this disease, although he mentions no attempt to reproduce the lesions by exposing the patient to light. The disease was not described again for about thirty years, but since that time a relatively large number of cases have been reported which resemble that of Bazin, with an attendant confusion of nomenclature which still persists.

In 1889 Hutchinson, who was apparently not acquainted with the description given by Bazin, reported such a case under the name summer eruption. He considered it somewhat like several cases which he had previously described as summer prurigo (1878), but which were not bullous in character. Berliner (1890) recognized that Hutchinson's summer eruption was probably the same as Bazin's Hydroa vacciniforme, but preferred to translate Hutchinson's term to Eruptio aestivalis, dividing this into Eruptio aestivalis pruriginosa and Eruptio aestivalis bullosa to represent summer prurigo and summer eruption respectively. Berliner's nomenclature exemplifies a tendency to regard the two conditions described by Hutchinson as different manifestations or degrees of the same disease. This was unfortunate, since it is quite certain that they have entirely different etiologies if, in fact, they do not represent a number of etiological groups. Brooke (1892) introduced another term for the bullous type of eruption by combining Berliner's and Bazin's terminology to form Hydroa aestivale; this term is much used today. From that time on such terms were employed in various combinations, and with frequent additions. There was a tendency to separate the bullous type of eruption (Hydroa) by using the term *vacciniforme* only when scarring occurred; thus Buri (1891) suggests Eruptio aestivale vacciniforme, and later we find Hydroa aestivale and Hydroa vacciniforme separated on this basis (Senear and Fink, 1923). Perutz (1917a) has also suggested a separation on the basis of the presence or absence of porphyrinuria.

Today the terms *Hydroa aestivale* and *Hydroa vacciniforme* are used quite interchangeably for the most part, and the others have been discarded. In examining the older literature one may be confused, and the foregoing account has been inserted to show the general state of affairs. Since none of the terms which have been employed seems particularly suited from a descriptive standpoint, it might be preferable to employ Bazin's original Hydroa vacciniforme, but in the following pages the term *Hydroa* will be used for brevity without a qualifying adjective. This should lead to no confusion because the term has lost most of its general meaning, although it is sometimes used to describe conditions in which light sensitivity could have no part, *e.g.*, Hydroa gravidarum.

Symptoms. Dermatologically, Hydroa manifests itself as bullae (blisters) of various sizes up to one or two centimeters in diameter, confined to the exposed parts of the body; in a few cases placed in this category there have been lesions on other parts. The lesions may be preceded by erythema and are sometimes accompanied by edema. They contain clear or discolored fluid, and are sometimes purulent; they often become depressed in the center (umbilicated), and generally develop crusts. In most cases scars result which may bring about severe disfigurement and destruction of parts such as the ears, nose and fingers; the latter may become useless because of scarring in the region of the phalangeal joints. Sometimes scarring is absent, however. Pigmentation of the skin, peculiar fine downy hair growth and abnormal sensitivity to trauma are sometimes found (see Table 22); scleroderma and lesions of the cornea also have been reported.

There are certain inconstant accompaniments of the skin lesions, which should be mentioned. Porphyrinuria is the most interesting of these and will receive more attention below. Enlargement of the spleen and liver, and changes in the blood picture are frequently but not always found. Discoloration of the teeth and bones by porphyrin is sometimes observed. Gottron and Ellinger (1933) and Schreus and Carrié (1931) report a reduced number of capillaries in the skin of individuals with porphyrinuria. All these symptoms have been given various clinical significance by different writers, but their conclusions cannot be discussed here. Attention should be called to the variability of these findings which in itself suggests that the cases which have been described as Hydroa may have more than one etiology.

Porphyria. After the discovery of the association of porphyrin with Hydroa lesions and the recognition of its possible role as a photosensitizer, attempts were made to separate different types of porphyrinuria. Günther made careful reviews of the known cases of porphyrinuria at various times (1911, 1919, 1922) and proposed the following classifications:

I. Hematoporphyria * acuta. Sudden onset of porphyrin in the urine associated with acute central nervous system disorders such as Landry's paralysis, with acute infections, etc. Not associated with sensitivity to light.

II. Hematoporphyia acuta toxica. Sudden appearance of porphyrin in the urine with chronic poisoning such as by sulfonal, veronal, trional, lead, etc. Not associated with sensitivity to light.

III. Hematoporphyria chronica. Porphyria persists over a long time and is not associated with any acute nervous system or intestinal disorders. Associated with sensitivity to light and tendency to hydroaform lesions.

IV. Hematoporphyria congenita. Porphyria present from birth or early childhood. Associated with sensitivity to light, and tendency toward hydroaform lesions.

Günther's classification is more or less used at present (but see Carrié, 1936); it indicates that sensitivity to light is not associated with the acute, but only with the chronic or congenital forms of porphyrinuria. However, it is sometimes reported that the former conditions are accompanied by photosensitivity, and there are a number of interesting observations in this connection. For the present only the chronic or congenital forms will be considered, as these seem to bear the closest relationship to Hydroa.

Hydroa is not associated with all cases of chronic or congenital porphyria, nor does porphyria appear to be a constant accompaniment of Hydroa. Porphyria has been demonstrated in only a relatively small number of cases diagnosed as Hydroa, but this may be due in part to the

* At the time of Günther's first study (1911a), it was believed that the porphyrin found in such cases was hematoporphyrin, and in his later papers (1919, 1920, 1922) he continues to use the prefix hemato-. By porphyria (hematoporphyria in Günther's terminology), is meant the output of porphyrin by the body in quantities much in excess of normal. Porphyrinuria means the presence of abnormal proportions of prophyrin in the urine, and is sometimes used synonymously with prophyria, but the distinction is often made.

failure to detect porphyrins when present. They may exist as colorless reduction compounds (leucobases) called porphyrinogens, in which case they can be detected only if proper methods are employed to oxidize the porphyrinogens to porphyrins. The occurrence of porphyrins in the urine is more or less periodical (see Anderson, 1898; Perutz, 1917a), so that unless repeated examinations are made they may escape detection; Perutz (1917b) found that porphyrinogen could be found in the urine during the time when no porphyrin could be detected. Recently, van den Bergh and Grotepass (1933) have shown that large excess of porphyrin may be present in the body without being manifest in the urine; in such cases excess of porphyrin may be detected in the feces or elsewhere. Thus it is not surprising that the percentage of Hydroa cases in which porphyrinuria has been detected is so small. On the other hand, it must be recognized that some of the cases which have been diagnosed as Hydroa may belong to some other etiological group than that which is accompanied by porphyrinuria. It is even possible that the porphyrinuria reported in some of these cases may represent only the normal porphyrin excretion.

Recognizing this situation, Hausmann and Haxthausen (1929) separate the following categories:

Hydroa aestivale with porphyrinuria congenita.
Hydroa aestivale with porphyrinuria chronica.
Hydroa aestivale without porphyrinuria.

They state that the lesions are somewhat less severe when there is no porphyrinuria, but admit that this is a difficult distinction to make; and they suggest that Hydroa may have more than one etiology. Further clinical classifications have been made, but seem to shed little light on the problem and will not be discussed here.

Extensive studies of a few cases have been made, among which the famous Matthias Petry must be mentioned. This individual was discovered and first studied by Günther in 1911 (1911b). Later he was studied by Schum, and finally by Hans Fischer and collaborators in Munich. He died in 1925 and was immediately dissected and the pieces subjected to chemical as well as morphological study, porphyrins being found in almost all the organs; the findings provide the material for a monograph by Borst and Königsdörfer (1929). Rimington (1936) and Carrié (1936) may be consulted for a more detailed discussion of the physiology and pathology of porphyria.

Heredity of Porphyria and Hydroa. Another interesting approach to the problem of the etiological classification of this disease, or diseases, is by way of genetics, for in at least some cases the condition is inherited. Early studies of the heredity of Hydroa and porphyria were made by Siemens (1922), and by Hofman (1928), and more recently an extensive analysis has been given by Cockayne (1933), who separates the following groups:

Porphyrinuria congenita. Hydroa vacciniforme of Bazin.
- (1) A dominant type.
- (2) A recessive type which appears to be a single gene recessive with a preference for the male sex.

Hydroa vacciniforme without porphyria. Hydroa aestivale.
- (1) A dominant type.
- (2) A recessive type which appears to be a single gene recessive with a preference for the male sex.

This classification is to be considered as tentative. Cockayne has followed the suggestion of Perutz (1917a) in separating Hydroa vacciniforme and Hydroa aestivale according to whether porphyria accompanies the disease. The same hereditary types are found in both cases, suggesting that the separation is not a fundamental one. In the following analysis of experiments in which the attempt has been made to reproduce the lesions of Hydroa by exposure to light, the classification of Cockayne has been followed to see if any correlation is to be made between such findings and the genetic classification. In making this analysis isolated cases on which genetic data is not available are treated separately.

Experimental Studies of Photosensitivity. If Hydroa lesions result from sensitization to light by porphyrins, they should be produced by approximately the same wave-lengths to which normal skin is sensitized by injection of hematoporphyrin. Thus, from the evidence previously presented, it should be possible to produce the lesions of the disease by subjecting the skin to any wave-lengths between the lower limit of sunlight (about 2900 Å) and about 6500 Å, provided the intensity and time of exposure were great enough; but the greatest sensitivity should lie between about 3000 and 4500 Å.

I have undertaken (Blum and Hardgrave, 1936; Blum and Pace, 1937) to reproduce the lesions of Hydroa by exposure to the same sources in conjunction with the same filters used in the preceding experiments on skin sensitized by hematoporphyrin, so that the two sets of measurements should be comparable. The case of Hydroa on which these studies were made displayed severe porphyrinuria, hirsutus and sensitivity to slight trauma, being similar to a number of other cases in these respects (see Table 22). The lesions were bullous in character, and limited to the parts not covered by clothing; the bullae were followed by destructive scarring, the fingers being badly damaged. There can be little doubt that the case is comparable with a number of others which have been placed in this category, and it seems typical.

Table 20 presents the results of these experiments. In no case was it found possible to reproduce the bullous lesions by using any of the combinations of sources and filters with which whealing was elicited in normal skin after the injection of hematoporphyrin. It was not possible to reproduce the lesions of Hydroa or elicit any signs of abnormal sensitivity by frequently repeated irradiation with the mercury arc, but this is not evidence against the hypothesis that the lesions are the result of sensitization by porphyrins, since hematoporphyrin does not sensitize appreciably to this radiation (see p. 219). The other experiments show, however, that exposure to sunlight or carbon arc equal to or in excess of that which the patient would normally encounter in a single exposure to sunlight did not produce the lesions.

Table 20. Photosensitivity in Hydroa Vacciniforme.
(From Blum and Pace, 1937)

Source	Date	Filter	λ	*A*	*t* min.	Region	Result †
Sun	6.iv.36	None	All	1.00	30	Back	Blackish erythema.
Sun	9.iv.36	None	All	1.00	30	Back	Blackish erythema.
Sun	6.iv.36	W.G.	3150→	0.75	30	Back	Blackish erythema.
Sun	9.iv.36	W.G.	3150→	0.75	30	Back	Blackish erythema.
Sun	12.x.36	P.G.	3100→	0.89	90	Forearm	Blackish erythema.
Sun	6.iv.36	986	2600–4100	0.14	30	Back	Nothing.
Sun	9.iv.36	306	3800→	0.36	30	Back	Nothing.
Sun	9.iv.36	585	3100–4600	0.57	30	Back	Blackish erythema.
Sun	12.x.36	585	3100–4600	0.57	90	Forearm	Blackish erythema.
Sun	9.iv.36	352	3700→	0.21	30	Back	Nothing.
Sun	9.iv.36	243	6150→	0	30		Nothing.
Carbon arc	29.ix.36	585	3100–4100	..	45	Upper arm	Blackish erythema.
Therapeutic "C" carbons	15.x.36	None	All	..	3	Upper arm	Nothing.
Hg arc *	Bi-weekly	None	All	..	Up to 10	Entire body	Normal tanning.

λ = wave-lengths in Å passed by filter. *t* = period of irradiation. *A* = relative amount of energy of sunlight.

* "Therapeutic" type. † see p. 231.

These experiments themselves would seem to throw grave doubt on the hypothesis that porphyrins are photosensitizers in Hydroa, but there are a few possible explanations which may be offered for such negative results. It may be suggested that the patient happened to be insensitive to the radiation at the particular time that the experiments were performed. This argument might be supported by the fact that the quantity of porphyrin in the urine varies widely from time to time, so that one might expect that the quantity of porphyrin in the skin also undergoes wide variation. But against this is the fact that hematoporphyrin is very slowly removed from the skin (Meyer-Betz, 1913; Blum and Templeton, 1937; Blum and Pace, 1937), so that the porphyrin content of the skin need not undergo such wide variations as that of the urine. Furthermore, it was found on one occasion that the Hydroa patient whose arm had been exposed to sunlight for one and one-half hours without the production of hydroa lesions actually developed a few lesions on the face which were observable on the following day although this part had been shielded from direct sunlight throughout the experiment, and to a great extent from indirect sunlight. Thus the patient must have been susceptible to development of the Hydroa lesions at the time of exposure. This suggests the possibility that those parts of the skin which have been frequently exposed to light over long periods of time, and have developed lesions, are more sensitive to light than those which have not been previously exposed, a suggestion which has been frequently made. It will be best to examine the merits of this hypothesis after considering some other experimental studies.

Table 21 brings together data from published experimental attempts to elicit the lesions of Hydroa by exposure to light. Undoubtedly some experiments have been missed, in spite of extensive search, and a few have not been recorded because the description of the experiments was too indefinite; but those which have been assembled show a general lack of agreement among the results. The interpretations of the various individual experimenters, all of which there is not space to consider, are even more at variance. For convenience in discussion, the cases with and without porphyrinuria have been separated and a third group has been made, to contain those in which the presence of porphyrin is not definitely reported. Comparison with the classification of Cockayne may be more easily made with this arrangement. My interpretation of the results may not agree in all respects with those given by the authors.

The number of cases in which the reproduction of the lesions or even the existence of an abnormal reaction is reported is not great. This might be due to the fact that relatively small doses of radiation have been employed in many cases, but there can be no doubt that the results of a number of well conducted experiments have been entirely negative. It is often impossible to be sure that a lesion which is reported as abnormal is not within the limits of normal response to excessive dosage with the sunburn wave-lengths, or to excessive heating. Often the experiments are not controlled in any way, and at times the conclusions are equally free from restraint. As a whole, the data cannot be regarded as very reliable. One of the greatest difficulties is that with rare exceptions no attempt has been made to rule out the normal effects of the sunburn radiation.

Results with the Mercury Arc. If the lesions of Hydroa result from photosensitization by porphyrin they should not be produced by mercury-arc radiation, except with excessive doses which alone would produce severe sunburn, as is evident from the experiments discussed earlier in this chapter (p. 219). On the other hand, if the lesions result from hypersensitivity of the normal sunburn mechanism, they should be produced by mild doses of this radiation. In many of the cases shown in Table 21 the response to this type of radiation seems to be within normal limits, while in others it is difficult to judge because the dosage cannot be estimated. It is possible to produce severe blistering in normal skin by excessive exposures to the mercury arc, and many, if not all, the responses reported as abnormal in Table 21 may have resulted in this way. It is probable that some of the cases which have been reported as Hydroa really belong with those which are discussed in the next chapter as polymorphic light eruption, most of which seem to be hypersensitive to sunburn radiation. As a whole, the results indicate that Hydroa is not a condition of hypersensitivity of the sunburn mechanism, particularly

Table 21. Experimental Attempts to Elicit the Lesions of Hydroa with Light.

Key: (–) = no response; N = normal response; >N = response greater than normal; <N = response less than normal; A = abnormal response; H = hydroaform lesions. In some cases, attempts have been made to delimit the spectrum, but these show little consistency, and are not treated in the table for lack of space. The genetic arrangement of Cockayne has been followed. All cases were not diagnosed as Hydroa by the investigators.

Case	Sunlight	Carbon arc	Mercury arc	Other sources	Remarks
Cases Showing Porphyria					
Dominant type					
Jacoby (1928)			(–)	"visible light" (–)	
Radaeli (1911) case II		N	N		
Recessive type					
Arzt and Hausmann (1920) case II		<N	<N		
Ehrmann (1905, 1909)		A	A	carbon filament N	
Gray (1926)	(–)	(–)	N		
Martenstein (1922) cases I and II		A*	N	X-ray, radium (–)	*Immediate erythema and whealing
Schmidt La Baume (1927)		N	N	X-ray, radium N	
Schreus and Carrié (1931)	>N		>N	tungsten filament (–)	
Isolated					
Barber, Howitt and Knott (1926) case 14		A*		tungsten arc N	*No erythema; papules and vesicles
Barber, Howitt and Knott (1926) case 15		A*		tungsten arc >N	*No erythema; papules and vesicles
Blum and Pace (1937), Blum and Hardgrave (1936)	N*	N*	N		*Blackish discoloration (see p. 231)
Cappelli (1914)			N		
Cerutti (1935) case I			N		
Cerutti (1935) case II	A		N	"infrared" N	
Dowling (1935)		(–)	(–)	tungsten arc (–)	
Freund (1912), Perutz (1917)		H*	A†	X-ray N†	*Freund, †Perutz
Gottron and Ellinger (1931)		<N	<N	tungsten filament (–)	Hot and cold air (–)
Gottron and Ellinger (1933)			<N	"visible light" (–)	
Günther (1911) (Mathias Petry)		N*		X-ray, heat N	*Blackish discoloration (see p. 231)
Goeckerman, Osterberg and Sheard (1929)				source not stated	Eczematous lesions produced, maximum at 3000 Å
Haxthausen (1929) case I		(–)			
Linser (1906)		N		X-ray N	
Meineri (1931)	A*	N		radium (–)	*Increased sensitivity to trauma
Radaeli (1911) case I		A	A		
Sparacio (1926)	N		A	tungsten filament (–)	

Table 21.—*Continued.*

Case	Sun-light	Carbon arc	Mer-cury arc	Other sources	Remarks
Cases Showing Porphyria					
Isolated					
Sparacio (1928)	A*		N	tungsten filament A	*Through red and yellow filters
Urbach and Blöch (1934) case I	A	N	N		
Urbach and Blöch (1934) case II	>N		>N		
Weiss (1925)			<N		
Cases Without Porphyria					
Dominant type					
Funfack (1924) case III	A*	A*	A*		*Immediate erythema and edema
Werther (1924)				"short ultraviolet" >N	
Recessive type					
Scolari (1933) case I	H*				*After repeated exposure and dosing with sulfonal
Überschär (1926)		(−)			
Isolated					
Barber, Howitt and Knott (1926) cases 6, 8, 10, 16, 13		N		tungsten arc N	
Barber, Howitt and Knott (1926) cases 1, 3, 5		<N*		tungsten arc N*	Bulli formed in some cases
Buquiccio (1923)	A*	A	N		*Apparently sensitive to blue and violet
Carrié (1931)			N	tungsten filament (−)	
Epstein (1933) case I			N	thorium X. N	
Epstein (1933) case II			N	X-ray N, thorium and mesothorium. N	
Epstein (1933) case III			N	X-ray N; thorium X. H; mesothorium >N	
Funfack, (1924) case I	A*	A*	A*		*Immediate erythema, edema
Funfack (1924) case II	A*	A*	A*		*Immediate erythema, edema
Haxthausen (1929) case II		A*			*Perhaps sensitive to blue and violet
Scolari (1933) case II		A*	N	*"infrared lamp" N	*Immediate edema, possibly from blue and violet
Isolated: No Mention of Examination for Porphyrins					
Becker (1908)			A*		*Erythematous dermatitis
Greenbaum (1927)			>N	X-ray, infrared N	
Möller (1900)	H*	H*			*Through glass; only after repeated exposure
Moro (1906)		N		iron arc N	
Pautrier and Payenneville (1913)			(−)		
Scholtz (1907)		N		iron arc N	

those reported by Blum and Hardgrave (1936), where no abnormal response was produced by bi-weekly exposures to the mercury arc (see Table 20).

Results with Carbon Arc and Sunlight. Sunlight should be the most important source for such experiments, since this is the radiation to which the Hydroa patient is normally exposed, but as reference to the table will show, it has not often been employed. As the experiments on photosensitization with hematoporphyrin demonstrate, the carbon arc should be an excellent source for the production of lesions if porphyrins are the sensitizers, but the characteristics of carbon arcs vary widely (see p. 26) and it is difficult to compare experiments in which this source has been used. In many of the experiments listed in Table 21 the Finsen-Reyn arc was used, which, according to Johansen (1915), gives a much greater intensity in the visible than the arc which was used in the experiments of Blum and Pace. Some arcs give a much greater emission in the sunburn region than others, so that many of the experiments may be subject to the same criticism as those in which the mercury arc was used, namely, that lesions may be produced by the sunburn radiation before enough of the longer wave-lengths have been applied to elicit any lesions that might result from the action of these longer wave-lengths on porphyrins in the skin. This difficulty has been avoided in the experiments of Blum and Pace by the use of a window-glass filter to eliminate the sunburn radiation. There is also the danger of burning by excessive heat when the carbon arc is used.

In a number of cases the findings, like those shown in Table 20, are completely negative. In some of these repeated irradiation has produced no lesions, but they may not have covered sufficient time.

More abnormal responses are reported with the carbon arc and sunlight than with the mercury arc, but the number is small. Möller's experiments, conducted in 1900, were well performed, but somewhat variable and difficult to interpret; he seems to have got hydroaform lesions on some occasions by successive exposure to sunlight or carbon arc through window glass. Ehrmann (1905) found some apparently abnormal results when he irradiated with a carbon arc through cobalt glass or copper ammonium solution, which should allow the greater part of the wave-lengths absorbed by porphyrins to pass, while absorbing most of the sunburn radiation. He found somewhat abnormal results with the same patient (1909) when he irradiated with a mercury arc. It is difficult to evaluate these experiments, and he does not claim to have produced the lesions of Hydroa.

Freund (1912) states that he obtained hydroaform lesions after irradiating with a carbon arc through filters, but it is difficult to tell whether this was due to excessive sunburn radiation or to wave-lengths in the longer ultraviolet region where porphyrins have their principal absorption, since his filters seem to have passed both to a certain extent. The blackish discoloration which was observed by Günther (1911a), and by myself (Blum and Hardgrave, 1936) does not seem to be an abnormal response. I have since observed such a response in a number of normal skins; it seems to be more prevalent in pigmented skins, and may have been the response discussed on p. 186. Funfack's (1924) patients, in whom careful examination revealed no porphyria, seem to have been abnormally sensitive to sunlight, carbon arc and mercury arc, but he was not able to produce hydroaform lesions with any of these sources. Sparacio (1928) produced bullae in one of his cases which he exposed to sunlight, but since the lesions were produced either through red and yellow filters, or by the heat of a stove, it seems probable that this was a sensitivity to heat rather than light. At any rate, it would seem to be produced by wave-lengths outside the absorption spectrum of the porphyrins. Scolari (1933) claims to have produced hydroaform lesions in one case by exposure to sunlight, but since he dosed his patient with sulfonal the interpretation of his results is questionable. One of Urbach and Blöch's (1934) cases seems to have been sensitive outside the sunburn region, but Hydroa lesions were not produced. Sellei (1936) claims to have produced the lesions of Hydroa in one of two cases by exposure to sunlight, but not with a mercury arc. However, he gives no description whatsoever of the procedure, so that it is difficult to judge his results. The case of Goeckerman, Osterberg and Sheard (1929) seems to have been sensitive to the sunburn radiation; it will be discussed again in Chapter 20.

The experiments of Meineri (1931) are of special interest. This investigator exposed his patient to direct sunlight, day after day throughout the summer, without producing any other manifest response than the normal pigmentation which might be expected to result from such treatment. He states, however, that the patient, who responded to slight trauma with the formation of bullae, was more sensitive to such trauma after this exposure. Meineri's results and conclusions will be discussed again below.

Buquiccio (1923), who separated spectral regions by means of filters and with a spectroscope, concluded that his case was sensitive to blue and violet light. Such a result does not fit the action spectrum for sensitization by porphyrins, and his results suggest that the etiology of the condition which he studied may have been the same as that discussed in the preceding chapter. One of the cases studied by Haxthausen (Hausmann and Haxthausen, 1929) developed a large blister in an area irradiated with carbon arc through quinine sulfate solution, which cuts out most of the ultraviolet; in another case which was handled in an identical manner no lesions were produced. Haxthausen * suggests that the first case may have been sensitive to the same radiation as Buquicchio's, but is inclined to be skeptical of his result because of the possibility that the lesion might have been produced by heat.

* Personal communication.

Very few definite conclusions are to be culled from the data in Table 21. It seems very probable that more than one disease is described therein, but the hope that the experimental results might group themselves along the lines of Cockayne's genetic classification finds little to support it. The reader is advised to consult the original literature, but more is to be expected from careful experiments in the future than from reference to those which have been carried out in the past. It is hoped that the present treatment may point the way to obtaining better experimental evidence, particularly in the elimination of the sunburn radiation by employing a window-glass filter. If, in all cases, two areas of the skin had been simultaneously exposed, one to the direct rays of the source, the other to the same source but with a filter of window glass to remove the sunburn wave-lengths, the data might shed infinitely more light on the etiology of Hydroa.

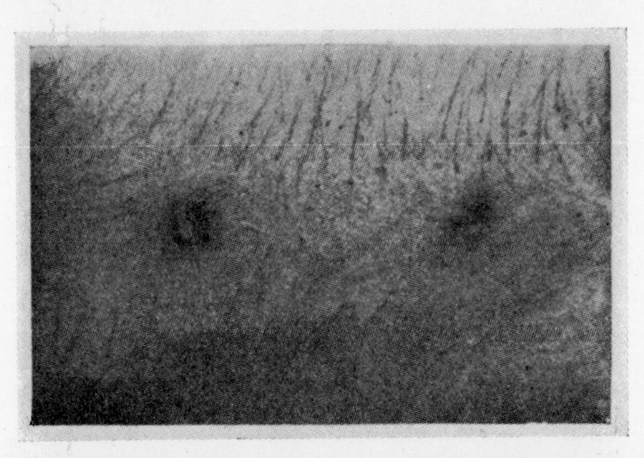

FIGURE 49. Two regions of papulo-vesicular eruption following photosensitization by intradermal injection of hematoporphyrin. The lesions appeared after exposure to sunlight nine weeks after the injection and original exposure. The legions appear on circumscribed pigmented areas, the pigmentation dating from a few days after the first exposure to light. [From H. F. Blum and N. Pace, *British Journal of Dermatology and Syphilis,* **49,** 465 (1937).]

Some Further Experiments: *Delayed lesions.* A chance incident which occurred in the course of the experiments of Blum and Pace (1937) may have significance. Three normal subjects were used in these experiments, in whom local sensitization of the skin with hematoporphyrin was produced. The immediate reaction was the same in all: an itching wheal appeared upon exposure to sunlight, which was followed in a few days by pigmentation persisting for months and marking the place at which the wheal had been produced. On a number of occasions this region was exposed to sunlight within a few weeks after the original experiment, and signs of persisting sensitivity were indicated by the appearance of wheals similar to the original, or by mere itching or erythema. On one occasion, nine weeks after the original experiment, one of the individuals who had had four areas on the forearm sensitized, exposed himself to the sun for a considerable time. On the next day he discovered that small papulo-vesicles had formed on the four pigmented spots which marked the sensitized areas. A photograph of two or these areas is shown in Figure 49. One such area which was excised and sectioned showed mild changes in the epidermis and in the papillary layer; the other lesions disappeared in a few days leaving very slight scarring. Whether this may be taken as an example of experimental Hydroa is open to question; it is difficult to understand why similar lesions did not appear in the other individuals. Possible significance of this finding will be discussed below.

Irradiation plus Pressure. Gottron and Ellinger (1931) found that their patient, although having a much higher threshold to sunburn radiation than normal, responded to such radiation with bullae when a quartz compression lens was used. Such a lens is designed both to concentrate the radiation and to apply pressure to the skin. They concluded that Hydroa lesions result from the combined action of mechanical pressure and light. It is difficult to see how such a condition could arise in normal exposure to the sun's rays, and the finding is not confirmed by all investigators. While Schmidt-LaBaume (1927), Martenstein (1922) and Gottron and Ellinger (1931 and 1933) found reactions greater than normal when a mercury arc or carbon arc was used in conjunction with pressure; Epstein (1933), Urbach and Blöch (1934), Scolari (1933), Moro (1906) and Cerutti (1935) found the same or greater sensitivity in the control. It seems probable, however, that mechanical trauma may play an important part in the production of Hydroa lesions, as will be pointed out below.

Passive Transfer of Photosensitivity. Stein (1928a, b) stated that he was able to produce local photosensitivity in the skin of a normal individual by the injection of serum from a Hydroa patient. Scholz (see Stein, 1928b) could not confirm this.

Relationship between Hydroa and Epidermolysis Bullosa. Epidermolysis bullosa is characterized by sensitivity to trauma such that slight injury to the skin results in bullous lesions of the injured part, a symptom which has also been reported in some cases diagnosed as Hydroa. Porphyrinuria is sometimes associated with Epidermolysis bullosa as with Hydroa, and relationship between the two conditions has been suggested (Ehrmann, 1909; Grosz, 1914; Meineri, 1931; Gottron and Ellinger, 1931; Cockayne, 1933; Carrié, 1936; and Blum and Pace, 1937.* Data have been gathered in Table 22 which suggest that sensitivity to trauma, excessive hair growth, and excess pigmentation are often associated together in

* Since this chapter was written there has appeared an excellent experimental paper by Turner and Obermayer (1938), who arrive independently at much the same point of view adopted herein. Their case, diagnosed as porphyria accompanied by epidermolysis bullosa, hypertrychosis and melanosis, has not been included in the discussion or tables because their paper merits separate reading.

patients showing porphyrinuria. In some of the cases represented in this table there were apparently no bullous lesions. Diagnosis varies but includes both Epidermolysis bullosa and Hydroa. It is probable that these symptoms are more often associated together than is reported in the literature.

Table 22. Symptoms Associated with Porphyrinuria.

Cases diagnosed as Hydroa				
Case	Excessive pigmentation of skin	Hypertrichosis	Hypersensitivity to trauma	Porphyrinuria
Artz, Hausmann (1920 Case I		+	+	+
Artz, Hausmann (1920) Case II		+		+
Ashby (1926)	?		+	+
Blum, Hardgrave (1936)	?	+	+	+
Blum, Pace (1937)				
Cappelli (1914)		+		+
Gottron, Ellinger (1931)	+		+	+
Gray (1926)		+	+	+
Linser (1906)		+		+
Mackey, Garrod (1926)			+	+
Martenstein (1922)			+*	+
Meineri (1931)	+	+	+	+
Sparacio (1928)	+	+		+
Urbach, Blöch (1934) Case I			+	+
Urbach, Blöch (1934) Case II			+	+

Cases diagnosed as other than Hydroa					
Case	Diagnosis	Excessive pigmentation of skin	Hypertrichosis	Hypersensitivity to trauma	Porphyrinuria
Cerutti (1935) Case III†	Epidermolysis bullosa			+	+
Cerutti (1935) Case I	?			+	+
Gottron, Ellinger (1933)	Porphyria	+		+	+
Hegler, Fraenkel, Schum (1913)	Porphyria	+	+		+
Hudelo, Montlaur (1919)	Pemphigus etc.	+	+	+	red urine
Marcozzi (1929)	Epidermolysis bullosa distrofica		+	+	+
Michelli, Dominici (1931) Case I	Porphyria	+	+		+
Michelli, Dominici (1931) Case II	Porphyria	+	+		+
Radaeli (1911)‡	Pemphigus traumaticus or Hydroa (Case III)		+	+	red urine
Schreus, Carrié (1931)	Congenital porphyria	+		+	+

* More sensitive to trauma in summer.
† Also described by Cerutti (1933).
‡ The father of Radaeli's case II (see Table 21).

Relationship between Epidermolysis bullosa and Hydroa is indicated in other ways. In the former disease, there is a deficiency in the elastic tissue of the skin, and Meineri has pointed out that this was the case with his patient whose lesions resembled Hydroa. The mode of inheritance is similar, there being dominant and recessive groups (Cockayne, 1933).

There seems to be a separation in both cases into two groups, one in which the bullous lesions result in scarring, and one in which this is not the case.

These resemblances and differences render any etiological study difficult, and numerous questions arise which are not easily answered. Is it not possible, for example, that certain of the cases listed in Table 21, and perhaps many others diagnosed either as Hydroa or Epidermolysis bullosa, belong to a single etiological group which is characterized by porphyria and sensitivity to trauma, and that other cases diagnosed as Hydroa should be placed with the conditions described in the following chapter, or possibly in other etiological groups? It seems plausible that sensitivity to slight trauma represents a condition of abnormality of the skin which is frequently, if not always, accompanied by porphyrinuria; but that photodynamic action of porphyrin in the skin is an essential factor in the production of the bullous lesions on the exposed parts is certainly open to question.

Meineri (1931) suggests that the Hydroa lesions do not result directly from the photosensitizing action of porphyrins, but that the skin is altered and made more sensitive to trauma after exposure to light, due to photodynamic action of these substances. This would explain his experiments in which successive exposure to sunlight did not result in lesions, but in increased sensitivity to trauma. This attractive hypothesis might also explain the experiment of Blum and Pace just described (see p. 232), in which papulo-vesicles appeared on normal skin which had been photosensitized with porphyrin, for the lesions may have resulted from slight trauma to a region which was subjected intermittently to the photodynamic action of porphyrin over a considerable time. The observation of Schreus and Carrié (1931) that lesions could be produced on the hands by rubbing, but not on the unexposed parts, might be cited in support of Meineri's hypothesis. Others of Meineri's experiments show, as do those of Urbach and Blöch (1934), that protection of the hand by a glove may prevent the lesions of Hydroa; such treatment might prevent trauma of the part as well as exposure to light. Altogether, Meineri's hypothesis seems the most consistent with the existing evidence, but requires further testing.

Porphyrias not Related to Hydroa

Porphyrinuria in Cattle. In 1936 Fourie reported that a herd of cattle in South Africa had been discovered to contain a number of animals which consistently voided red urine. Rimington (1936) was able to show that these animals were producing abnormal quantities of porphyrin. The herd which contained these animals was served by a single bull, which itself did not show porphyrinuria. A back cross of one of the porphyrinuric cows with the same bull has produced a porphyrinuric calf (Fourie and Rimington, 1937), which indicates that the porphyrinuria is a recessive character. Thus it seems probable that this porphyria among cattle is an inherited recessive as in most porphyrinuria of man. It seems not improbable that the inherited metabolic abnormality which results in the excretion of large quantities of porphyrin by these cattle is the same, or similar, to that found among human beings. Thus it may be expected that studies of these porphyrinuric cattle will result in better knowledge of human porphyria as well as of that found in cattle. Members of the herd have been preserved at Onderstepoort, in South Africa, where they will be studied.

Some of the porphyrinuric cattle showed signs which might be attributed to sensitivity to sunlight, *i.e.*, lesions of the more exposed parts of the skin. It is to be hoped that this phase of the problem will be carefully studied in the South African herd.

Difference in Photosensitizing Action of Porphyrins. Much of the evidence presented above suggests that porphyrinuria may often be present without photosensitivity. This in itself seems rather puzzling, since the porphyrin output is very great in many instances. It has been suggested that the porphyrins are often present in some form in which they are not photoactive, and it is also stated that some porphyrins are much more effective photosensitizers than others. It was pointed out in Chapter 7 that photodynamic effectiveness is in great part dependent upon the ability of the dye to be taken up by the cell, and thus a direct comparison of the photodynamic effectiveness of different dyes is hardly possible, except for a given set of conditions. The experiments of Dankemeyer, Table 19, illustrate such differences for the photosensitization of rats by various porphyrins. The metal complexes of the porphyrins seem as a rule to lack photodynamic activity, and it has been proposed from time to time that porphyrins exist in the body as metal complexes and therefore do not act as photosensitizers (see Carrié, 1936). Lignac (1925) claimed that Ca^{++} injections protected rats against the photosensitizing action of hematoporphyrin, possibly by combining with the pigment. Dhéré has suggested that the porphyrin in the earthworm is not photodynamically active because the porphyrin is in colloidal suspension (see p. 123).

Porphyrinuria from Drugs, etc. Abnormally high excretion of porphyrin may follow the administration of barbiturates (see Hausmann, 1923), and has been found recently after the administration of sulfanilimide (see Rimington and Hemming, 1938). Photosensitivity has been reported to follow luminal (Haxthausen, 1927), and recently there have been frequent reports of abnormal sensitivity to light after sulfanilimide therapy. It has been supposed that such sensitivity results from the photosensitizing action of the porphyrins, but there is strong experimental evidence against this, in the case of sulfanilamide at least (see p. 270).

Porphyrinuria has also been associated with lead poisoning (see Hausmann, 1923), and sensitivity to light has been claimed in this condition, though the writer has been unable to find any report of experimental demonstration in man. Götzl (1911) treated rabbits with an organic lead compound and reported porphyrinuria and photosensitization, but these experiments should certainly be repeated in the light of present knowledge.

Porphyria and Photosensitivity in General. The idea is current, unfortunately, that porphyrins are responsible for all cases of abnormal sensitivity to light. The action spectra of the conditions discussed in the preceding and following chapters will not permit such an interpretation, as may be seen by comparing them with the absorption spectrum of porphyrins.

Porphyrinuria has been reported in many instances of abnormal sensitivity to light, or in association with lupus erythematosus, pellagra, etc., the relationship of which to light is doubtful (see Chapter 24). Small quantities of porphyrin are present in normal urine and this may provide the basis for some of these reports. Careful quantitative studies by McFarland and Strain (1938) and Brunsting, Brugsch and O'Leary (1939) show no correlation between porphyinuria and these various conditions.

It has been suggested by Goeckerman, Osterberg and Sheard (1929) and by Templeton and Lunsford (1932) that the porphyrinuria of Hydroa is a result of the skin damage resulting from exposure to light rather than a causal factor. Templeton and Lunsford thought that they found increases of porphyrin in the urine of normal individuals exposed to mercury-arc radiation, but Scolari (1935) did not find any increase in coproporphyrin after such treatment.

Numerous hypothesis have been formulated to account for porphyrinuria without sensitivity to light, and sensitivity to light without porphyrinuria. Such speculation seems futile, without much more adequate experimental study of the action of light in such cases, particularly wavelength studies.

General Conclusions

The idea that sensitization by porphyrins is the basis of pathological photosensitization in man is so firmly intrenched that it requires a certain degree of temerity to attack it. However, it seems quite clear that the photosensitizing action of these compounds has not been experimentally demonstrated in the condition which is most often attributed to such action, namely, Hydroa vacciniforme of Bazin. Moreover, the experimental evidence supports the contention that porphyrins play little if any part in producing the lesions of this disease.

On the other hand, there can be no doubt that porphyrins may act as photosensitizers in man, and that the presence of abnormally high concentrations of porphyrins provides a condition in which photosensitivity might be expected.

There is a certain correlation between skin lesions of a particular bullous type, and the presence of abnormally high quantities of porphyrin. There is also correlation between porphyrinuria and skin lesions in a certain group of cattle having similar heredity, and there is similarity in the inheritance of the tendency to produce excess of porphyrin in both these animals and man.

In man, the lesions accompanying excess of porphyrins are often produced by slight trauma, and it is possible that this is the usual stimulus for their formation. On the other hand, it is possible that light may play either a direct or an indirect part in their production. If light is a factor, it seems probable that it must act in an indirect manner, since experiment has not proved a direct effect.

Decision of the many questions which arise from these considerations must await further experimental study.

Chapter 20

Abnormal Sensitivity to the Sunburn Spectrum—Polymorphic Light Eruption

It is probable that most abnormal photosensitivity in man represents abnormal reaction to the sunburn radiation. Although the character of the lesions may vary rather widely, it is logical to group them together if all are produced by the same exciting agent. The factor which evokes the lesions is one to which normal skin responds, a condition widely different from that described in Chapter 18, for example, where the skin responds to a stimulus which is not normally effective. The sequence of changes which follow the irradiation of normal skin with the sunburn spectrum are rather complex, and it might be expected that abnormality of any part or function of the skin which comes within the sweep of these changes would result in an abnormal response of one kind or another when the skin is subjected to such radiation. Thus a variety of abnormal responses and lesions should be expected. It is possible that in many cases sunburn radiation simply exacerbates a dermatological condition whose origin is quite independent of the action of that agent, while in others the lesions appear only after exposure to such radiation.

Polymorphic Light Eruption. Haxthausen (Hausmann and Haxthausen, 1929) groups together certain fairly common dermatological conditions which result from exposure to sunlight, under the term *chronic polymorphic light eruptions* (chronischen polymorphen Lichtausschläge). He includes under this name the conditions commonly diagnosed as Hutchinson's summer prurigo, Eczema solare etc. According to Haxthausen the lesions may be papulous, eczematous or a mixture of the two; in some of his patients the morphology was different during separate attacks. The lesions are generally accompanied by severe itching, and are confined to the exposed parts of the body. Other names applied to conditions which seem to fall within this group are prurigo aestivale and summer acne.

The variability of the lesions renders any purely morphological classification of little value, and shows the necessity for further etiological study. Evidence presented below indicates that most of these cases respond abnormally to the sunburn spectrum, but in a few cases one must conclude that the sensitivity is due to other wave-lengths. It seems certain, therefore, that more than one etiology is included within the group, but that there is no clear cut relationship between etiology and morphology. This does not simplify the problem, and makes it all the more important to determine the wave-lengths which provoke the lesions in individual cases. Until more is known of these conditions, Haxthausen's category may be employed as a tentative expedient, and I shall use the term *polymorphic light eruption* in his sense.

The Active Wave-lengths. The experimental data for the determination of the wave-lengths which evoke these abnormal skin eruptions are quite fragmentary. There is only one study, that of Schaumann and Lindholm (1932), which approaches a quantitative determination of the action spectrum. The experiments of these investigators seem to offer quite conclusive evidence that the sunburn spectrum alone was responsible for the production of the lesions in the case they studied, although they give a somewhat different interpretation. Their patient suffered from lupus of the face which was cured by local "Finsen" treatment (exposure of the affected parts to carbon arc through a quartz compression lens), but attempts to give general treatments with carbon or mercury arc on parts of the body not affected by lupus produced erythemato-papulous eruptions. They conclude that the patient was suffering from summer prurigo, as well as lupus, and it seems logical to include this case with the polymorphic light eruptions.

Schaumann and Lindholm exposed the arm of the patient to the lines of the mercury arc dispersed by the prism of a spectrograph, the skin of the patients being placed in the position ordinarily occupied by the photographic plate. Two different spectrographs, a large and a small, were used on different occasions, the results with both instruments being in agreement. The 3022 and 2967 Å lines produced erythema in the patient and in a control individual three to four hours after exposure for seventy to ninety minutes. After a four-hour exposure with the larger spectrograph, a feeble erythema also appeared in the position of the 3130 Å line, again in both control and patient. These findings are easily explained by reference to Figure 50. The 3022 and 2967 Å lines are relatively strong, and both fall in the region of the greatest erythemic effect, whereas although the intensity of the 3130 Å line is greater than either of these, it falls in a region of the sunburn spectrum where its effectiveness is not great. Allowing for unknown factors, such as differences in the spreading of the lines by the prisms and its effect upon the intensity, these results may be considered to be in agreement with the sunburn spectrum. In one experiment the authors attempted to equalize the intensity of the 3022 and 2967 Å lines, measuring the intensities by means of a thermopile and adjusting the time of exposure so as to give equal amounts of energy; but they found that the 3022 Å line was still the most effective in producing erythema. This seems to disagree with the sunburn spectrum, since the activity at 2967 Å should be considerably greater than at 3022 Å; but the measurement of the intensity of the lines under such conditions is beset with difficulties and a considerable error may have been introduced. The important thing in all their experiments is that, as far as erythema production is concerned, the patient behaved exactly as the normal control. The only difference in the behavior of the two was that the patient developed itching papules at the same time as the erythema, in the position of the 3022 Å line only. The papules showed, upon histological examination, the same appearance as those elicited by exposure to sunlight.

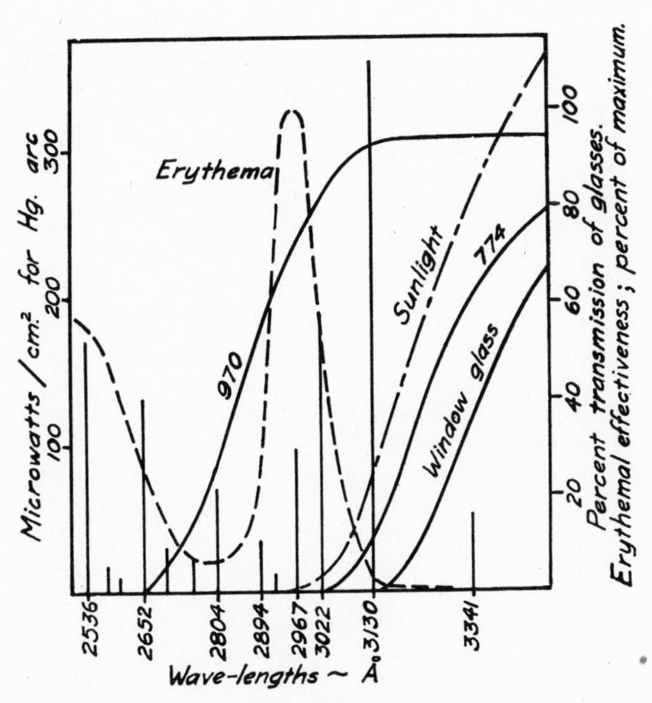

FIGURE 50. Vertical lines represent wave-lengths and intensities of the lines emitted by a "high-intensity" mercury arc (after McAllister, 1933). The curve marked Window glass is the percentage transmission for this glass; those marked 970 and 774 are for Corning glasses designated by these numbers. The curve marked Erythema represents the action spectrum of sunburn. The curve marked Sunlight represents the spectral distribution of sunlight as measured by Brian O'Brian (from Forsythe and Christison, 1929). The ordinates for this curve are arbitrarily chosen, and are not to be directly compared with those for the mercury arc.

Schaumann and Lindholm explain their experiments by assuming that only wave-lengths in the region of 3022 Å were effective in producing the lesions exhibited by their patient. However, since the erythema produced by the 3022 Å line was always more intense in both patient and control than that produced by the other lines, it seems probable that the erythemato-papular lesions shown by their patient would have been produced by any wave-lengths of the sunburn spectrum, provided there was sufficient dosage and penetration. We must assume in this case that the threshold for the production of the erythemato-papular lesions is higher than that for erythema, since erythema sometimes resulted without the appearance of any abnormal lesions. Schaumann and Lindholm's experiments may best be interpreted as showing that the lesions were produced in their case by the action of sunburn radiation, and hence represent an abnormal response to that agent. More recent studies of the same patient by these investigators (1938) further demonstrates the absence of sensitivity to visible radiation, since there was no response to sunlight passing through red or yellow filters.

Other studies are less complete than those of Schaumann and Lindholm, but present rather convincing evidence that polymorphic light eruptions may be produced by sunburn radiation. Numerous investigators have found the lesions to appear after exposure to the mercury arc (*e.g.*, Haxthausen, 1919; Sellei and Leibner, 1926; D'Amato, 1926; Flarer, 1930; Touraine and Dupperat, 1935; Gilfillan, 1937) which in itself is good evidence that it is the sunburn radiation which produces the lesions. This is true because the mercury-arc spectrum, as compared to sunlight,* is strong in sunburn radiation and weak in longer wave-lengths (see Figure 50). Thus, if the mercury arc elicits the same lesions as sunlight, it is probable that it is the sunburn radiation to which the skin is abnormally sensitive. If it were the longer wave-lengths of the mercury spectrum which produced the lesions, the patient must be extremely sensitive to sunlight. Hence the fact that the lesions may be reproduced by exposure to the mercury arc furnishes presumptive, although not definite, evidence that a patient is abnormally sensitive to the sunburn spectrum.

Haxthausen (1919) exposed a number of his thirty patients with polymorphic light eruption to sunlight, carbon arc, and mercury arc, and found that all three sources elicited papular lesions. Sometimes these appeared as primary lesions, while in other cases the first reaction was a diffuse, intensely itching erythema upon which papular lesions appeared as the result of scratching. Haxthausen (1929) states that the reactions did not occur when the "ultraviolet" portion of the radiation was removed, but gives no description of his method of delimiting this region.

D'Amato (1926), Goekermann, Osterberg and Sheard (1929), and Lomholt (1930) all made some attempt to delimit the wave-lengths producing lesions of the polymorphic light eruption type, with results which strongly suggest the sunburn spectrum.

The writer has made a few studies of cases of this type, using filters to delimit the effective wave-lengths, all of which are best explained by the assumption that sunburn radiation is the provoking factor. The transmissions of the more important filters are shown in Figure 50, and the results are summarized in Table 23.

A brief glance at this table shows that in none of the patients could the

* Presumably the "therapeutic" type mercury arc was used in all such studies.

Table 23. Spectral Sensitivity in Polymorphous Light Eruption.

Case	Source	Filter	Shortest wave-length trans-mitted (Å)	Expo-sure time (min.)	Response	Remarks
I*	Hg arc	None		3	Severe erythema	Developed into erythematous dermatitis
	Hg arc	970	2650	3	Mild erythema	
	Hg arc	774	3000	15	None	
	Hg arc	Window glass	3150	15	None	
	Hg arc	385	3500	15	None	
	Hg arc	038	4300	15	None	
	Sun	None		20	None	
II†	Hg arc	None		4	Erythema	Erythema followed normal course; September 1, 1935
	Hg arc	None		3	Erythema	
	Sun	None		30	None	September 10, 1935
	Sun	Window glass	3150	30	None	
	Sun	None		75	Erythema	May 1, 1936. Erythema developed into a scaly eruption
	Sun	Window glass	3150	75	None	
	Sun	038	4300	75	None	
	Sun	348	5700	75	None	
	Hg arc	None		3	Erythema	
	Sun	None		70	Erythema	May 7, 1936. Erythema developed into a scaly eruption
	Sun	None		30	Erythema	
	Sun	Window glass	3150	70	None	
III‡	Hg arc	None		3	Erythema	Erythema became eczematous
	Hg arc	774	3000	3	None	
	Sun	None		40	Erythema	
	Sun	774	3000	40	Erythema	
	Sun	038	4300	40	None	
IV	Hg arc	None		4	Erythema	Erythema followed normal course; no papules
	Hg arc	None		6	Erythema	
	C arc	None		$1\frac{1}{2}$	Erythema	Developed wheal and itching papules
	C arc	Window glass	3150	5	None	
	C arc	None	All	1	Erythema	All developed a "red rash"
	C arc	970	2600	1	Erythema	(All developed a "red rash")
	C arc	774	3000	$1\frac{3}{4}$	Slight erythema	(All developed a "red rash")

* Case studied by Blum and Schumacher (1937).

† Studied in collaboration with Dr. Lloyd Hardgrave.

‡ Studied in collaboration with Dr. Herman Allington.

lesions be produced when all wave-lengths shorter than 3150 Å were excluded, *i.e.*, radiation passing through window glass was not effective. On the other hand, the lesions were reproduced in all cases, at one time or another, by radiation rich in the sunburn wave-lengths, *i.e.*, those shorter than about 3200 Å. Apparently all these cases may be grouped among the polymorphic light eruptions, and in all probability represent abnormalities involving the sunburn mechanism. The cases will be considered individually, since they illustrate a number of points to be kept in mind in making such tests.

The general procedure in these tests was similar to that employed in studies described in Chapters 18 and 19. The region of the skin to be subjected to the tests is covered with a sheet of cardboard in which a number of holes are cut; some of these are covered by filters, but at least one of them is directly exposed. The individual holes are a few square centimeters in area, as a rule; it is convenient to make them of different shapes so that areas of erythema or abnormal lesions may be identified later, since such areas will have the same shape as the hole through which they were irradiated. This allows the simultaneous irradiation of a number of test areas, so that errors due to differences in the intensity of the radiation from a given source on different occasions are avoided. This is particularly important when sunlight is used as the source because of its great spectral variability, particularly in the sunburn region. Because of the possibility of encountering very sensitive patients it is advisable to begin a series of tests by a short exposure and to increase this if the first attempt is without result. In taking this precaution, one must reckon with two human characteristics: if severely burned on one occasion, the patient may not return for observation, or if the tests are too prolonged the patient may become bored and terminate them. Examination of the test areas was usually made about twenty-four hours after the exposure. The occasional appearance of immediate erythema due to heat was disregarded, and only persistent changes were considered in the analysis.

The first case, a woman of forty, was observed during the early stages of a neurological condition which terminated fatally in ascending paralysis resembling Landry's syndrome. On entering the hospital the patient displayed a fiery red erythematous eruption on the face, hands and arms. The lesions were distributed in such a manner as to suggest that they resulted from exposure to light; the submental triangle and the interior surfaces of the hands and arms which are ordinarily protected from light were free from lesions. A complete description of the case is presented by Blum and Schumacher (1937). The tests made to determine the action of light are summarized in the first part of Table 23; exposure to a mercury arc for periods of time which produced only a mild sunburn response in a control caused an erythema in the patient which appeared at about the same time as in the control, *i.e.*, a few hours after the irradiation, but persisted many days. This erythema changed its appearance to that of the erythematous, somewhat eczematous eruption which characterized the lesions on the face and hands. Exposure to such radiation passing through filters which removed wave-lengths longer than 3150 Å did not produce the lesions even though the irradiation was continued for a much longer time. Exposure to the sun was made for a period which was not long enough to provide dosage of sunburn radiation sufficient to elicit the lesions, but was long enough to indicate that there was no sensitivity to the longer wave-lengths of sunlight.

The second case, a woman twenty years old, showed an eczematous eruption covering the face, hands, and parts of the back exposed when wearing a bathing suit. The lesions appeared during the summer after a sun bath at the beach. The patient applied a mixture of vinegar, olive oil and common salt after the sunbath, believing this to be efficacious in the development of tanning. This may have exacerbated the condition, but probably was not a fundamental factor in the development of the lesions, since they reappeared the following summer without further application of the mixture, after having disappeared during the winter. At the time the first tests were made, in September of the year of the first outbreak, the patient still showed eczematous lesions although they were clearing up. As will be seen from the table, these first tests were entirely negative. Mercury-arc radiation produced a mild degree of erythema which followed the normal course of sunburn, and there was no response to a moderate dose of sunlight. However, during the following May, similar tests were performed with quite different results, as can be seen from the table. Exposure to the mercury arc or to sunlight for relatively short periods resulted in the appearance of erythema which developed a scaly eczematous surface, resembling the original lesions to a certain extent. No lesions were developed when the radiation was filtered through window glass or other filters which removed wave-lengths shorter than 3150 Å. These tests indicate that the lesions were produced by the sunburn spectrum, and that the patient was quite sensitive to this radiation. This was further shown by the fact that a few lesions had begun to appear on the face when the last tests were made at the beginning of summer, although the patient had been very careful to prevent more than slight occasional exposure to direct sunlight, and had been free from the lesions during the previous winter. No other pathological condition was found in this patient. This case is of particular interest because it was possible to reproduce the lesions at one time but not at another, although the same procedure was followed. Templeton and Lunsford (1932) had a similar experience with their patient, sunlight being known to elicit the lesions, but no lesions developing when the individual was subjected to sunlight for experiment. This should emphasize the importance of making more than one trial if first experiments yield negative results.

The third case displayed an eczematous eruption of the exposed parts of the skin, but no other pathology. The tests are particularly interesting because of the fact that "Pyrex" glass (Corning 774), which transmits wave-lengths longer than about 3000 Å, protected the patient against the action of mercury-arc radiation, but did not completely prevent the development of lesions when the patient was exposed to direct sunlight for forty minutes. This apparent discrepancy is explained by reference to Figure 50. This glass cuts out all but a small part of the sunburn radiation from the mercury arc; only a small proportion of the 3130 Å line passes, and this has very little sunburn action. However, it allows a considerable part of the sunburn radiation in sunlight to pass. Window glass was not used in this case, because it was thought that the "Pyrex" glass would remove the sunburn radiation from sunlight, since it was effective in removing it from the mercury-arc spectrum. This offers a good example of the necessity of knowing the transmissions of filters quite exactly. This patient was very sensitive to sunlight, sometimes developing lesions on parts covered by light clothing.

While the first three cases might be considered as belonging to the eczematous type of polymorphic light eruption, and might all be called Eczema solare, the fourth is of the papular type which has been called summer prurigo. The lesions were small, severely itching papules limited to the areas exposed, particularly the backs of the hands. The tests show that the mercury arc produced only a normal sunburn response, whereas the carbon arc reproduced the lesions. This might seem to indicate that the lesions were produced by longer wave-lengths than the sunburn spectrum since the former are abundantly present in the emission spectrum of the carbon arc which was used (see Figure 48, p. 218), except for the fact that no lesions developed when the carbon-arc radiation was filtered through window glass even though the period of exposure was relatively long. This apparently anomalous situation may be explained on the basis of the greater penetrating power of the longer wave-lengths of the sunburn spectrum. The sunburn radiation from the mercury arc is largely made up of shorter wave-lengths which, while producing erythema, do not penetrate very deeply (see Figure 41, p. 202). If the production of papules involves deeper layers, mercury-arc radiation should have relatively little effect; but carbon-arc radiation, being much richer in the longer wave-lengths, would be more effective. That the longer wave-lengths are more effective in the production of papular lesions in abnormally sensitive skin is indicated by the experiments of Schaumann and Lindholm (see p. 240), who found that the 3022 Å line of the mercury arc produced papules, while the 2967 Å line produced only erythema. This is also indicated in a recent study by Turner (1939) who found that sunlight produced the lesions of a polymorphic light eruption, whereas the mercury arc did not.

The foregoing studies illustrate a number of points which might lead to confusion when testing patients for spectral sensitivity. They illustrate the importance of using more than one source of radiation, of knowing the transmission of filters with some accuracy, and of careful consideration of all possibilities when the tests are interpreted. In the present state of the problem it does not seem wise to formulate a standard technique for the performance of such tests, as our knowledge is too scanty, and too great reliance on a standardized method might lead to erroneous conclusions. It may be possible after more complete study to devise a set of tests which can be used by dermatologists in separating different types of diseases caused by light.

Although most of the experimental studies of polymorphic light eruptions indicate that the sunburn spectrum is responsibe for the lesions, a few appear to demonstrate sensitivity to longer wave-lengths; and it is probable that at least a small number of cases which present clinical pictures similar to those just described are abnormally sensitive to wave-lengths other than those of the sunburn spectrum. The case described by Veiel in 1887 and diagnosed as Eczema solare displayed sensitivity to sunlight which had passed through window glass; in a room with closed windows only the side of the face toward the window developed the lesions. The patients of Wucherpfennig (1928) and Liebner (1931) seem to have been sensitive to longer wave-lengths, and that of Urbach and Konrad (1929) was definitely sensitive to visible wave-lengths, but not to the sunburn spectrum. It is thus necessary to admit that in some cases, which from their general clinical picture belong to the polymorphic light eruption group, the lesions are produced by wave-lengths outside the sunburn spectrum. These must have a different etiology from the majority of such cases. Hence it is important to make at least rough tests to delimit the spectrum which produces the lesions.

Separation of Different Etiologies within the Group. Numerous attempts have been made to separate different etiologies which might fall within the polymorphic light eruption group of Haxthausen. Among these, that of Sellei and Liebner (1926) is of particular interest since it has some experimental basis. These writers separate two sets of symptoms. The first—Eczema solare—is confined entirely to the parts exposed to light; it is eczematous in character, and is unaffected by changes in temperature. The second, Prurigo simplex chronica recidivans, appears on certain predisposed areas, *e.g.*, the forehead, the outer side of the upper arms and the ankles; exposure of any region of the body to sunlight exacerbates the lesions on these parts although the sunburn response of the exposed region may be normal. This conclusion is backed by experiments on four patients. Unaffected areas of the skin were exposed to mercury-arc radiation; the exposed regions responded with the normal sunburn sequence, but the lesions were exacerbated on the unexposed parts. Mercury-arc radiation applied to the affected areas likewise produced exacerbation. In this condition the lesions tend to be of itching papular type; they improve in cold weather, and are worse on hot days.

The first of these conditions certainly falls within the scope of Haxthausen's polymorphic light eruption group, but it may be necessary to place the second in a separate category. The second condition seems to be elicited by the sunburn spectrum since it is produced by the mercury arc, and this is probably true also of the first. It is of particular interest, because it indicates that substances formed in normal sunburn enter the blood-stream and cause exacerbation of lesions on other parts of the body. The first type appears, on the other hand, to be an entirely local reaction to sunburn radiation.

Transfer of Sensitization. Flarer (1930) studied two cases which showed more or less bullous lesions when exposed to light. He injected the serum from these bullae into the skin of three subjects. When the injected area was exposed, 48 hours later, to mercury-arc radiation, erythema and intense vesiculation was produced. Irradiation of other areas of skin produced normal sunburn. The sensitization of the injected area persisted for as long as eight days. Blood serum, on the other hand, did not produce these effects. If the serum from the bulli was irradiated and then injected, it produced the same vesiculation.

Haxthausen (1919) found that injection of irradiated blood serum from his patients into their own skin or that of normal controls produced the same effects as the injection of blood serum from normal individuals. Mühlmann and Akobjan (1930) claim to have rendered rats sensitive to mercury-arc radiation by injecting blood serum from an individual suffering from Prurigo aestivalis. D'Amato (1926) found that urine from his patient did not sensitize rats to the radiation from a tungsten filament lamp. All such experiments need repetition and further study before their significance can be estimated.

Relation to Other Conditions. It has been suggested in Chapter 19 that some cases reported as hydroa may belong to the group of diseases which has been discussed in the present chapter. The idea that porphyrins act as photosensitizers in all diseases produced by light frequently appears in discussion of the polymorphic light eruptions, and there are a few cases on record in which porphyrinuria has been reported in association with such conditions (Goeckerman *et. al.,* 1929; Barber, Howitt and Knott, 1926; Delbanco, 1920; Urbach, 1931). Pick (1924) would make the presence of porphyria a distinguishing characteristic of hydroa as contrasted with summer prurigo. The inexactitude of the nomenclature and description of all these diseases makes it difficult to place individual cases. The case of Goekermann, Osterberg and Sheard (1929) and those of Barber, Howitt and Knott (1926), which showed porphyrinuria, have been reviewed, together with other cases of porphyrinuria, in Chapter 19, although they might on any other basis be grouped with the cases described in this chapter.

The urticarias described by Beinhauer (1925) and Weiss (1932) were undoubtedly produced by sunburn radiation, and should be grouped etiologically with the conditions discussed in this chapter although the lesions had different morphology. They have been discussed in Chapter 18 because of their morphological resemblance to the other cases described therein. The patient described by Keller (1924b), who responded to mercury-arc radiation with pemphigus-like bullae, seems also to belong to this group.

Weinberger (1929) describes what appears to have been a case of combined summer prurigo and vitiligo, in which the lesions developed only on the white vitiliginous spots.

It is possible that sometimes, as has been suggested by Jesionek (1912), lesions diagnosed as summer prurigo are not produced by the action of light. In a number of the cases described by Hutchinson (1878) 1891) they seem to have been produced by heat or some other agent, since they were not restricted to the exposed areas of the skin, and this seems to be true of a number of other cases. However, if the conclusions of Sellei and Liebner (1926) are correct, this is not evidence against the participation of light, since products resulting from the irradiation of one part of the body may exacerbate lesions on another part. This possibility indicates once more the importance of producing lesions, or exacerbation of lesions, by exposure to light, before a diagnosis of photosensitivity is made.

Quantitative vs. Qualitative Differences. The wide quantitative variation in sensitivity to sunburn radiation among normal persons has been discussed in Chapter 17. It is very difficult to say just where normal sensitivity ends and hypersensitivity begins; for example, a given dose of sunlight or of mercury-arc radiation which would produce only a mild degree of erythema and pigmentation in one individual, might produce rather severe lesions in another. Very severe cases of acute sunburn with dangerous sequellae following overdosage with sunburn radiation have been described by numerous authors (*e.g.*, Lipschitz 1925; MacCormac and MacCree, 1925; Urquhart, 1933). It is reasonable to assume that such sequellae are to be expected most frequently in those individuals who are quantitatively most susceptible to sunburn radiation, but this can hardly be classed as abnormal photosensitivity. Again, certain changes which Haxthausen (Hausmann and Haxthausen, 1929) classes together as the chronic effects of light on normal skin (see Chapter 21), might be more prominent in persons with low sunburn threshold.

Abnormal sensitivity in the qualitative sense seems far more important in these conditions than quantitative variations. The lesions of the polymorphic light eruptions are abnormal responses which are qualitatively different from those of normal sunburn. It is probable that the dermatological changes which occur in this group bear little or no relationship in many cases to the erythemic threshold of the skin, but depend upon the qualitative reaction which the skin makes to this type of stimulus. This is clearly shown in Turner's case (1939). Certainly, comparisons of quantitative sensitivity to sunburn radiation between normal individuals and patients showing abnormal lesions have provided little information.

We may assume with justification that the initial photochemical process initiates a set of reactions, or responses in the various tissues of the skin, all of which are not directly related. Any one of these semi-independent responses may differ quantitatively from the normal, and these quantitative variations may result in the qualitative differences which are recognized as polymorphic light eruption. Thus the final response may be only indirectly related to the intensity of the initial photochemical process.

The possibility of different types of abnormal response which might result from exposure to such radiation is borne out by the variability of the morphology of the lesions and other symptoms which have been discussed above. The complexity of the problem is not lessened by the possibility that radiation from other regions of the spectrum may in some cases produce clinical pictures similar to those brought about by sunburn radiation (as for example in the case described by Urbach and Konrad, 1929) although the etiology must obviously be quite different.

Chapter 21

Effects of Long Exposure to Sunlight; Skin Cancer and the Sunburn Spectrum; Xeroderma Pigmentosum

Effects of Prolonged Exposure to Sunlight. Prolonged exposure to sunlight may produce changes in normal skin which are more pronounced in some individuals than in others. In such cases it may be difficult to distinguish between the extreme of normal variation and pathological conditions. It is presumably the sunburn radiation which is responsible for such effects, but it is not feasible to determine the active wavelengths experimentally. Haxthausen (Hausmann and Haxthausen, 1929) lists a number of these conditions, which will be discussed briefly.

Some persons tend to deposit pigment in splotches or freckles when they are exposed to sunburn radiation. The same individuals often tend to pigment in a similar manner as the result of rubbing or heat, freckles often being observed on other parts of the body than those exposed to sunlight. Thus, freckling is probably an abnormality of the pigment-producing mechanism rather than a specific abnormal response to light. The tendency to form freckles varies greatly among individuals, ranging from a very mild degree to what Sequiera (1911) terms permanent freckles. The tendency to develop patches of pigment, which is found in senile skins, may be increased by the action of sunlight.

Repeated exposure tends to produce a reddening of the skin due to a more or less permanent dilation of the minute vessels of the corium, as was pointed out by Möller (1900). An observation of my own shows that a severe sunburn may render the affected area sensitive not only to light, but to other types of stimulus. A blond individual developed a moderately severe sunburn as a result of sunbathing. An irregular region on the back, several square centimeters in area, blistered slightly, but healed leaving no changes which could be observed under ordinary conditions. This occurred eight years ago, but even up to the present year, this area tends to develop a more pronounced erythema than the surrounding skin when the individual is exposed to sunlight, so that it can be readily distinguished under these circumstances, though not at other times. The same occurs when the individual indulges in exercise sufficient to produce a generalized flushing.

Chronic reddening of a triangular area of the skin of the neck, sometimes accompanied by pigmentation, telangiectases and thickening of the corneum, is frequently observed in women when this area is habitually exposed. This has been called by Brocq (1916) "Dermatose du triangle sternoclaviculaire."

Haxthausen also lists degenerative changes in the skin of aged persons who have been habitually exposed to light; degeneration of the elastic and collagen fibers may occur. These changes give an appearance to the face which is recognized as typical of the aged. He believes that light is a factor in producing these changes because the more shaded parts are least affected.

Malignant Tumors of the Skin

It has been suggested from time to time that prolonged exposure to sunlight may stimulate the production of malignant tumors of the skin (Unna, 1894; Shield, 1899; Hyde, 1906; Dubreuilh, 1907; and many others). More recently, Roffo has been particularly active in defense of this thesis. His argument is based on three principal sets of evidence: (1) malignancy of human skin appears most frequently on those parts of the body most exposed to light; (2) malignancy of the skin may be produced in laboratory animals by prolonged exposure to light; and (3) he reports a direct relationship between cholesterol content of the skin, exposure to light and the incidence of malignancy. Most of Roffo's original studies have appeared in a journal published by the institute in Buenos Aires which he directs—Boletin del Instituto de Medicina Experimental para el Estudio y Tratamiento del Cáncer—but have been reviewed elsewhere (Roffo, 1933b, 1934a, 1935a, 1936a).

Roffo (1932b) presents an analysis of cases of malignancy of the skin, showing that this condition occurs most frequently on those areas ordinarily exposed to light, and indeed that its incidence may be closely correlated with the regions which receive the greatest amount of sunlight, for example, the great majority are limited to the face, 61 per cent occurring on the nose. Like Dubreuilh (1907), he points out that these same parts show a tendency to hyperkeratosis even when there is no malignancy, and believes that this may be considered as a pre-cancerous condition, at least in the aged.

Experimental Production of Malignant Tumors. The first demonstration of cancer production by light in laboratory animals is due to Findlay, who discovered in 1928 that white mice developed papillomata and malignant epithelioma after eight months' daily exposure to mercury-arc radiation. In 1930 he reported a similar experiment on a rat, and in the same year Putschar and Holtz reported that they could produce skin cancer in rats with great regularity by such treatment. This has been confirmed by a number of later workers (see Table 24). Huldschinsky (1933) produced sarcoma of rats' eyes by exposure to mercury-arc radiation.

The primary lesions which these experimenters describe are usually confined to the parts which are devoid of or sparsely covered with hair,

Table 24.

Investigator	Source	Period of exposure	Approximate time to appearance of tumors
Findlay, 1928 (mice)	Hg arc	1 to 2 minutes daily*	8 months
Findlay, 1930 (rats)	Hg arc	1 minute 3 times weekly	21 months
Putschar, Holtz, 1930 (rats)	Hg arc	24 hours daily†	27 weeks
Herlitz, Jundell, Wahlgren, 1931 (mice)	Hg arc	1 and 2 hours daily	Up to 7½ months
Roffo, 1933c (rats)	Hg arc	20 hours daily†	8 to 10 months
Roffo, 1934a (rats)	Sun	5 hours daily	7 to 10 months
Beard, *et al.*, 1936 (rats)	Hg arc‡	24 hours daily†	5 to 12 months
Rusch, Baumann, 1939a (mice)	Hg arc	10 to 60 minutes daily	3½ to 14 months

* For 5 months, then 3 minutes four times weekly.

† Graduated doses were given at first, reaching this value after the first few weeks.

‡ In "Pyrex" glass envelope.

e.g., the ears, eyes, and paws, or to parts from which the hair has been removed for the purpose of the experiment. They may be either epithelial tumors (cancer) or sarcomas. Roffo has shown (1935b) that the tumors may be artificially transplanted to normal rats.

The fact that mercury-arc radiation produces these malignant growths is strong evidence that the exciting wave-lengths lie in the region of the sunburn spectrum, and some of Roffo's experiments lend further support to this.* Using various filters he obtained the results listed in Table 25.

Table 25.

(From Roffo 1934a)

Source	Filter	Per cent of rats developing tumors
Sun	None	70
Sun	Transparent glass	0
Sun	Blue glass	0
Sun	Red glass	0
Sun	Yellow glass	0
Sun	Orange glass	0
Sun	Green glass	0
Sun	Violet glass	0
Hg arc	None	100
Hg arc	Transparent glass	0
Neon lamp		0
Tungsten filament lamp		0
Short radio waves (0.3 to 20 meters)		0

He gives no further description of the glasses which he used than to state the color which they presented to the eye, but it can be safely assumed that all had a common glass base, and did not transmit any wave-lengths shorter than about 3200 Å. It appears that only wave-lengths shorter than this produce the malignant tumors, since only sources rich in such radiation (sunlight and mercury arc) were effective, and that these were

* This has been further demonstrated by Dr. H. P. Rusch (personal communication).

not effective if filtered through glass. This is not definite proof that sunburn radiation is the sole noxious agent, since the shape of the action spectrum for the production of these malignant tumors has not been determined. Sterols absorb in the same general region of the spectrum, and according to Roffo these substances play a role in the development of malignancy; their relationship to the problem will be discussed below.

Various investigators have employed different dosages of radiation. It is impossible to make an exact comparison because the type of lamp, the operating conditions, and the distance from the animals is not always stated, but Table 24 provides some bases of comparison.

The large numbers of animals used in some of these experiments permits little doubt of the reality of the effect. Roffo (1934b) exposed six hundred rats to sunlight, of which three hundred and sixty-five died from excessive insolation within the first few days, but 70 per cent of the survivors developed tumors. Rusch and Baumann exposed several hundred mice to mercury-arc radiation, and obtained relatively high incidence of tumors.

Beard, Boggess and von Hamm (1936) state that the dosage employed by Roffo is so excessive that his experiments have no significance with regard to the development of skin cancer in human beings because the latter are not subject to so much radiation. This argument is not too convincing. While Roffo's dosages with mercury-arc radiation are certainly greater than human beings normally encounter, the tumors appear after a relatively short time, and the incidence is very high. Findlay's dosage was certainly not excessive, yet his rat developed a malignant tumor in 21 months, and his mice within a much shorter time. The dosage of sunlight which Roffo employed in his later experiments (1934b), five hours per day, does not seem greater than that to which agricultural laborers or enthusiastic "nudists" may be subjected, yet 70 per cent of the rats exposed to this radiation developed malignancy within a year. Beard, Boggess and von Hamm argue that, according to their estimate, the average time required to produce tumors in their rats corresponds to a period of eighteen years of such irradiation for man. Miescher (1937) also minimizes the importance of Roffo's experiments as regards man. He exposed guinea pigs and mice to mercury-arc radiation daily for several months, and observed pronounced thickening of the epidermis, but no cancer. Rothman and Bernhardt (1929) likewise could not produce cancer, although they produced what they thought to be a precancerous condition.

While it is unwise to apply too directly to man the finding that skin malignancy can be produced in laboratory animals by exposure to light, it is still less wise to minimize the possibility that malignancy of human skin may result from excessive insolation. This is particularly true in a public convinced of the hygienic and esthetic value of sunbathing.

Cholesterol and Skin Cancer. Roffo, who has studied the cholesterol* content of malignant tumors for many years, finds that there is a direct relationship between cholesterol in various regions of the skin and the incidence of skin cancer of these parts. He reports (1935c) that the cholesterol content of human skin can be increased by irradiation with a mercury arc. He finds that skin cholesterol increases with age, being relatively low in the fetus, and relates this to exposure to sunlight. Kawaguchi (1930) reported increased cholesterol in the skin of guinea pigs by irradiation with mercury arc or sunlight. Roffo (1932b, 1933c) and Knudson, Sturges and Bryan (1938) obtained similar results with rats. Baumann and Rusch (1939) found such increase in rats but not in guinea pigs. The interpretation of all these findings on cholesterol content of skin seems uncertain (see Bergmann, 1939).

Assuming, with Roffo, a relationship between exposure to sunlight, cholesterol content, and incidence of malignancy, the exact role of cholesterol in the development of malignant tumors is open to various interpretations. One may ask whether the accumulation of cholesterol in the skin is due to the sunburn spectrum; *e.g.*, is it another of the sequellae, of which pigmentation is one, which follow changes in the prickle cells caused by these wave-lengths (see page 184), or is it formed from some precursor by direct photochemical action? The former explanation receives some support from Knudson, Sturges and Bryan's finding that blood cholesterol decreases as skin cholesterol rises. That total cholesterol content of the body does not influence tumor production is indicated, however, by experiments on animals fed a high-cholesterol diet. Wahlgren (1934) and Baumann and Rusch (1939) attempted in this way to increase the incidence of tumors in rats exposed to mercury-arc radiation, but without success.

Once the cholesterol is present in the skin, does it act directly to produce malignancy, or does it require activation by radiation to exert its carcinogenic effect? Cholesterol absorbs in the same spectral region which produces sunburn, and the interposition of window glass would remove the wave-lengths which might activate it, so that the latter hypothesis would be in accord with the experimental evidence in this respect (see Table 25). This is the opinion of Roffo, who has attributed importance to the fact that cholesterol which has been irradiated blackens a photographic plate brought near it; but Stavely and Bergmann (1937) have shown that this is an error in interpretation. The blackening is due to hydrogen peroxide, not to radiation emitted by the cholesterol. Cholesterol is closely related to a number of hydrocarbons which are known to produce cancer, and it is conceivable that a photochemical transformation might take place in the skin. However, Strong and Bergmann (reported by Bergmann, 1939) have attempted to produce cancer by application of irradiated cholesterol to the skin without success. Bergmann (1939)

* Presumably total sterols.

presents a number of strong arguments against the hypothesis that cholesterol is changed into a carcinogenic substance in the skin.

Roffo refers at times to the supposed changes in cholesterol by light as "photodynamic," but the relationship of these changes to photodynamic action in the accepted sense employed in this book is doubtful. Körbler (1935) suggested that porphyrins might be photosensitizing agents for tumor production in the skin, but this finds little to support it. The photodynamic action of the carcinogenic hydrocarbons will be discussed in Chapter 22.

Indirect Action of Sunburn Radiation. Recently, Ellinger (1939a) has offered evidence for an indirect action of mercury-arc radiation in the production of tumors by showing that the incidence of development of cancer in mice inoculated with a suspension of cancer cells is greater in irradiated animals. He has shown (1938, 1939b) that mercury-arc radiation inhibits the growth of mice, which he credits to increased action of the thyroid gland induced by increased blood flow resulting from histamine liberated by the skin. This condition would appear to favor tumor development.

Xeroderma Pigmentosum

Xeroderma pigmentosum is a rare condition of the skin appearing in the early years of life, developing into malignancy, and almost always resulting in early death. Photophobia indicated by the child's avoidance of light, is often described as the earliest symptom (but see p. 278). Intense erythema and reddening of the conjunctiva usually appear at the time the child is first exposed to sunlight, and recur with subsequent exposures, the erythema being confined to the exposed parts of the body. Diffuse pigmentation and small or large freckles appear later. Atrophy and telangiectases appear after a time, as a rule on the pigmented spots, the telangiectases often surrounding atrophic areas. The atrophy may be superficial or produce deep scars. Hyperkeratosis and the appearance of multiple tumors which may be either carcinomas or sarcomas, involving both the epidermis and the corium, follows sooner or later. All these changes are most pronounced on the regions exposed to light, but in some cases the lesions are more generally distributed. Hausmann and Haxthausen (1929) or Rich, Davison and Green (1924), may be consulted for further details of the clinical and histological picture of this disease, or accounts may be found in most treatises on dermatology.

Many of the victims of this disease die between the ages of 5 and 10 years, and two-thirds before the age of 15. Rarely do they live beyond 20, although one case is reported to have died at 53 (Hausmann and Hauxthausen, 1929), and there are rare instances of marriages resulting in normal children (see Cockayne, 1933). Abortive cases have been described.

Xeroderma pigmentosum is an inherited abnormality. Siemens (1925), Cockayne (1933), and Macklin (1936) discuss its mode of inheritance, showing that it is clearly a single gene recessive equally distributed between the sexes. Relatives of individuals suffering from this disease often show a tendency to more or less intense freckling. Ordinarily freckling shows a high correlation with skin and hair color, being most common in the red blonds, but within the families of those suffering from Xeroderma this is not the case. This has led to the suggestion that the condition is not completely recessive to the normal and that the freckled relatives of xerodermatous individuals are heterozygous for this character.

Unna (1894) seems to have been the first to suggest that Xeroderma pigmentosum is a disease produced by light. Numerous investigators have attempted to demonstrate abnormal reactions to light (Weik, 1907; Rothman, 1923; Martenstein, 1924; Heiner, 1925; Martenstein and Bobowitsch, 1926; Greenbaum, 1927; Juon, 1928; Dörffel, 1929; MacCormac, 1933; Birnbaum, 1933; Lynch, 1934, Heft, 1935), but the results give little information which may be regarded as conclusive. Rothman found that mercury-arc radiation produced a persistent erythema followed by telangiectases, and Heiner, Juon, and Lynch found hyperpigmentation to follow such radiation. Others have found no reaction to light which might be considered abnormal in a qualitative sense, and there is disagreement as to the quantitative reaction, some reporting a low erythema threshold and others a high threshold, while the majority find no deviation from the normal. Martenstein and Lynch found no reaction to mercury-arc radiation when a window-glass filter was used, as did Weik and Martenstein when a carbon arc was employed. No abnormal reactions to tungsten filament lamps or carbon arc are reported by investigators who have used these sources (Martenstein, Juon, Rothman, Martenstein and Bobowitsch, and Weik). Both normal sensitivity and hypersensitivity to x-rays, α-rays, γ-rays, and chemical irritants have been reported. These studies leave open the possibility that sunburn radiation precipitates the occurrence of the lesions of Xeroderma. It seems most probable that the abnormal response is the result of some abnormality of the skin which is exacerbated by the action of the sunburn radiation, but that the condition might develop independently of this radiation although its appearance may be accelerated by it. Recently, Zoon (1938), using a series of filters, found an increased sensitivity to the long wavelength sunburn spectrum (around 3000 Å) in two Xeroderma patients. This might indicate an increased transparency of the epidermis.

Haxthausen (Hausmann and Haxthausen, 1929) lists Xeroderma pigmentosum tardivum as a condition closely resembling that which has just been discussed, but appearing late in life. It is interesting to speculate that such cases might represent heterozygous inheritance of Xeroderma pigmentosum, the incompletely recessive character becoming manifest later in life after more exposure to sunlight. Haxthausen groups this together with a cancerous or precancerous condition of the skin of sailors or agricultural workers who have been long exposed to sunlight, which was described by Unna (1894), and with skin cancer produced by light. Unna called this precancerous condition "Seemannshaut," and it has since been variously called, "seaman's skin," "peasant's skin," "farmer's skin," "tropical skin," etc. The exact relationship between true Xeroderma pigmentosum and the malignancies of the skin which have been discussed in the preceding section is of considerable interest. The discussion of these conditions has been placed in the same chapter because of their malignant nature, and because sunburn radiation seems to be concerned in their development.

Chapter 22

Photosensitization by Substances Coming Into Contact with the Skin

Guillaume (1926a, b, c; 1927) has shown that a photosensitizer must reach the malphigian layer of the epidermis in order to produce observable effects, no response to light occurring if only the dead corneum is invaded. If an aqueous solution of one of the common photodynamic dyes is applied to the surface of the skin it penetrates only the latter layer. However, if the corneum is carefully removed by scraping with a razor or rubbing with pumice stone before applying such a solution, the dye penetrates the malphigian layer, and there is evidence of photosensitization, *i.e.*, erythema appears upon exposure to light.

Thus, photosensitization by external contact is not the general rule, since most substances do not penetrate sufficiently; but there are photosensitizing substances which penetrate below the corneum and produce sensitization to light, *e.g.*, bergamot oil (see p. 262). It should also be possible to carry a photosensitizer into the skin by dissolving it in a substance soluble in the corneum, and Hickel (1927) was able to produce photosensitization by rubbing in a "pommade" containing a fluorescein dye, and wiping away the excess material with alcohol. Applied in this manner, a part of the dye should be taken up in the corneum where it would act as a light filter to prevent the active radiation from reaching the malphigian layer; wiping off with alcohol serves the double purpose of removing the excess light absorbing material from the surface, and of washing it out of the corneum. The writer has been able to confirm these findings, applying eosin in a fatty cosmetic base. This was kept in contact with the skin for twenty-four hours, then wiped off with alcohol immediately before exposure to sunlight. A few minutes exposure resulted in whealing, which subsequently disappeared leaving little or no trace. The whealing was not accompanied by itching, in contrast to that following intradermal injection and exposure to light (see p. 104).

Guillaume showed that the wave-lengths which produce the photosensitized effects are those which are absorbed by the particular photosensitizer, and are independent of the sunburn radiation. In some cases he found them to act as light filters to reduce the intensity of the sunburn radiation while at the same time sensitizing to another region of the spectrum. This is easily explained by differences in absorption of the dye, and penetration of the skin by radiation. Nevertheless, the belief that the dye increases the sunburn response has vitiated many experiments. For instance, Marceron (1925) and Ambrogio (1936a, b) both studied the effect of dyes in reducing the erythema threshold to mercury-arc radiation; obviously a mixed effect is studied in this case, which is difficult to evaluate.

Dermatitis Associated with Contact with Plants. Many instances of dermatitis following sunbathing in meadows or fields, where the skin has come into contact with green plants, have been reported. It has been called *bullöse Freïbad-und-Wiesendermatitis,* and *dermatose bulleuse des bains de soliel dans les prés,* which indicate the usual conditions under which the dermatitis appears.

Oppenheim first described the disease in 1926 (see 1932). The usual history is that the patient develops itching blisters which are more or less striated in their arrangement; they appear, as a rule, the day after the sun bath. The blisters usually clear up without difficulty, but leave pigmentation which persists for months. A most important clue to the etiology of the dermatitis is that it has not been reported from the seashore or other regions where sun-bathing is indulged in, but where there is no opportunity for contact with plants. Oppenheim (see 1932) first thought of an insect parasite, which came from plants in the particular locality in Austria in which he observed the condition, but he could find no such parasite. Like many later investigators, he then turned to the idea of sensitivity to a particular plant. From the striated arrangement of the lesions, it appears as though they might follow scratches received by rubbing against plant leaves and stems while lying in a grassy meadow, and the fact that only a limited number of those who are exposed exhibit the lesions indicates a specific sensibility of the individual. A number of workers occupied themselves with the search for a specific plant that was responsible for the lesions, but reached varying conclusions (Hartman and Briel, 1927; Siemens, 1927; Gans, 1929; Oppenheim and Fessler, 1928; Phyladelphy, 1931). The latter two writers suspected that silicious plants were responsible, the effect not being confined to a single genus.

Phyladelphy (1931) considered the possibility that sunlight might play a role, and tried the effect of plant extracts applied to the skin in conjunction with sunlight or mercury-arc radiation, but Kitchevatz (1934a, b) seems to have been the first to make extensive experiments. The latter arrived at the conclusion that the lesions arise from photosensitization by chlorophyll which is rubbed into the skin by lying upon crushed plant leaves, and has supported this explanation with experiments. He rubbed into the skin extracts of spinach and also commercial chlorophyll, following this with exposure to mercury-arc radiation, or to red light from a "Pannexol" lamp (probably a tungsten filament lamp with filter). In the former case, he obtained a more marked erythema on the area into which the chlorophyll had been rubbed, and in the latter, erythema on the sensitized area only. It is difficult to judge from these experiments alone whether chlorophyll acted as a sensitizer or not. He demonstrated (1934b) the penetration of chlorophyll into the epidermis, a necessary prerequisite for its photosensitizing action, by histological study of biopsy material. Fine granules of chlorophyll were found in the corneum and granulosum, but particularly in the basal cell layer; some granules were found even in the dermis.

Kitchevatz concludes that the bullous lesions are brought about in susceptible individuals in the following way: The individual lies in the grass for a considerable time so that chlorophyll from crushed leaves is rubbed into the skin, scarification of the epidermis by plant stems assisting in this process. He then rolls over so that the impregnated area is exposed to the sun, and the lesions are produced through the photosensitizing action of chlorophyll. Why this occurs only in susceptible individuals—many persons indulge in such sunbathing without untoward effects—is a question which might be more closely related to variations in the physiology of the skin than to photodynamic action.

While Kitchevatz' general explanation of the mode of production of the lesions is reasonable enough, more recent experiments render it questionable that chlorophyll is the photosensitizing substance. Jausion, Jacowski and Kouchner (1937) have carried out experiments which, while somewhat complicated, indicate that chlorophyll alone does not produce lesions such as those which are typical of this condition, and the same was found by Behçet *et. al.* (1939).

Hirschberger and Fuchs (1936) found that contact with parsnip plants *(Pastinaca sativa)* or extracts from various parts of these plants, followed by exposure to sunlight or mercury-arc radiation, resulted in bullous lesions. Extracts exposed to light before application to the skin were not effective. Jensen and Hansen (1939) find the wave-lengths which produce the lesions to lie in the ultraviolet longer than about 3200 Å. This spectral region does not correspond to the absorption spectrum of chlorophyll (see Figure 18, p. 73), nor with the sunburn spectrum.

Kuske (1938) presents evidence that the photosensitizing substances responsible for this condition, and for photosensitization by figs (see below) are members of the furocumarin group. He obtained characteristic lesions with such substances which he extracted from various plants. Comparison of action spectrum with absorption spectrum was not made.

The etiology of this condition does not seem clearly established as yet, although it seems clear that light plays an important part. The lesions are quite different morphologically and in their time relationships from those produced by photodynamic dyes (see pp. 104, 259), and it is not at all certain that the abnormal response to light arising from contact with green plants is an example of photodynamic action. There may be numerous complicating factors; *e.g.*, Hirschberger and Fuchs believe that the skin must be moist at the time of contact with the parsnip plant.

Of particular interest at this time is the suggestion of Hirschberger and Fuchs that this condition may have importance in military medicine. Their studies were made after observing the lesions in soldiers who had been occupied in the construction of camouflage with green plants, including parsnip. The similarity of the lesions to those produced by mustard gas suggests that confusion of diagnosis might occur under conditions of modern warfare.

Photosensitization by Contact with Figs. Kitchevatz (1934c, 1936) describes an erythemato-bullous dermatitis occurring among individuals handling figs. He found that alcoholic extracts of fig leaves, the skins of figs, or the white part of figs act as photosensitizers when applied directly to the skin. In Kitchevatz' experiments the "milk" from figs did not seem to contain the photosensitizing substance, but Vigne and Ponthieu (1936) describe a case of dermatitis after contact with fig branches and exposure to the sun, in which they believe the "milk" from the fig branches produced photosensitzation. This is confirmed by recent experiments of Behçet, Ottenstein, Lion and Dessauer (1939). The experiments of these investigators seem to indicate that the active wave-lengths lie in the near ultraviolet and violet. The lesions were also produced by heat in susceptible individuals. This suggests a somewhat complex etiology.

Photosensitization by Perfume. *(Berlock dermatitis, Dermite pigmentée en forme de coulée).* In 1916, E. Freund reported a number of cases characterized by peculiar localized patches of pigmentation which appeared on the skin after exposure to the sun. Freund traced their origin to the application of cologne water previous to exposure to sunlight, the peculiar shape of the lesions corresponding to the manner in which the liquid had run over the skin. Rosenthal, who was unaware of Freund's paper and had quite different ideas as to the etiology, used the name *Berlock dermatitis* in describing cases which he reported in 1924, because the pigmented patches reminded him of a locket or other ornament suspended on a chain.* Hoffman and Schmidz (1925) rediscovered Freund's observation, but employed Rosenthal's term which has since been generally applied to dermatitis caused by perfume. The lesions often lack the breloque form, *e.g.*, when perfume is applied behind the ear (see Gross and Robinson, 1930).

Freund showed in his original paper that cologne water, or the ingredient, bergamot oil, produced erythema and subsequent pigmentation if applied to the skin and followed by exposure to sunlight. Szántó (1929) thought that all the essential oils in cologne water might act as photosensitizers, but Del Vivo (1930) found that only bergamot oil and citron oil (olii de Cedro) were effective. Both Del Vivo (1930) and Giradeau and Acquaviva (1934) found that cologne water or bergamot oil, applied

* Use of the corresponding word, *breloque,* seems rare in this sense in English.

to the skin without friction, did not produce any reaction without exposure to light.

Bergamot oil, a common ingredient of perfumes, is extracted from the skin of a citrus fruit. It varies in constitution, but usually contains chlorophyll (see Rogin and Sheard, 1935) so that that compound has been suspected as the photosensitizer. Action spectrum studies do not seem to support this, however. Del Vivo (1930) found the radiation from a tungsten filament lamp (Sollux) which passed through a red filter was without effect, but that which passed through a blue filter produced erythema and pigmentation. Giradeau and Acquaviva, using an arc as the source of radiation, found that the effect was strongest behind blue and violet filters, less behind yellow, and was not produced behind red. If chlorophyll were the sensitizer, red light should be effective, although the lack of quantitative measurements may leave some doubt as to whether photosensitization by this substance can be eliminated. Both the last named investigators found that bergamot oil or cologne water acted as a filter for the sunburn radiation, so that the mercury arc produced less erythema on skin to which these substances were applied. This accounts for the failure of numerous other writers to produce sensitization to mercury-arc radiation with these substances. It seems clearly established that bergamot oil sensitizes the skin to blue and violet light, and that this is the cause of Berlock dermatitis.

Szántó (1929) found that, although Berlock dermatitis is more often found in women than in men, there is no difference between the sexes in susceptibility to photosensitization by cologne water. He found a large proportion of both sexes to be susceptible.

Urbach (1928) observed a case of breloque like dermatitis, due to the application of benzine and exposure to sunlight. It seems more probable that this was due to lowering of the erythema threshold to sunburn radiation, perhaps by "clearing" the corneum, than that the benzine acted as a photosensitizer. It is important to recognize that dermatitis having the characteristic breloque shape may result from causes other than photosensitization by perfume.

Photosensitization by Green Soap. Baeyer and Dittmar (1928) and Proft (1931) studied the photosensitizing action of green soap (soap prepared from linseed oil and KOH) finding that this material increased the erythemic and pigment producing action of sunlight, and of mercury-arc radiation. Baeyer and Dittmar found no sensitization to a tungsten lamp. Hence it is probable that this effect is due to increased effectiveness of the sunburn radiation, rather than to a true photosensitization. Both authors examined the possibility that one of the ingredients of green soap might act as a photosensitizer (chlorophyll is usually present), but their experiments give no evidence of such action. It is possible that green soap acts as a "clearing" agent to increase light transmission by the corneum. Baeyer and Dittmar did not find such increased photosensitivity when sodium soap was employed.

Photosensitization by Coal-tar and Petroleum Products. It has long been recognized that coal-tar products and products from some petroleums are active in producing skin affections which may terminate in malignancy. Lewin in 1913, however, was the first to recognize that such materials may also render the skin sensitive to sunlight. He observed a large group of workers employed in the fabrication of cables in an electrical factory where they came regularly into contact with coal-tar products used in the manufacture of insulation. He found that most of these workers developed dermatitis and itching when exposed to sunlight. Kistiakowski (1930) has reported extensive observations on workers in a large factory in the Ukraine dealing with petroleum products, in whom a similar clinical picture is developed, and Foerster and Schwartz (1939a, b) have described similar observations in this country.

During the last war, such dermatoses seem to have been particularly common. In Germany this was attributed to the use of impure products, and of adulterants, because of the blockade of the Central Powers by the Allies (*e.g.*, see Habermann, 1933b), but the fact that there was a high incidence in other countries as well (Thibierge, 1925a) indicates that the blockade alone was not responsible.

Herxheimer and Nathan (1917) describe photosensitization by Carboneol, a coal-tar product used as a medicament, and Frieboes (1917) found photosensitization by Vaseline. Oppenheim (1921) points out that the Vaseline used during the war was of very poor quality (again the blockade), and it is probable that impurities and not petrolatum itself accounted for Frieboes observation. This view is confirmed by Baeyer and Dittmar's (1928) and Proft's (1931) finding that pure Vaseline has no photosensitizing action.

Fleischhauer (1930) found that a coal-tar preparation, Lianthal, sensitized skin to light. She noted that this preparation actually protected the skin from light when spread on in the usual manner and allowed to remain on the surface, but if the excess was wiped off after a short time, say fifteen minutes, the skin was found to be sensitive to light, displaying erythema followed by pigmentation if exposed to the sun's rays. The photosensitization lasted as long as 72 hours after the application.

Although erythema appeared after as little as two minutes' exposure to sunlight, there was little or no sensitization to mercury-arc radiation. Experiments with filters indicate that the action spectrum lies somewhere beween 3300 and 5000 Å, which accounts for the failure of the mercury arc to produce the response (see Figure 46, p. 217). Foerster and Schwartz (1939a) set the spectral limits of the action spectrum for light-evoked lesions after contact with tar products used in the manufacture of electric insulation, between 3900 and 5000 Å. This is in general agreement with Fleischauer's findings.

Lewin (1913) and others have regarded acridine compounds as the most probable sensitizers, since these compounds are active photosensitizers and are found in coal-tar residues. The absorption spectrum of such compounds would correspond well enough to the limits of the action spectrum found by Fleischhauer and Foerster and Schwartz. On the other hand, such residues contain anthracene and phenanthrene compounds which are photosensitizers; but the latter absorb principally in the ultraviolet (see Mayneord and Roe, 1935).

Doniack and Mottram (1937) found that white mice painted with benzpyrene, di-benzanthracene or tar were rendered photosensitive, as shown by scratching followed by erythema and edema when they were exposed to direct sunlight. Dermatitis appeared a day later on the parts which had been exposed. They state that these effects were produced by blue and violet light, which might be considered to be in agreement with the findings of Fleischauer and of Foerster and Schwartz; but they give the wave-lengths active in photodynamic killing of paramecia as lying between 3500 and 4100 Å. All these limits were determined with filters and may not be very exact. The absorption spectrum of benzpyrene lies principally in the ultraviolet, but extends into the violet; one isomer of dibenzanthracene is yellowish in color, but the absorption of this type of compound is characteristically in the ultraviolet. It is obvious from the above that it is dangerous to credit the photosensitizing action of coal-tar residues too definitely to any one group of compounds.

Relationship between Carcinogenic Action and Photosensitization. The nature of the carcinogenic (cancer-producing) hydrocarbons in coal-tar products need not be discussed here; the reader is referred to Fieser (1937) and Voegtlin (1938) for discussions of this subject. However, there has been much recent interest in the possible relationship between photosensitization and the production of malignancy, both of which are associated with coal-tar and its constituent compounds. As we have seen, some of the photosensitizing activity of coal-tar residues is probably due to substances which are not carcinogens, *e.g.*, acridine, but at least some of the carcinogenic substances are active photosensitizers. Both benzpyrene and di-benzanthracene, which have been mentioned above, are carcinogens. In addition, Lewis (1935) has sensitized chick embryo cells to light with methylcholanthrene, the most active of the carcinogens, and Hollaender (1939) has shown that this compound is a powerful photosensitizer for yeast. Mottram and Doniach (1938) have added others to the list. I have recently been able to produce hemolysis with methylcholanthrene and light, and to show that O_2 is required for this effect, so that it seems clear that this is an example of photodynamic action. Mottram and Doniach (1938) believe that water-soluble impurities are responsible for the photodynamic action of these compounds and that these impurities are increased in quantity by previous irradiation.

The fact that these compounds produce photodynamic action does not indicate that light increases their carcinogenic action, although this view has been entertained by numerous workers. As has been pointed out (p. 123), photodynamic activity may be unrelated to other biological properties of a compound. A parallelism has been pointed out by Mottram and Doniach (1938) between photodynamic action and "blastogenic" action (presumably related to carcinogenic action) for a number of carcinogenic hydrocarbons, showing that those compounds which are most effective photodynamically are likewise the most effective blastogens. This recalls the parallelism among the fluorescein dyes between apparent photodynamic effectiveness, lytic effectiveness and ability to bring about the disk sphere transformation in red cells (see p. 88); and it is probable that both correlations have the same basis, namely that the principal factor in determining the apparent effectiveness of a given compound in producing any of these effects is its ability to be taken up by the cell.

Maizin and De Jonghe (1934) report that a greater number of tumors occurred among rats treated with benzpyrene and kept in lighted rooms than among such animals kept in the dark; the incidence of tumors seemed to parallel the amount of light in the rooms. They concluded that the light produced a more active carcinogen within the skin of the animal by oxidation of the original compound, an idea which has been accepted by some workers in the field. Assuming the adequacy of Maizin and De Jonghe's experiments, their conclusions are not implicit in the findings. There is no evidence that the benzpyrene was oxidized in the animals' skin, and it is more probable that it acted as a photosensitizer for the oxidation of cellular constituents (see Chapter 6). They actually found that benzpyrene oxidized with ozone was less carcinogenic than the original compound, but argued that in this case the oxidation had been carried farther than that which they supposed to occur in the skin.

Other attempts to demonstrate increased carcinogenic action of tar or carcinogenic hydrocarbons by light have given varying results, some workers finding increased tumor formation, others no effect (see Seelig and Cooper, 1933; Vlès, de Coulon and Ugo, 1935; Schorr and Ssobolewa, 1930; Taussig, Cooper and Seelig, 1938; and Rusch, Baumann and Kline, 1939.* Proper consideration of the important questions of light intensities, and particularly of wave-lengths, seems to have been neglected in a number of these studies, and there is also a possibility that some of the differences in conclusions result from different statistical treatment of the data.

Rideal has recently stated (1939) that photo-oxides of carcinogenic hydrocarbons are more strongly active in destroying protein monolayers and in killing paramecia than are the parent compounds. Fluorescein dyes destroy serum films and produce lysis after they have been irradiated, but

* Dr. Luce-Clausen and Dr. G. B. Mider have recently found that the incidence of tumors was less in benzpyrene-treated mice exposed to light than in similarly treated animals kept in the dark. The sunburn radiation was eliminated in these experiments (personal communication).

as has been pointed out in Chapter 9 (p. 97) this does not explain their photodynamic action. One must be very careful in interpreting the phenomena which suggest that light plays a part in carcinogenic action, until the question is more adequately examined.

It is possible that tumor production by the carcinogenic hydrocarbons may be increased by an additional effect due to destructive photodynamic action. This might be regarded as an addition of effects rather than an increase in the inherent carcinogenic action of these compounds, and if this were the case we might expect to find that non-carcinogens, such as the fluorescein dyes, would produce tumors in animals which were exposed to light. Büngeler (1937a, b) reports tumor production in white mice by photodynamic action with eosin and hematoporphyrin, as well as with a coal-tar product, Anthrasol. He does not report any difference between the action of the latter, which presumably contained carcinogenic hydrocarbons, and the two former, which are non-carcinogenic. All his animals showed severe skin damage. Preliminary studies by the writer, using rose bengal as a photosensitizer have not shown cancer production.

Chapter 23

Photosensitization by Substances Applied Internally

Photodynamic substances have been frequently used as therapeutic agents, either inadvertently without regard to their photosensitizing properties, or deliberately with the belief that photosensitization has curative powers. In the latter case it has been often assumed that the results of photodynamic action are the same as those produced by ultraviolet radiation, a fallacy which has been pointed out in earlier chapters (see *e.g.*, p. 116).

The first recorded instance of photosensitization by a substance employed as a therapeutic agent seems to be that described by Prime (1900) who administered eosin to twenty-six patients suffering from epilepsy. Prime used eosin because it is an organic compound containing bromine, which element has a more or less specific effect in that disease. He was unaware of the photosensitizing action of this substance, for his observations were contemporary with Raab's discovery of photodynamic action. He noted that his patients displayed erythema of the exposed parts; in his own words, "Cette rougeur peut occuper tout la face, le cou, la partie supérieure du thorax, la face dorsale des mains et des doigts, envahir le pharynx, c'est à dire, en somme, toutes les parties et rien que les parties exposées a l'air." Prime could only think of contact with the air as the precipitating factor, but obviously those regions which he mentions as exposed to air are the same which are exposed to light, with the exception of the pharynx, which showed lesions in only one case. There is little doubt that the erythema which his patients displayed was due to photodynamic action.

After Raab's discovery there were attempts to apply photodynamic action in light therapy. The first seems to have been that of Dreyer (1903), in the Finsen Medicinske Lysinstitut in Copenhagen, who attempted to increase the curative effect of the Finsen treatment of lupus vulgaris (carbon-arc radiation through a quartz compression lens) by the application of eosin. Success was claimed at first, but Neisser and Halberstaedter (1904) could not confirm his results, and in 1906 Reyn says that the use of eosin at the Finsen Institute was finally abandoned as not worth continuing. Bouveyron (1922), however, reported good results from bathing the skin with eosin or erythrosine before exposure to light. Lupus vulgaris is a tuberculous condition, and Bouveyron considered his finding that tuberculin was destroyed by photodynamic action to be of significance in this respect. Lange, Bamatter, Gruninger and Löwenstädt (1933), however, could find no evidence that tuberculous guinea pigs were improved by the injection of eosin in conjunction with exposure to sunlight. Duplicate experiments were carried out at Berlin, Geneva, and Davos, to rule out any possible climatic effect.

Jesionek and Tappeiner (1905) claimed some success in the treatment of skin carcinoma with this method, but this does not seem to have been highly effective, for it was apparently abandoned.

The therapeutic use of photodynamic action seems to have been neglected after the earlier efforts until interest was revived by attempts to use it in the treatment of rickets, the first of which was that of György and Gottlieb in 1923. This subject has been discussed already in Chapter 11 (p. 118), where it was pointed out that the value of such treatment is extremely doubtful.

About the same time, Jausion and Marceron (1925) accidently rediscovered the photosensitizing action of acridine compounds. Of twenty soldiers who were injected with trypaflavine (diamino-methyl acridine), for treatment of gonorrhea, nine became very sensitive to sunlight, displaying erythema and edema of the exposed parts, followed by intense pigmentation and heavy hair growth. This discovery led to widespread use of trypaflavine and other photodynamic dyes in conjunction with light. The rationale for this seems to have been based on the belief that ultraviolet radiation (presumably the sunburn radiation) has important general therapeutic powers, which are enhanced by the use of a photosensitizing agent. Guillaume (1926b, c) showed that the action spectrum for trypaflavine sensitization of skin is different from that of sunburn, and hence that the mechanisms are independent. Important fundamental differences between photodynamic action and the action of ultraviolet radiation on unsensitized living systems have been pointed out in Chapter 11. The practical results of such treatment have been disputed, and references can only be given to two of the numerous articles which have appeared on this subject. Jausion and Pagès (1933) may be cited as pro, and Videbech (1931) as con.

Haxthausen (1933) reports the case of an individual who developed a persistent photosensitivity after treatment with trypaflavine and mercury-arc radiation. Previous to the treatment, the patient was not abnormally sensitive to sunlight, but shortly after it, he noticed that either diffuse or direct sunlight produced swelling of the hands and face. The patient was not sensitive to sunlight passing through window glass, indicating that the sensitivity was due to sunburn radiation, and not to photosensitization by trypaflavine. The observations of Marceron (1925) and Ambrogio (1936) show that trypaflavine can be detected in the blood stream only for a few days after injection, at most, and that photosensitivity disappears about the same time. Haxthausen's patient, however, remained photosensitive for over six months. It is probable that this patient suffered from a hypersensitivity to sunburn radiation, which was set off by the photodynamic action of trypaflavine.

Within the last few years, hematoporphyrin has been used rather extensively in the treatment of melancholia. It is administered by intramuscular injection, or sometimes by mouth, in spite of the fact that it is not apparently absorbed from the digestive tract. Whether there is a marked effect upon the sensitivity of the patient to light over the whole body is doubtful when the size of the dose used is considered, but local photosensitivity at the point of injection may occur, as shown by the incident described by Blum and Templeton (1936). An individual received intramuscular injections of hematoporphyrin while in Germany in December, 1935. In February of the following year, after returning to the United States, she exposed herself to bright California desert sunlight, with the result that two large wheals appeared on each arm in the region at which the hematoporphyrin had been injected. The wheals subsided within a few hours after the end of the exposure, but were followed by persistent pigmentation of the same areas. The reaction, as described by the patient, was comparable in every way to that resulting from intradermal injection of hematoporphyrin followed by exposure to sunlight (see p. 232), in which case photosensitivity of the injected area may persist for two months or longer.

Rose bengal is quite generally used in testing liver function. The dye is injected, and its rate of disappearance from the blood-stream estimated. Normal individuals excrete the dye so rapidly that it does not accumulate in the skin in sufficient quantity to produce photosensitivity. Any liver dysfunction which results in retention of the dye in the blood-stream may cause accumulation of dye in the skin, so that the patient becomes sensitive to light. Fiessinger, Albeaux-Fernet and Aussannaire (1937) have recently described a number of cases in which sensitivity to light, manifested by severe erythema and edema, has followed this test. All those individuals in which such photosensitivity occurred had liver disturbances of one kind or another. Thus, patients who show retention of the dye should be kept out of sunlight after the test, until time has been allowed for the disappearance of the dye from the blood stream. Photosensitization of local areas of the skin by intradermal injection of rose bengal shows that photosensitivity of these areas lasts only about forty-eight hours.

Since the introduction of sulfanilamide as a therapeutic agent, sensitivity to light has been reported as sequel to the use of this drug in numerous instances (*e.g.*, see Tedder, 1939). It has been frequently assumed that this is due to the photosensitizing action of porphyrin. Apparent support for this assumption was given by Rimington and Hemming (1938), who found that increased porphyrin excretion and increased sensitivity to mercury-arc radiation follow dosage with this drug. However, as shown in Chapter 19 (p. 219) porphyrins do not sensitize strongly to radiation from this source.

Epstein (1939) was able to demonstrate local sensitization to mercury arc and to sunlight after intradermal injection of solutions of sulfanilamide, and this has been confirmed by the writer (1940). If an area of the skin including such a sensitized region is irradiated with a mercury arc, the skin in the immediate neighborhood of the injection develops a sunburn response greater than that of the surrounding area. The latent period is reduced, erythema and pigmentation are greater, and there may be desquamation of the sulfanilamide-treated area when this does not occur in the surrounding skin which has received the same dose of radiation. The general picture is the same as that of sunburn (except in some of Epstein's cases where inflammatory symptoms developed) and is very different from that obtained with hematoporphyrin or other photodynamic dyes, both in the morphology of the response and in the time relationships (compare pp. 104 and 175). I have found that the response is not inhibited by cutting off the O_2 supply, in contrast to photodynamic action (p. 116). It is thus highly improbable that sulfanilamide acts as a photodynamic sensitizer.

The reaction does not occur if window glass is used as a filter; and if filters are used which allow a part of the sunburn spectrum to pass (similar to those whose transmission is shown in Figure 50), both the sulfanilamide response and the sunburn response of the surrounding area are proportionately decreased. This indicates that the action spectrum of the sulfanilamide response is identical with or very near that of sunburn. This is incompatible with photosensitization by porphyrins, and suggests that sulfanilamide acts by increasing the sensitivity of the epidermal cells to the normal sunburn mechanism. The absorption spectrum of sulfanilamide is such that it might have an action spectrum very much like that of sunburn, and it is thus possible that the drug itself is the light absorber for a photochemical reaction leading to this abnormal response. This seems the less probable explanation, however. That sulfanilamide does not act by altering the response of the blood vessels of the skin, causing them to respond more readily to the products of the sunburn reaction, is indicated by the fact that no sulfanilamide response is obtained if the drug is injected into the skin after the radiation has been applied (Epstein 1939, Blum 1940).

Barbiturates have been known to produce sensitivity to light (see p. 236), and since porphyrin excretion may be increased by these drugs, a relationship between these symptoms has often been postulated. Perutz (1910, 1917a) reported increased sensitivity to mercury-arc radiation in rats dosed with sulfonal, which parallels the finding of Rimington and Hemming in the case of sulfanilamide. As pointed out above, however, this is evidence against, rather than for, photosensitization by porphyrins.

The possibility that microörganisms in the blood-stream may be killed by photodynamic action has been frequently entertained, particularly with regard to the animal parasites. Busck and Tappeiner in 1906 carried out *in vitro* experiments on killing of trypanosomes with this possibility in mind. More recently, Jancsó (1931a, b, c), Roskin (1929, 1930a, b), Roskin and Romanowa (1929a, b), Roskin and Levinson (1930) and others have occupied themselves with this question. This is a definite possibility, since visible radiations which might be used in such treatment penetrate appreciably to the minute blood vessels in the corium (see Figure 41, p. 202), in contrast to the sunburn radiation which is virtually all absorbed in the epidermis. The problem seems to be that of finding a photodynamic dye which will be taken up by the microörganism in preference to the blood cells or the serum. Jancsó has indicated that this may be the case with trypaflavine, which seems to be taken up by trypanosomes in the bloodstream. It has been claimed, and disputed, that quinine acts more effectively in malaria when the patient is exposed to light, allegedly due to photosensitization of the plasmodium of that disease by this drug.

Another possible therapeutic application of photodynamic action lies in the treatment of parasitic skin infections.

Although the results of therapeutic application of photodynamic action are not encouraging thus far, the question has not often been studied with a clear picture of the mechanism of photodynamic action in mind. It may deserve more thorough investigation.

Chapter 24

Other Diseases Attributed to the Action of Light

A number of diseases other than those which have been discussed have been thought to be caused or exacerbated by light. Experimental data is meager in most instances.

Pellagra. It would be beyond the scope of this book to discuss the many theories regarding the etiology of pellagra, and I shall mention only the alleged role of light in this disease. In the original description of the disease by Casal in 1762 there seems to have been no mention of light as a possible etiologic factor (see Major, 1932). This seems to have been first suggested by Strambio in 1784 (see Hausmann, 1923). Such terms as *mal del sole,* and *scottura di sole,* which have been used by Italian peasants to describe the disease (see Bouchard, 1862), suggest belief in a relationship with light. The pellagra patient often shows dermatological changes after exposure to sunlight.

While there are some who claim that sunlight is an important factor in the development of pellagra, others think that the pellagrous individual is hypersensitive to light, but believe this to be only a minor manifestation of the disease. Bouchard found the same state of affairs in 1862. He quotes numerous bizarre contemporaneous theories, but himself believes that sunlight elicits an abnormal skin response in the hypersensitive pellagrous individual: "l'erythem pellagreux ne serait autre chose qu'un coup de soleil developpé chez un sujet pellagreux."

While the hands, face and neck, which are ordinarily exposed to light, are regions of predilection for the pellagrous lesions, other areas, *e.g.,* the pubic region, are common sites (*e.g.,* see Spies, 1935). An interesting observation is that of Neusser (1887, quoted from Hausmann, 1923) who noted that gypsy children that wandered half naked in the sun showed pellagrous lesions on the hands and feet only, while other parts of the body which were sufficiently exposed to sunlight to receive a dark tan were not affected. It seems that the lesions typical of pellagra appear on certain susceptible areas of the skin rather than on those which are exposed to light. Hausmann (1923) cites other evidence to indicate that pellagra is not directly related to light.

The belief that pellagra results from the excessive use of maize in the diet is an old one, which at one time dominated thought in this field (see Sambon, 1905). High incidence of the disease was often observed in regions where maize was a large factor in the diet, *e.g.*, the high incidence among Italian peasants was linked with the eating of large quantities of polenta, a cooked maize meal. But it has been pointed out that pellagra occurs where maize is entirely absent from the diet (*e.g.*, Enright, 1920; Voegtlin, 1920). Feeding experiments on laboratory animals in which diets high in maize content were used have yielded conflicting results. A number of workers, Lucksch (1908), Raubitschek (1910), Horbaczewski (1910, 1912), Lode (1910), and Umnus (1912), have reported that animals fed on such diets remain relatively unharmed if kept in the dark, but show symptoms similar to photodynamic sensitization if exposed to light. These experiments, linked with the supposed importance of light in pellagra, led to the hypothesis that this disease is due to photodynamic sensitization. In a more recent study, Suaréz (1916) extracted a fluorescent substance from maize which photosensitized red blood cells, paramecia, and rabbits when injected intravenously; but mice, guinea pigs and pigeons were not photosensitized by a diet of maize meal. This diet, however, produced symptoms of vitamin deficiency. Pfeiffer (1911), and Gay and McIver (1922) failed likewise to produce photosensitization by feeding maize.

Jobling and Arnold (1923) suggested that photodynamic sensitization originates from a quite different source. These writers isolated strains of bacilli, which produced fluorescent substances, from the feces of five out of nine acute, one out of six subacute, and two out of twenty-three chronic cases of pellagra. They were unable to find such bacilli in the feces of fifty non-pellagrous individuals. They produced photosensitivity in white mice and paramecia with extracts containing the fluorescent material. This investigation has not been continued.

Massa (1932) who has isolated uro- and coproporphyrins from the urine of pellagrous patients, suggests that such substances play the role of photosensitizers in this disease. Dobriner, Strain, and Localio, (1938) have found somewhat increased porphyrin output in a pellagran (but see p. 237).

Little effort has been made to delimit the spectral region which might be responsible for the pellagrous lesions. Mook and Weiss (1925) exposed a patient to sunlight for two hours and produced a rather severe sunburn which may have been somewhat abnormal in character; but they made no attempt to discover the exciting wave-lengths. Gougerot and Meyer (1933) studied three cases; the erythema threshold to mercury-arc radiation varied widely, but there was no abnormal response to this source. One case, studied with a monochromator, responded to yellow and red with erythema, but no lesions; the second responded to sunlight with bullous lesions; and the third seemed abnormally sensitive to heat. Spies (1935) found that mercury-arc radiation, although increasing pigmentation, did not provoke pellagrous lesions and had little or no effect on the development or healing of such lesions.

Recently (1937) Smith and Ruffin described experiments on pellagrans in which they produced lesions with an electrical heater quite as effectively as with sunlight. From this it seems probable that the pellagrous skin is sensitive to heat rather than to light. Hypersensitivity to heat would explain the fact that pellagrous lesions are produced by exposure to sunlight, and also in the protected areas such as the pubic region. All in all there seems little to substantiate the belief that sunlight *per se* plays an important role in pellagra, although J. H. Smith (1931) and Smith and Ruffin (1937) show a convincing relationship between pellagra incidence and the season of greatest insolation.

Lupus Erythematosus. This condition seems to be on the increase in this country at present (see Ludy and Corson, 1938). Its etiology is quite uncertain (*e.g.*, see Pels, 1933). An increasing number of cases are reported as appearing subsequent to sunburn or therapeutic treatment with mercury-arc radiation (Feit, 1927; Audry, 1930; Brain, 1933; Bechet, 1934; Gougerot and Burnier, 1934). Haxthausen (Hausmann and Haxthausen, 1929) found that the appearance of the lesions showed marked increase during the spring and summer; he dealt with clinic patients in Copenhagen, where there is a great seasonal variation in ultraviolet radiation.

It seems probable that this condition is at least exacerbated by light, but since no wave-length studies are available, it is impossible to know what part of the spectrum is concerned. The fact that the lesions are sometimes reported as appearing after irradiation with the mercury arc indicates, however, that sunburn radiation is responsible. Thus the disease may resemble polymorphic light eruption, and there are perhaps instances of confusion in diagnosis.

It is generally believed at present that there is no relationship between lupus erythematosus and lupus vulgaris, the latter a tuberculous condition. However, such a relationship has been frequently suggested, and Kren and Löewenstein (1931) claimed to have isolated tubercle bacilli from the lesions of lupus erythematosus. It is interesting, therefore, to mention the experiments of Grosz and Volk (1914), who found that if guinea pigs were given intradermal injections of killed tubercle bacilli, the areas injected became hypersensitive to mercury-arc radiation. This effect was not observed when other bacilli were used, except in a single instance with *B. coli*. Kitchevatz and Bugarski (1938) have found a similar effect in a human patient.

Ludy and Corson (1938) indicated a possible relationship to lead poisoning, and also implicate porphyrins, but there seems little reason to suspect that the disease results from photosensitization by porphyrins.

Granuloma Annulare. Kitchevatz and Bugarski recently (1938) made an interesting study of a patient suffering from this disease. They were able to show that sunlight exacerbated the lesions, and that the reaction to tuberculin injections or scarifications was much more severe when exposed to light. By applying the tuberculin on bandages which remained in contact with the skin for twenty-four hours, they were able to reproduce the lesions when the skin was later exposed to light.

Riehl's War Melanosis. During the first World War, Riehl (1917) described a number of cases of peculiar pigmentation of the face. While the disease seems to have been most common in Germany and Austria, and has been credited to conditions, possibly dietary, resulting from the blockade of the Central Powers, it has been reported from France (Thibierge, 1925b), from Denmark (Rasch, 1926), and from Greece (Photinus, 1933).

There have been two theories regarding its etiology: (1) it results from faulty diet—Rhiel thought it might come from war bread which contained bean flour—and (2) from contact with oil or tar, in which case it should be classed with conditions discussed in Chapter 22. Sensitization to light has been suggested in connection with both hypotheses, but there is no experimental or other direct evidence that the condition is associated with exposure to light. Most recent cases described under this name have a history of contact with oil or coal-tar, and it is possible that Rhiel's melanosis is really not separable from photosensitization by such substances (see p. 264). For a more complete discussion, see Thibierge (1925a, b) and Habermann (1933b, c).

Other Diseases of the Skin. Light is suspected as the cause, or an exacerbating factor, in a number of diseases of the skin other than those which have been mentioned thus far. Since there is little or no experimental evidence to support such conclusions, an extensive discussion would be beyond the scope of this book. In many cases the localization of the lesions on those parts which are the most frequently exposed to light is the only evidence that light is a factor in their etiology. Sometimes there is record of the lesions following exposure to sunlight, or occurring after treatment with mercury-arc radiation. Wave-length studies are virtually lacking. Although much more evidence is necessary before definite conclusions can be reached, it seems probable that sunlight plays a minor part in many skin diseases, and that it is the sunburn radiation which is usually responsible.

Hausmann and Haxthausen (1929) list *acne vulgaris, eczema, pityriasis streptogenes faciei, psoriasis, rosacea,* and certain other unclassified dermatoses. Jausion and Pagès (1933) list a number of conditions in which they believe light to have a part, but their ideas as to the complicated relationship involved can be obtained only by reference to their book. Stokes and Callaway (1937) report that patients being treated with therapeutic doses of mercury-arc radiation, after having suffered infections, suddenly became hypersensitive to the radiation. Ayers (1923) described "chronic actinic cheilitis," chronic irritation of the lip; in one case, similar lesions occurred after exposure to mercury-arc radiation, which would indicate that sunburn radiation is the cause.

Acute Exanthemata. It has been reported that light exacerbates the lesions of smallpox and other acute exanthemata (see Hausmann, 1923), but this does not seem to be too clearly demonstrated. Protection of smallpox patients from light has been practiced at various times and places, but there has never been complete agreement as to the benefits derived. Finsen suggested that all but red light should be excluded from the rooms of such patients by the use of red glass or drapes.* Unfortunately this lead in some way to a belief which has not entirely disappeared, that red light has a curative effect.

Diseases of the Eye. *Photophthalmia.* This is a condition of inflammation of the cornea which seems comparable to sunburn. The wave-lengths which produce it lie in the same spectral region as the sunburn radiation, *i.e.,* wave-lengths shorter than about 3300 Å and it seems probable that the inflammation is a result of photochemical changes in proteins of the cornea. The inflammation has a long latent period, except when it results from very high dosage of radiation, as may result from exposure to electrical discharge, and is comparable with sunburn in this respect.

This condition occurs in normal individuals exposed to artificial sources of ultraviolet radiation, or prolonged exposure to intense sunlight. As in the case of sunburn, the presence of reflecting surfaces may increase the amount of incident radiation, and "snow blindness" and "glacier blindness" are terms which have been used to describe this condition when occurring in particular environments (see p. 189). Individuals may display hypersensitiveness as in the case of sunburn (see Duke-Elder, 1926a).

Lens Cataract. The lens is protected by the screening action of the cornea, so that the wave-lengths which its proteins absorb must be considerably reduced before reaching the lens. Nevertheless, there is good evidence that ultraviolet radiation may produce cataract (see Duke-Elder, 1926b). It seems probable that this is due to gradual changes in the proteins of the lens, which this tissue is unable to repair rapidly enough. Clark (1935b) has shown that lens protein may be denatured without coagulating when exposed to ultraviolet radiation, but that coagulation occurs when the denatured protein is treated with calcium or heated. She suggests that such denatured protein which has accumulated in the lens may coagulate in senile eyes, forming cataract, and that excess calcium or heat may play an important part in the coagulation.

Changes in the Retina. Duke-Elder (1926c) describes changes in the retina produced by radiation. The wave-lengths which reach the

* The history of such handling of smallpox patients goes back many centuries (see Hausmann, 1923).

retina lie between about 3300 and 12,000 Å, the short wave-length limit moving toward longer wave-lengths as the age of the individual increases. It does not seem probable that enough of that ultraviolet radiation which is generally destructive to living tissues (wave-lengths shorter than 3300 Å) reaches the retina to cause serious damage. Destructive changes in the retina by the specific action of visible wave-lengths may occur but is not definitely proven. Heat resulting from the action of intense radiation, regardless of wave-length, may be an important cause of retinal damage as *e.g.,* in sun blindness. In this and the preceding conditions there is little definite evidence as regards abnormal sensitivity.

Photophobia. Pain, accompanied by attempts to avoid light when there is a sudden increase in intensity, is an accompaniment of numerous pathological conditions of the eye. Normal eyes may react in this way if the stimulus is sufficiently severe, so that this condition must be regarded as one of hypersensitivity of a normal mechanism. The retina is apparently the receptor of the stimulus, since photophobia does not occur in blind eyes, and since the stimulating wave-lengths are those which affect the retina. Siegwart (1920) found that blind eyes did not manifest photophobia, but in cases of unilateral blindness where the consensual iris reflex was preserved, illumination of the normal eye resulted in pain associated with the blind eye. Siegwart and Vogt (see Siegwart, 1920) found that ultraviolet and infrared were ineffective, and that yellow, which has maximum effect on the retina (see Figure 8, p. 30), is the most effective in eliciting photohobia.

Siegwart (1920) and Lebensohn (1934) present quite convincing evidence that the pain is associated with reflex movement of the iris muscles, and reflex vasodilation of the vessels of the conjunctiva. The reflex changes which lead to the pain involve the trigeminal nerve, since extirpation of the Gasserian ganglion abolishes it.

It would seem that photophobia cannot be classed as a disease caused by light, but is a symptom which may accompany inflammatory conditions of the eye. It may occur in Xeroderma pigmentosum and Hydroa vacciniforme, but in these conditions it occurs, according to Lebensohn (1934), only after corneal lesions have appeared. It should not, therefore, be taken as indicating an essential relationship of these diseases to light.

The inflammatory condition described as vernal catarrh, or conjunctivitis vernalis, has often been regarded as abnormal sensitivity to light because of its recurrence in spring and summer, but its etiology seems very uncertain (see Hausmann, 1923).

Light and Rhinitis. Belief that hay fever is associated with sunlight was once current, but seems to have gone out of fashion with the advent of newer knowledge of allergy to pollens, etc. Cases of severe rhinitis occurring immediately upon exposure to sunlight are reported, but the writer knows of no experimental study of this condition. It may possibly result from a reflex vasodilation comparable to that which Lebensohn (1934) describes in photophobia, in which the retina is the point of reception.

Recently, Leonard Hill (1931, 1932) has published accounts of blocking of the nasal passages by the action of long wave-length infrared radiation upon the skin, particularly in individuals with deviated nasal septa. He has also claimed that red, or short wave-length infrared, has an antagonistic "nose opening" effect. Recent careful experiments by Dufton and Bedford (1933) and Winslow, Greenburg and Herrington (1934) have shown that the nose-closing effect results from heating of the skin in any manner, and not from a specific effect of infrared. They were unable to demonstrate the "nose-opening" effect.

Jaundice and Photosensitization. Photodynamic properties of bile have been reported (Hausmann, 1908), and denied (Meyer-Betz, 1913). Ruminant bile should have photosensitizing properties when the animals feed on chlorophyll-containing plants, because the bile then contains phylloerythrin (see Chapter 14). According to Saeki (1932a, b, c), bilirubin is a photosensitizer. The possibility has been suggested that photosensitivity should accompany jaundice, but so far as I know this has not been demonstrated. The relationship of icterus and photosensitization in sheep suffering from geeldikkop has been discussed in Chapter 14; phylloerythrin, which does not seem to be ordinarily an important constituent of human bile, is the photosensitizer in this condition. Since chlorophyll is not normally a very high dietary constituent for man, it is not probable that geeldikkop will ever be listed as a disease of human beings.

Tropical Sunlight. There is a persistent belief, firmly held in the last century and the beginning of the present, that tropical sunlight has peculiarly mysterious and dangerous properties. It allegedly affects white races more than colored. This assumption has been used as an argument that the white man undergoes degeneration in the tropics, and hence cannot compete with the colored races in that environment. Woodruff (1905) discusses the peril at length, and Jack London credited his peculiar nervous upset in the tropics to the effect of tropical sunlight (see the "Cruise of the Snark"), thus disseminating this belief.

Paul (1918) stresses the common occurrence of skin malignancy in Australia to support this contention; he states, "The common occurrence of these cancerous and precancerous diseases of the skin in Australia is to be regarded as one of the penalties to be paid for inhabiting a country normally destined to be occupied by a colored race." He would apparently blame the sunburn radiation of sunlight for this. As has been pointed out in Chapter 17, this is a very small and variable part of total sunlight, which cannot be estimated from measurements of total insolation, and it is probable that there is less sunburn radiation incident in many parts of the tropics than in some regions of higher latitudes. Severe sunburn is common in northern regions where snow and ice serve to reflect the sun's rays, and one might expect that the incidence of skin cancer might be higher among blonds in Scandinavia than in the tropics. Roffo (1932b) has found that the darker-skinned individuals in Argentina show a lower incidence of skin malignancy than blonds, but this could not have a very important differential effect upon the white and dark races as regards their ability to inhabit tropical climates.

Ideas of the great penetrating power of tropic sunlight are common. This leads, for instance, to the use of spine pads—thick cloth pads to cover the spine and protect against the penetration of sunlight. These pads are by preference red, perhaps a survival of the red light treatment of past centuries (see p. 277). A glance at Figure 41 (p. 202), will show that there is very little danger that radiation may reach the spinal cord. I am told that such pads were used by British troops in Mesopotamia during the last war, and are common in the tropics. Laymen, at least, are beginning to realize the futility of such apparatus, and I understand that sunbathing is as popular at Singapore as in more northern climates. A recent popular article by a medical missionary (Davis, 1938) shows that he, at least, was able to defy the tropical sun of the Congo without such protection, and I know of a young scientist who upset a British community in Nigeria a few years ago by playing tennis without a sun helmet, and survived the ordeal. To be sure, his resistance was attributed to the fact that he was an American (a Canadian by birth), since all Americans, presumably, have Indian blood.

Wedding (1887), related his work on buckwheat (see p. 154) to the tropical effects of sunlight, and went so far as to suggest that the German government send a ship to the tropics, one-half the crew to be painted with walnut stain, and the different reaction of the two halves studied. Needless to say the buckwheat experiments have nothing to do with the problem (see Chapter 15), but Wedding goes so far as to suggest that the blond races must mix with the darker Spanish or Italians if they are successfully to inhabit the tropics. Wedding's remarks were published in an anthropological journal, and it is interesting to speculate as to how much influence they had in establishing the rather common belief that white races cannot successively inhabit the tropics because of the sunlight.

Buckwheat Poisoning. The following editorial comment was appended to an article by the writer (1933), in which it was stated that photosensitization by buckwheat does not occur in man. "The well established opinions of the laity that 'buckwheat rash' develops in the spring, and that buckwheat should not be eaten during the summer months are explainable with difficulty unless buckwheat has a photodynamic effect in man as well as in the lower animals." The writer can find no published account to substantiate such photosensitization in human beings, and Blumenstein (1935), who has examined the subject carefully, is of the opinion that it does not occur. It seems improbable that the quantities of buckwheat ingested by human beings would be sufficient to bring about a condition similar to fagopyrism in domestic animals (see Chapter 15). A few cases of allergy to buckwheat have been reported (Smith, 1909; Schmidt and Peemöller, 1920; Blumenstein, 1935), but there was no evidence of sensitivity to light in any of these.

Mason and Mason (1927) observed a fall in the basal metabolic rate of a man who had been fed a diet high in buckwheat, upon exposure to mercury-arc radiation. This individual did not show a drop in basal metabolic rate when exposed to this radiation until after he had been fed the high-buckwheat diet, but some other individuals did. The evidence is too fragmentary to be accepted as demonstrating photosensitization.

The writer continues to enjoy his buckwheat cakes.

Bibliography

A

Adler, L., *Arch. exp. Path. Pharmakol.*, **85,** 152 (1919).
Ambrogio, A., *Archivio italiano di dermatologia, sifilografia e venereologia,* **12,** 602 (1936) a.
Ambrogio, A., *Arch. ist. biochim. ital.*, **8,** 205 (1936) b.
Amsler, C., and Pick, E. P., *Arch. exp. Path. Pharmakol.*, **82,** 86 (1918).
Anderson, T. M., *British Journal of Dermatology,* **10,** 1 (1898).
Andrade, E. N. daC., "The Structure of the Atom," London, G. Bell and Sons, 1924.
Anonymous, *Atti del Real Istituto d'Incorragiamento alle Scienze Naturali di Napoli.*, Vol. 2, Naples, Angelo Trani, 1818.
Anonymous, "Beetles and St. John's Wort," *Nature,* **132,** 525 (1933).
Arnow, L. E., *Science,* **86,** 176 (1937).
Arzt, L., and Hausmann, W., *Strahlentherapie,* **11,** 444 (1920).
Ashby, H. T., *Quart. Journ. Med.*, **19,** 375 (1926).
Audry, C., *Bulletin de la société française de dermatologie et de syphiligraphie,* **37,** 607 (1930).
Auger, D., and Fessard, A., *Ann. physiol. physicochim. biol.*, **9,** 873 (1933).
Awoki, T., *Biochem. Z.*, **158,** 337 (1925).
Ayres, S., *J. Am. Med. Assoc.*, **81,** 1183 (1923).

B

Bachem, A., *Am. J. Physiol.*, **91,** 58 (1929).
Bachem, A., *Strahlentherapie,* **39,** 30 (1930).
Bachem, A., and Reed, C. I., *Am. J. Physiol.*, **97,** 86 (1931).
Baeyer, H., and Dittmar, O., *Münch. med. Wochschr.*, **75,** 428 (1928).
Bailey, L. H., "The Standard Cyclopedia of Horticulture," Vol. 3, p. 1629, New York, Macmillan, 1915.
Baldwin, W. M., *Biol. Bull.*, **39,** 59 (1920).
Barber, H. W., Howitt, F. D., and Knott, F. A., *Guys Hosp. Rep.*, **76,** 314 (1926).
Barnes, T. C., *J. Cellular Comp. Physiol.*, **14,** 83 (1939).
Barnes, T. C., and Golubock, H. L., *Science,* **87,** 556 (1938).
Barron, E. S. G., and Hoffman, L. A., *J. Gen. Physiol.*, **13,** 483 (1929).
Bateman, T., "Cutaneous Diseases," pp. 258-261, Philadelphia, Collins and Croft, 1818.
Baumann, C. A., and Rusch, H. P., *Am. J. Cancer,* **35,** 213 (1939).
Baumberger, J. P., Bigotti, R. T., and Bardwell, K., *Am. Journ. Physiol.*, **90,** 277 (1929).
Bazin, E., "Leçons Théoriques et Cliniques sur les Affections Cutanées de Nature Arthritique et Dartreuse," pp. 192-201, Paris, Adrien Delahaye, 1860.
Bazin, E., Leçons Théoriques et Cliniques sur les Affections Cutanées de Nature Arthritique et Dartreuse, 2nd edition, p. 460, Paris, Adrien Delahaye, 1868.
Beard, H. H., Boggess, T. S., and Von Haam, E., *Am. J. Cancer,* **27,** 257 (1936).
Bechet, P. E., *Arch. Dermatol. Syphilol.*, **29,** 221 (1934).
Beck, L. V., and Nichols, A. C., *J. Cellular Comp. Physiol.*, **10,** 123 (1937).
Becker, *Arch. Dermatol. Syphilis.*, **91,** 376 (1908).
Behçet, H., Ottenstein, B., Lion, K., and Dessauer, F., *Annales de Dermatologie et de Syphiligraphie,* **10,** 7th series, 32, 125 (1939).
Beinhauer, L. G., *Arch. Dermatol. Syphilol.*, **12,** 62 (1925).
Bergmann, W., *Z. Krebsforsch.*, **48,** 546 (1939).
Berliner, C., *Monatshefte für praktische Dermatologie.*, **11,** 449, 480 (1890).
Bertholf, L. M., *J. Agri. Research,* 42, 379 (1931) a.
Bertholf, L. M., *J. Agri. Research,* 43, 703 (1931) b.
Bertholf, L. M., *Z. vergleich. Physiol.*, **18,** 32 (1933).
Bichlmaier, *Monatshefte für praktische Tierheilkunde,* **23,** 305 (1912).
Bie, V., *Mitteilungen Finsens Medicinske Lysinstitut,* **9,** 5 (1905).
Bier, O., and Rocha e Silva, M., *Compt. rend. soc. biol.*, **118,** 911 (1935) a.
Bier, O., and Rocha e Silva, M., *Compt. rend. soc. biol.*, **118,** 914 (1935) b.
Birkeland, J. M., *Science,* **80,** 357 (1934).
Birnbaum, *Zentralblatt für Haut und Geschlechtskrankheiten,* **44,** 140 (1933).
Black, W. A., *Science,* **82,** 495 (1935).
Bloch, B., *Z. physiol. Chem.*, **98,** 226 (1916).
Blum, H. F., *Biol. Bull.*, **58,** 224 (1930) a.
Blum, H. F., *Biol. Bull.*, **59,** 81 (1930) b.
Blum, H. F., *Ann. Internal Med.*, **6,** 877 (1933).
Blum, H. F., *Cold Spring Harbor Symposia Quant. Biol.*, **3,** 318 (1935) a.
Blum, H. F., *Cold Spring Harbor Symposia Quant. Biol.*, **3,** 210 (1935) b.
Blum, H. F., *Science,* **86,** 285 (1937) a.
Blum, H. F., *Am. Naturalist,* **71,** 350 (1937) b.
Blum, H. F., *J. Cellular Comp. Physiol.*, **9,** 229 (1937) c.
Blum, H. F., *J. Am. Vet. Med. Assoc.*, **93,** 185 (1938).
Blum, H. F., *Am. J. Physiol.*, **129,** 312 (1940).
Blum, H. F., Allington, H., and West, R. J., *J. Clin. Investigation,* **14,** 435 (1935).
Blum, H. F., and Fox, D. L., *Univ. Calif. Pub. Physiol.*, **8,** 21 (1933).
Blum, H. F., and Gilbert, H. W., *J. Cellular Comp. Physiol.*, **15,** 75 (1940) a.

Blum, H. F., and Gilbert, H. W., *J. Cellular Comp. Physiol.*, **15**, 85 (1940) b.
Blum, H. F., and Hardgrave, L. E., *Proc. Soc. Exp. Biol. Med.*, **34**, 613 (1936).
Blum, H. F., and Hyman, C., *J. Cellular Comp. Physiol.*, **13**, 281 (1939) a.
Blum, H. F., and Hyman, C., *J. Cellular Comp. Physiol.*, **13**, 287 (1939) b.
Blum, H. F., and McBride, G. C., *Biol. Bull.*, **61**, 316 (1931).
Blum, H. F., and Morgan, J. L., *J. Cellular Comp. Physiol.*, **13**, 269 (1939).
Blum, H. F., and Pace, N., *British Journal of Dermatology and Syphilis*, **49**, 465 (1937).
Blum, H. F., Pace, N., and Garrett, R. L., *J. Cellular Comp. Physiol.*, **9**, 217 (1937).
Blum, H. F., and Schumacher, I. C., *Ann. Internal Med.*, **11**, 548 (1937).
Blum, H. F., and Scott, K. G., *Plant Physiol.*, **8**, 525 (1933).
Blum, H. F., and Spealman, C. R., *J. Phys. Chem.*, **37**, 1123 (1933) a.
Blum, H. F., and Spealman, C. R., *Copeia*, p. 150 (1933) b.
Blum, H. F., and Spealman, C. R., *Proc. Soc. Exp. Biol. Med.*, **31**, 1007 (1934) a.
Blum, H. F., and Spealman, C. R., *Am. J. Physiol.*, **109**, 605 (1934) b
Blum, H. F., and Templeton, H. J., *J. Am. Med. Assoc.*, **108**, 548 (1937).
Blum, H. F., Watrous, W. G., and West, R. J., *Am. J. Physiol.*, **113**, 350 (1935).
Blum, H. F., and West, R. J., *J. Clin. Investigation*, **16**, 261 (1937)
Blumenstein, G. I., *J. Allergy*, **7**, 74 (1935).
Bohn, G., and Drzewina, A., *Compt. rend. soc. biol.*, **89**, 386 (1923).
Borst, M., and Königsdörfer, H., "Untersuchungen über Porphyrie mit besonderer Berücksichtigung der Porphyria Congenita," Leipzig, Hirtzel, 1929.
Bourbon and Chollet, *Recueil de médecine vétérinaire*, **103**, 530 (1927).
Bouchard, C., "Recherches Nouvelles sur la Pellagre," pp. 86-87, Paris, F. Savy, 1862.
Bourdillon, R. B., Gaddum, J. H., and Jenkins, R. G. C., *Proc. Roy. Soc. London (B)*, **106**, 388 (1930).
Bouveyron, A., *Compt. rend. soc. biol.*, **87**, 1018 (1922)
Bowen, E. J., and Steadman, F., *J. Chem. Soc.*, p. 1098 (1934).
Brain, R. T., *Proc. Roy. Soc. Med.*, **26**, 748 (1933).
Brocq, L., *Annales de dermatologie et de syphiligraphie*, **5**, *series 3*, 1133 (1894).
Brocq, L., *Annales de dermatologie et de syphiligraphie*, **6**, *series 5*, 113 (1916).
Brooke, H. G., *British Journal of Dermatology*, **4**, 128 (1892).
Brown, R. H., *Journal of General Psychology*, **14**, 62 (1936).
Bruce, E. A., *J. Am. Vet. Med. Assoc.*, **52**, 189 (1917).
Brunner, O., and Kleinau, W., *Sitzber. Akad. Wiss. Wien. Math. natur. Klasse Abt. II b.*, **145**, 464 (1936).
Brunsting, L. A., Brugsch, J. T., and O'Leary, P. A., *Arch. Dermatol. Syphilol.*, **39**, 294 (1939).
Bull, L. B., *Australian Vet. J.*, **3**, 53 (1927).
Bull, L. B., and Macindoe, R. H. F., *Australian Vet. J.*, **2**, 85 (1926).
Büngeler, W., *Klin. Wochschr.*, **16**, 1012 (1937) a.
Büngeler, W., *Z. Krebsforsch.*, **46**, 130 (1937) b.
Bunker, J. W. M., and Harris, R. S., *New England J. Med.*, **216**, 165 (1937).
Buquicchio, A., *Giornale italiano delle malattie veneree e della pelle*, **64**, 474 (1923).
Buri, T., *Monatshefte. fur praktische Dermatologie*, **13**, 181 (1891).
Burmeister, *Magazin für die gesamte Thierheilkunde*, **10**, 112 (1844).
Buruiană, L., *Biochem. J.*, **31**, 1452 (1937).
Busck, G., *Mitteilungen Finsens Medicinske Lysinstitut*, **9**, 193 (1905).
Busck, G., *Biochem. Z.*, **1**, 425 (1906).
Busck, G., and Tappeiner, H., *Deut. Arch. Klin. Med.*, **87**, 98 (1906).
Byrne, K. V., *Australian Vet. J.*, **13**, 74 (1937) a.
Byrne, K. V., *Agri. Gaz. N. S. Wales*, **48**, 214 (1937) b.

C

Campbell, A., and Hill, L., *Brit. J. Exper. Path.*, **5**, 317 (1924).
Cappelli, J., *Giornale italiano delle malattie veneree e della pelle*, **55**, 481 (1914).
Carn, K. G., *Agr. Gaz. N. S. Wales*, **47**, 608 (1936).
Carrié, C., *Arch. Dermatol. Syphilis*, **163**, 523 (1931).
Carrié, C., *Strahlentherapie*, **46**, 697 (1933).
Carrié, C., "Die Porphyrine," Leipzig, G. Thieme, 1936.
Carter, C. W., *Biochem. J.*, **22**, 575 (1928).
Caspersson, T., *Skand. Arch. Physiol.*, **73**, *supplement no. 8* (1936).
Castle, E. S., *J. Gen. Physiol.*, **14**, 701 (1931).
Cěrný, C., *Z. Physiol. Chem.*, **73**, 371 (1911).
Cerutti, P., *Giornale italiano di dermatologia e sifilologia*, **74**, 335 (1933).
Cerutti, P., *Archivio italiano di dermatologia, sifilografia e venerologia*, **11**, 1 (1935).
Chaffee, E. L., and Hampson, A., *J. Optical Soc. Am.*, **9**, 1 (1924).
Chakvararti, D. N., and Dhar, N. R., *Z. anorg. allgem. Chem.*, **142**, 299 (1925).
Charcot, *Compt. rend. soc. biol.*, **5**, series 2, 63 (1858).
Chollet, E., *Recueil de médecine vétérinaire*, **109**, 97 (1933).
Clar, E., and Haurowitz, F., *Ber.*, **66**, 331 (1933).
Clark, J. H., *Physiol. Rev.*, **2**, 277 (1922).
Clark, J. H., *J. Gen. Physiol.*, **19**, 199 (1935) a.
Clark, J. H., *Am. J. Physiol.*, **113**, 538 (1935) b.
Clark, J. H., *Am. J. Hyg.*, **24**, 334 (1936).
Clark, J. H., and Chapman, J., *Am. J. Hyg.*, **15**, 755 (1932).
Claus, G., *Biochem. Z.*, **204**, 456 (1929).
Clawson, A. B., and Huffman, W. T., *The National Wool Grower*, **27**, 13 (1937).
Clementi, A., and Condorelli, F., *Arch. fisiol.*, **28**, 174 (1930).
Clifton, C. E., *Proc. Soc. Exp. Biol. Med.*, **28**, 745 (1931).
Coblentz, W. W., Dorcas, M. J., and Hughes, C. W., *Scientific Papers of the Bureau of Standards*, **21**, 535 (1926).
Coblentz, W. W., and Stair, R., *Bur. Standards J. Research*, **12**, 13 (1934).
Coblentz, W. W., Stair, R., and Hogue, J. M., *Proc. Nat. Acad. Sci.*, **17**, 401 (1931).
Coblentz, W. W., Stair, R., and Hogue, J. M., *Bur. Standards J. Research*, **8**, 541 (1932).
Cockayne, E. A., "Inherited Abnormalities of the Skin," Oxford, Univ. Press, 1933.

Commoner, B., *J. Cellular Comp. Physiol.*, **12**, 171 (1938).
Crew, W. H., and Whittle, C. H., *J. Physiol.*, **93**, 335 (1938).
Crozier, W. J., *J. Gen. Physiol.*, **6**, 647 (1924).
Cummins, *Arch. Dermatol. Syphilol.*, **13**, 419 (1926).
Curtis, W. E., Dickens, F., and Evans, S. F., *Nature*, **138**, 63 (1936).
Cuzin, J., *Bull. soc. chim. biol.*, **12**, 745 (1930) a.
Cuzin, J., *Bull. soc. chim. biol.*, **12**, 1401 (1930) b.
Cuzin, J., *Compt. rend. soc. biol.*, **104**, 15 (1930) c.

D

Daley, C. J., *Agr. Gaz. N. S. Wales*, **48**, 301 (1937).
D'Amato, G., *Policlinico*, **33**, 1750 (1926).
Daniels, F., "Chemical Kinetics," Ithaca, Cornell Univ. Press, 1938.
Dankemeyer, W., "Ueber die photosensibilizierende Wirkung von Porphyrinen," *Inaug. Diss.*, Hamburg, Riefling und Weingärtner, 1930.
Darrow, K. K., "Electrical Phenomena in Gases," Baltimore, Williams and Wilkins, 1932.
Dartnall, H. J. A., and Goodeve, C. F., *Nature*, **139**, 409 (1937).
Davis, W. E., *The Saturday Evening Post*, **211**, 16 (1938).
Davson, H., and Ponder, E., *J. Cellular Comp. Physiol.*, **15**, 67 (1940).
Dejust, L. H., Verne, J., Combes, R., Parat, M., Urbain, A., Dujarric de la Riviere, R., de Saint Rat, L., "Etudes sur la Chimie de la Peau," Paris, Legrand, 1928.
Delbanco, *Dermatol. Wochschr.*, **70**, 201 (1920).
Del Vivo, G., *Giornale italiano di dermatologia e sifilologia*, **71**, 467 (1930).
Dempsey, T. F., and Mayer, V., *J. Comp. Path. Therap.*, **47**, 197 (1934).
Dhéré, C., *Compt. rend.*, **195**, 1436 (1932).
Dhéré, C., *Compt. rend.*, **197**, 948 (1933).
"Dispensatory of the United States of America," *19th ed.* Philadelphia, J. B. Lippincott, 1907.
Djourno, A., and Piffault, C., *J. physiol. path. gen.*, **34**, 746 (1936).
Dobriner, K., *J. Biol. Chem.*, **113**, 1 (1936).
Dobriner, K., Strain, W. H., and Localio, S. A., *Proc. Exp. Soc. Biol. Med.*, **38**, 748 (1938).
Dodd, S., *J. Comp. Path. Therap.*, **29**, 47 (1916).
Dodd, S., *Agri. Gaz. N. S. Wales*, **31**, 265 (1920) a.
Dodd, S., *J. Comp. Path. Therap.*, **33**, 105 (1920) b.
Dognon, A., *Compt. rend. soc. biol.*, **97**, 1590 (1927).
Dognon, A., *Compt. rend. soc. biol.*, **98**, 21 (1928) a.
Dognon, A., *Compt. rend. soc. biol.*, **98**, 283 (1928) b.
Dognon, A., *Compt. rend. soc. biol.*, **98**, 374 (1928) c.
Doniach, I., and Mottram, J. C., *Nature*, **140**, 588 (1937).
Dörffel, *Zentralblatt für Haut und Geschlechtskrankheiten*, **30**, 294 (1929).
Dowling, G. B., *Proc. Roy. Soc. Med.*, **29**, 98 (1935).
Dreyer, G., *Dermatol. Z.*, **10**, 578 (1903).
Dreyer, G., and Campbell-Renton, M. L., *Proc. Roy. Soc. (London) B*, **120**, 447 (1936).
Dubreuilh, W., *Annales de dermatologie et de syphiligraphie*, **8**, *series 4*, 387 (1907).
Dufton, A. F., and Bedford, T., *J. Hyg.*, **33**, 476 (1933).
Duggar, B. M., "Biological Effects of Radiation," 2 vols., New York, McGraw-Hill Book Co., 1936.
Duke, W. W., *J. Am. Med. Assoc.*, **80**, 1835 (1923).
Duke, W. W., *J. Am. Med. Assoc.*, **83**, 3 (1924).
Duke, W. W., *J. Am. Med. Assoc.*, **84**, 736 (1925) a.
Duke, W. W., "Allergy, Asthma, Hay Fever, Urticaria and allied Manifestations of Reaction," St. Louis, C. V. Mosby and Co., 1925 (b).
Duke, W. W., "The Practitioner's Library of Medicine and Surgery," Vol. 8, p. 561, New York, D. Appleton-Century Co., 1935.
Duke-Elder, W. S., *Lancet*, **210**, 1137 (1926) a.
Duke-Elder, W. S., *Lancet*, **210**, 1188 and 1250 (1926) b.
Duke-Elder, W. S., *Lancet*, **211**, 16 (1926) c.

E

Efimoff, W. W., *Biochem. Z.*, **140**, 453 (1923).
Efimoff, A., and Efimoff, W. W., *Biochem. Z.*, **155**, 376 (1925).
Ehrmann, S., *Arch. Dermatol. Syphilis*, **77**, 163 (1905).
Ehrmann, S., *Arch. Dermatol. Syphilis*, **97**, 75 (1909).
Ehrismann, O., and Noethling, W., *Z. Hyg. Infectionskrank.*, **113**, 597 (1932).
Eidinow, A., *Brit. J. Radiol.*, **3**, 112 (1930) a.
Eidinow, A., *J. Path. Bact.*, **33**, 769 (1930) b.
Eidinow, A., *British Journal of Dermatology and Syphilis*, **47**, 277 (1935).
Ellinger, F., *Arch. exp. Path. Pharmakol.*, **136**, 129 (1928).
Ellinger, F., *Biochem. Z.*, **215**, 279 (1929).
Ellinger, F., *Strahlentherapie*, **38**, 521 (1930).
Ellinger, F., *Biochem. Z.*, **248**, 437 (1932) a.
Ellinger, F., *Strahlentherapie*, **44**, 1 (1932) b.
Ellinger, F., *Z. Klin. Med.*, **122**, 272 (1932) c.
Ellinger, F., *Archiv. Gynäkol.*, **156**, 471 (1934).
Ellinger, F., "Die biologischen Grundlagen der Strahlenbehandlung," Strahlentherapie, XX Sonderband., Berlin, Urban and Schwartzenberg, 1935.
Ellinger, F., *Radiologica*, **3**, 195 (1938).
Ellinger, F., *Radiologica*, **4**, 181 (1939) a.
Ellinger, F., *Radiology*, **32**, 157 (1939) b.
Emmons, C. W., and Hollaender, A., *Am. J. Botany*, **26**, 467 (1939).
Enright, J. I., *Lancet*, **198**, 998 (1920).
Epstein, S., *Arch. Dermatol. Syphilis*, **168**, 67 (1933).
Epstein, S., *The Journal of Investigative Dermatology*, **2**, 43 (1939).
Erdt, *Magazin für die gesammte Thierheilkunde*, **6**, 303 (1840).

F

Fardon, J. C., Carroll, M. J., and Sullivan, W. A., *Studies of the Institutum Divi Thomae*, **1**, 117 (1937).
Feit, H., *Journal of the Medical Society of New Jersey*, **24**, 226 (1927).
Fessler, K., *Z. physiol. Chem.*, **85**, 148 (1913).
Fieser, L. F., "The Chemistry of Natural Products Related to Phenanthrene," New York, Reinhold Publishing Corp., 1937.
Fiessinger, N., Albeaux-Fernet, M., and Aussannaire, M., *Bulletin de la société française de dermatologie et de syphiligraphie*, **44**, 1768 (1937).
Findlay, G. M., *Lancet*, **215**, 1070 (1928).
Findlay, G. M., *Lancet*, **218**, 1229 (1930).
Fischer, H., "Handbuch der Biochemie," 2nd ed., Erganzungsband, p. 72, edited by C. Oppenheimer, Jena, Gustave Fischer, 1930.
Fischer, H., *Chem. Rev.*, **20**, 41 (1937).
Fischer, H., and Herrle, K., *Z. physiol. Chem.*, **251**, 85 (1938).
Fischer, H., and Hess, R., *Z. physiol. Chem.*, **187**, 133 (1930).
Fischer, H., and Hilmer, H., *Z. physiol. Chem.*, **143**, 1 (1925).
Fischer, H., Moldenbauer, O., and Süs, O., *Liebigs Ann.*, **485**, 1 (1931).
Fischer, H., and Niemann, G., *Z. physiol. Chem.*, **146**, 196 (1925).
Fischer, H., and Orth, H., "Die chemie des Pyrrols," Vol. 2, 1st half, Leipzig, Akademische Verlagsgesselschaft M.B.H., 1937.
Fischer, H., and Riedmair, J., *Liebigs Ann.*, **490**, 91 (1931).
Fischer, H., and Riedmair, J., *Liebigs Ann.*, **497**, 181 (1932).
Flarer, F., *Archivio italiano di dermatologia sifilografia e venereologia*, **5**, 542 (1930).
Fleischhauer, L., *Strahlentherapie*, **36**, 144 (1930).
Flint, L. H., *Science*, **80**, 38 (1934).
Flint, L. H., and McAlister, E. D., *Smithsonian Misc. Collections*, **94**, no. 5 (1935).
Flint, L. H., and McAlister, E. D., *Smithsonian Misc. Collections*, **96**, no. 2 (1937).
Foerster, H. R., and Schwartz, L., *Arch. Dermatol. Syphilol.*, **39**, 55 (1939) a.
Foerster, H. R., and Schwartz, L., *Arch. Dermatol. Syphilol.*, **39**, 955 (1939) b.
Forsythe, W. E., "Measurement of Radiant Energy," New York, McGraw-Hill, 1937.
Forsythe, W. E., and Christison, F., *Gen. Elec. Rev.*, **32**, 662 (1929).
Fourie, P. J. J., *Onderstepoort J. Vet. Sci. Animal Ind.*, **7**, 535 (1936).
Fourie, P. J., and Rimington, C., *Nature*, **140**, 68 (1937).
Franck, J., and Levi, H., *Z. physik. Chem. B.*, **27**, 409 (1934).
Franck, J., and Wood, R. W., *J. Chem. Phys.*, **4**, 551 (1936).
Franck, J., and Herzfeld, K. F., *J. Phys. Chem.*, **41**, 97 (1937) a.
Franck, J., and Herzfeld, K. F., *J. Chem. Phys.*, **5**, 237 (1937) b.
Frankenburger, W., *Naturwissenschaften*, **21**, 116 (1933).
Frei, W., *Arch. Dermatol. Syphilis*, **149**, 124 (1925).
Frei, W., *Arch. Dermatol. Syphilis*, **151**, 67 (1926).
French, C. S., *J. Gen. Physiol.*, **21**, 71 (1937).
Freund, E., *Dermatol. Wochschr.*, **63**, 931 (1916).
Freund, L., *Wien. Klin. Wochschr.*, **25**, 191 (1912).
Frieboes, W., *Dermatol. Z.*, **24**, 641 (1917).
Friedli, H., *Bull. soc. chim. biol.*, **6**, 908 (1924).
Fröhner, E., "Lehrbuch der Toxicologie für Tierärtzte," Stuttgart, Enke, 1927.
Fruitman, H. L., *Proc. Soc. Exp. Biol. Med.*, **32**, 610 (1935).
Fuchs, C. J., "Handbuch der allgemeinen Pathologie der Haussäugethiere," p. 145, Berlin, Veit und Comp., 1843.
Funfack, M., *Arch. Dermatol. Syphilis*, **146**, 303 (1924).

G

Gaffron, H., *Biochem. Z.*, **179**, 157 (1926).
Gaffron, H., *Ber.*, **60**, 755 (1927) a.
Gaffron, H., *Ber.*, **60**, 2229 (1927) b.
Gaffron, H., *Biochem. Z.*, **264**, 251 (1933).
Gaffron, H., *Ber.*, **68**, 1409 (1935).
Gaffron, H., *Biochem. Z.*, **287**, 130 (1936) a.
Gaffron, H., *Naturwissenschaften*, **24**, 81 (1936) b.
Gaffron, H., *Z. physikal. Chem. B.*, **37**, 437 (1937).
Galloway, I. A., *Brit. J. Exp. Path.*, **15**, 97 (1934).
Gans, O., *Deut. med. Wochschr.*, **55**, 1213 (1929).
Gassul, R., *Strahlentherapie*, **10**, 1162 (1920).
Gates, F. L., *J. Gen. Physiol.*, **14**, 31 (1930).
Gates, F. L., *J. Gen. Physiol.*, **17**, 797 (1934) a.
Gates, F. L., *J. Exp. Med.*, **60**, 179 (1934) b.
Gaviola, E., *Z. Physik*, **42**, 853 (1927).
Gay, D. M., and McIver, M. A., *Am. J. Trop. Med.*, **2**, 115 (1922).
Gicklhorn, J., *Sitzber. Akad. Wiss. Wien. Math. naturw. Klasse. Abt. I.*, **123**, 1221 (1914).
Giese, A. C., and Leighton, P. A., *J. Gen. Physiol.*, **18**, 557 (1935).
Gilfillan, W., *British Journal of Dermatology and Syphilis*, **49**, 241 (1937).
Giradeau and Acquaviva, *Bulletin de la société française de dermatologie et de syphiligraphie*, **41**, 973 (1934).
Goeckerman, W. H., Osterberg, A. E., and Sheard, C., *Arch. Dermatol. Syphilol.*, **20**, 501 (1929).
Gottron, H., and Ellinger, F., *Arch. Dermatol. Syphilis*, **164**, 11 (1931).
Gottron, H., and Ellinger, F., *Arch. Dermatol. Syphilis*, **167**, 325 (1933).
Götzl, A., *Wien klin. Wochschr.*, **24**, 1727 (1911).
Gougerot and Burnier, *Bulletin de la société française de dermatologie et de syphiligraphie*, **41**, 1656 (1934).
Gougerot, H., and Meyer, J., *Annales de l'Institut d'Actinologie*, **7**, 237 (1933).
Graham, C. H., and Hartline, H. K., *J. Gen. Physiol.*, **18**, 917 (1935).
Graham, C. H., and Riggs, L. A., *Journal of General Psychology*, **12**, 279 (1935).

Graham, N. P. H., and Gordon, H. M., *Australian Vet. J.*, **13**, 125 (1937).
Gray, A. M. H., *Quart. J. Med.*, **19**, 381 (1926).
Greenbaum, S. S., *Am. J. Diseases Children*, **34**, 81 (1927).
Griffith, R. O., and McKeown, A., "Photo-processes in gaseous and liquid systems," London, Longmans Green and Co., 1929.
Gross, P., and Robinson, L. B., *Arch. Dermatol. Syphilol.*, **21**, 637 (1930).
Grosz, S., and Volk, R., *Arch. Dermatol. Syphilis.*, **120**, 301 (1914).
Grosz, *Arch. Dermatol. Syphilis.*, **119**, 151 (1914).
Grundfest, H., *J. Gen. Physiol.*, **15**, 307 (1932) a.
Grundfest, H., *J. Gen. Physiol.*, **15**, 507 (1932) b.
Guerrini, G., *Boll. soc. ital. biol. sper.*, **5**, 567 (1930) a.
Guerrini, G., *Boll. soc. ital. biol. sper.*, **5**, 635 (1930) b.
Guerrini, G., *Boll. soc. ital. biol. sper.*, **5**, 1098 (1930) c.
Guerrini, G., *Arch. fisiol.*, **29**, 356 (1931) a.
Guerrini, G., *Boll. soc. ital. biol. sper.*, **6**, 401 (1931) b.
Guerrini, G., *Boll. soc. ital. biol. sper.*, **7**, 835 (1932).
Guerrini, G., *Boll. soc. ital. biol. sper.*, **9**, 816 (1934).
Guillaume, A. C., *Bull. mém. soc. méd. hôp. Paris*, **50**, 730 (1926) a.
Guillaume, A. C., *Bull. mém. soc. méd. hôp. Paris*, **50**, 1426 (1926) b.
Guillaume, A. C., *Bulletin médicale (Paris)*, **40**, 559, 589, 621 (1926) c.
Guillaume, A. C., "Les Radiations Lumineuse en Physiologie et en Therapeutique," Paris, Masson et Cie, 1927.
Günther, H., *Deut. Arch. klin. Med.*, **105**, 89 (1911) a.
Günther, H., *Deut. med. Wochschr.*, **37**, 1771 (1911) b.
Günther, H., *Dermatol. Wochschr.*, **68**, 177, 203, 213, 230, 243 (1919).
Günther, H., *Deut. Arch. klin. Med.*, **134**, 257 (1920).
Günther, H., *Ergebnisse der allgemeine Pathologie und pathologische Anatomie*, **20**, 608 (1922).
György, P., and Gottlieb, K., *Klin. Wochschr.*, **2**, 1302 (1923).

H

Haberlandt, F., *Strahlentherapie*, **29**, 161 (1928).
Haberlandt, L., *Pflügers Arch. ges. Physiol.*, **219**, 128 (1927).
Habermann, R., "Handbuch der Haut und Geschlechtskrankheiten," Vol. 4/2, p. 809, Berlin, Julius Springer, 1933.
Habermann, R., *Ibid.*, p. 833, 1933 b.
Habermann, R., *Ibid.*, p. 847, 1933 c.
Haig, C., *Biol. Bull.*, **69**, 305 (1935).
Hamilton, J. B., and Hubert, G., *Science*, **88**, 481 (1938).
Hammer, F., "Über die Einfluss des Lichtes auf die Haut," Stuttgart, F. Enke, 1891.
Hand, D. B., Guthrie, E. S., and Sharp, P. F., *Science*, **87**, 439 (1938).
Hannes, B., and Jodlbauer, A., *Biochem. Z.*, **21**, 110 (1909).
Harris, D. T., *Biochem. J.*, **20**, 271 (1926) a.
Harris, D. T., *Biochem. J.*, **20**, 280 (1926) b.
Harris, D. T., *Biochem. J.*, **20**, 288 (1926) c.
Harrison, G. R., *J. Optical Soc. Am.*, **19**, 267 (1929).
Harrison, G. R., *J. Optical Soc. Am.*, **24**, 59 (1934).
Hartmann, E., and Briel, I., *Dermatol. Z.*, **50**, 205 (1927).
Harzbecker, O., and Jodlbauer, A., *Biochem. Z.*, **12**, 306 (1908).
Hasselbalch, K. A., *Biochem. Z.*, **19**, 435 (1909).
Haubner, G. C., "Krankheiten der landwirtschaftlichen Haussäugethiere," 5th ed., p. 369, Anclan, W. Dietze, 1867.
Hauptmann, *Berliner tierärztliche Wochenschrift.*, **38**, 313 (1922).
Hausmann, W., *Biochem. Z.*, **14**, 275 (1908).
Hausmann, W., *Biochem. Z.*, **30**, 276 (1910).
Hausmann, W., *Biochem. Z.*, **67**, 309 (1914).
Hausmann, W., *Biochem. Z.*, **77**, 268 (1916).
Hausmann, W., "Grundzüge der Lichtbiologie und Lichtpathologie," Strahlentherapie, VIII Sonderband, Berlin, Urban and Schwartzenberg, 1923.
Hausmann, W., *Strahlentherapie*, **41**, 145 (1931).
Hausmann, W., and Haxthausen, H., "Die Lichterkrankungen der Haut," Strahlentherapie, XI Sonderband, Berlin, Urban and Schwartzenberg, 1929.
Hausmann, W., and Kolmer, W., *Biochem. Z.*, **15**, 12 (1908).
Hausmann, W., and Krumpel, O., *Biochem. Z.*, **186**, 203 (1927).
Hausmann, W., and Kuen, F. M., *Biochem. Z.*, **265**, 105 (1933).
Hausmann, W., and Loewy, A., *Biochem. Z.*, **173**, 1 (1926).
Hausmann, W., and Löhner, L., *Biochem. Z.*, **173**, 7 (1926).
Hausmann, W., and Pribram, E., *Biochem. Z.*, **17**, 13 (1909).
Hausmann, W., and Rosenfeld, P., *Strahlentherapie*, **45**, 125 (1932).
Hausmann, W., and Sonne, C., *Strahlentherapie*, **25**, 174 (1927).
Hausmann, W., and Spiegel-Adolf, M., *Klin. Wochschr.*, **6**, 2182 (1927).
Hausmann, W., and Zaribnicky, F., *Klin. Wochschr.*, **8**, 74 (1929).
Hausser, K. W., *Strahlentherapie*, **28**, 25 (1928).
Hausser, K. W., and Vahle, W., *Strahlentherapie*, **13**, 41 (1922).
Hausser, I., *Strahlentherapie*, **62**, 315 (1938).
Haxthausen, H., "Hysygdomme Fremkaldt af Lyset," Dissertation, Copenhagen, Pios, 1919.
Haxthausen, H., *Dermatol. Wochschr.*, **84**, 827 (1927).
Haxthausen, H., *British Journal of Dermatology and Syphilis.*, **45**, 16 (1933).
Hecht, S., *J. Gen. Physiol.*, **3**, 375 (1920).
Hecht, S., *J. Gen. Physiol.*, **11**, 657 (1928).
Hecht, S., and Williams, R. E., *J. Gen. Physiol.*, **5**, 1 (1922).
Heft, B. B., *Acta Dermato-venereologica*, **16**, 146 (1935).
Hegler, C., Fraenkel, E., and Schumm, O., *Deut. med. Wochschr.*, **39**, 842 (1913).
Heiman, M., *Wien. klin. Wochschr.*, **49**, 398 (1936).
Heiner, L., *Zentralblatt für Haut und Geschlechtskrankheiten*, **18**, 217 (1925).

Hemmingsen, A. M., and Krarup, N. B., *Biologiske Meddelelser*, **13**, no. 7, pp. 1-61 (1937).
Hendrickx, *Bull. acad. roy. méd. Belg.*, **12**, 224 (1932).
Henning, M. W., "Animal Diseases in South Africa," Chapt. 55, Vol. II, Johannesburg, South Africa, Central News Agency, Limited, 1932.
Henri, V., *Compt. rend. soc. biol.*, **73**, 323 (1912).
Henri, V., "Zangger-Festschrift," p. 792, Zürich, Rascher & Cie, 1934.
Henri, Mme. V., and Henri, V., *Compt. rend. soc. biol.*, **72**, 326 (1912).
Henri, Mme. V., Henri, V., and Wurmser, R., *Compt. rend. soc. biol.*, **73**, 319 (1912).
Henry, M., *Agr. Gaz. N. S. Wales*, **33**, 205 (1922).
Henschke, U., and Schulze, R., *Strahlentherapie*, **64**, 14 (1939) a.
Henschke, U., and Schulze, R., *Strahlentherapie*, **64**, 43 (1939) b.
Herlitz, C. W., Jundell, I., and Wahlgren, F., *Acta. Paediat.*, **10**, 321 (1931).
Hertel, E., *Z. allgem. Physiol.*, **5**, 95 (1905).
Hertel, E., *Z. allgem. Physiol.*, **6**, 44 (1906).
Herxheimer, K., and Nathan, E., *Dermatol. Z.*, **24**, 385 (1917).
Herzberg, K., *Z. Immunitäts*, **80**, 507 (1933).
Heupke, W., *Deut. Arch. klin. Med.*, **176**, 32 (1933).
Heusinger, *Wochenschrift für die gesammte Heilkunde.*, p. 277 (1846).
Hickel, R., *Annales de l'Institut d'Actinologie*, **2**, 212 (1927).
Hicks, C. S., and Holden, H. F., *Australian J. Exp. Biol. Med. Sci.*, **12**, 91 (1934).
Hill, L., *J. State Med.*, **39**, 683 (1931).
Hill, L., *J. Physiol.*, **74**, 1 P (1932).
Hilz, K., *Monatshefte für pracktische Tierheilkunde.*, **25**, 357 (1914).
Hinrichs, M. A., *J. Exp. Zoöl.*, **41**, 21 (1924).
Hinrichs, M. A., *Biol. Bull.*, **50**, 1 (1926).
Hirschberger, A., and Fuchs, H., *Münch. med. Wochschr.*, **83**, 1965 (1936).
Hoffman, E., and Schmidtz, H., *Münch. med. Wochschr.*, **72**, 1414 (1925).
Hofmann, E., *Dermatol. Z.*, **53**, 301 (1928).
Hollaender, A., *Biol. Bull.*, **75**, 248 (1938).
Hollaender, A., Cole, P. A., and Brackett, F. S., *Am. J. Cancer*, **37**, 265 (1939).
Hollaender, A., and Claus, W. D., *J. Gen. Physiol.*, **19**, 753 (1936).
Hollaender, A., and Emmons, C. W., *J. Cellular Comp. Physiol.*, **13**, 391 (1939).
Hollaender, A., Jones, M. F., and Jacobs, L., *J. Parasitol.*, **26**, 421 (1940).
Holtz, F., *Strahlentherapie*, **28**, 108 (1928).
Holz, P., *Arch. Exp. Path. Pharmakol.*, **175**, 97 (1934).
Honigmann, H., *Pflügers. Arch. ges. Physiol.*, **189**, 1 (1921).
Hoover, W. H., *Smithsonian Misc. Collections*, **95**, no. 21 (1937).
Hopkins, F. G., *Compt. rend. trav. lab. Carlsberg, Sér. chemique*, **22**, 226 (1938).
Hopkirk, C. S. M., *New Zealand J. Agr.*, **52**, 98 (1936).
Horbaczewski, J., *Oesterreichische Sanitätswesen*, **22**, *Beilagen*, 345 (1910).
Horbaczewski, J., *Oesterreichische Sanitätswesen*, **24**, 917 (1912).
Horsley, C. H., *J. Pharmacol.*, **50**, 310 (1934).
Howarth, J. A., *North American Veterinarian*, **12**, 29 (1931).
Howell, W. H., *Arch. intern. physiol.*, **18**, 269 (1921).
Hudelo and Montlaur, *Bulletin de la société française de dermatologie et de syphiligraphie*, **26**, 335 (1919).
Huldschinsky, K., *Deut. med. Wochschr.*, **59**, 530 (1933).
Hutchinson, J., "Lectures on Clinical Surgery," Vol. 1, p. 126, London, J. and A. Churchill, 1878.
Hutchinson, J., *Transactions of the Clinical Society of London*, **22**, 80 (1889).
Hyde, E. P., Forsythe, W. E., and Cady, F. E., *Astrophys. J.*, **48**, 65 (1918).
Hyde, J. N., *Am. J. Med. Sci.*, **131**, 1 (1906).

I

Inhoffen, H. H., *Liebigs Ann.*, **497**, 130 (1932).

J

Jacobson, R., *Z. Biol.*, **41**, 444 (1901).
Jacoby, *Zentralblatt für Haut und Geschlechtskrankheiten.*, **27**, 472 (1928).
Jadin, J., *Compt. rend. soc. biol.*, **110**, 124 (1932).
Jamada, K., and Jodlbauer, A., *Biochem. Z.*, **8**, 61 (1907).
Jancsó, N., *Zentr. Bakt. Parasitenk. I. Orig.*, **122**, 389 (1931) a.
Jancsó, N., *Zentr. Bakt. Parasitenk. I. Orig.*, **122**, 393 (1931) b.
Jancsó, N., *Zentr. Bakt. Parasitenk. I. Orig.*, **123**, 129 (1931) c.
Jausion, Jacowski, and Kouchner, *Bulletin de la société française de dermatologie et de syphiligraphie.*, **44**, 1756 (1937).
Jausion and Marceron, *Bulletin de la société française de dermatologie et de syphiligraphie*, **32**, 358 (1925).
Jausion, H., and Pagès, F., "Les Maladies de Lumière," Paris, Masson et Cie, 1933.
Jensen, T., and Hansen, K. G., *Arch. Dermatol. Syphilol.*, **40**, 566 (1939).
Jesionek, A., "Lichtbiologie und Lichtpathologie," in Praktische Ergebnisse auf dem Gebiete der Haut- und Geschlechtskrankheiten, vol. 2, Wiesbaden, J. F. Bergmann, 1912.
Jesionek, A., and Tappeiner, H., *Deut. Arch. Klin. Med.*, **82**, 223 (1905).
Jírovec, O., and Vácha, K., *Protoplasma*, **22**, 203 (1934).
Jírovec, O., and Ziegler, Z., *Z. ges. exp. Med.*, **90**, 651 (1933).
Jobling, J. W., and Arnold, L., *J. Am. Med. Assoc.*, **80**, 365 (1923).
Jodlbauer, A., *Deut. Arch. Klin. Med.*, **80**, 488 (1904).
Jodlbauer, A., and Busck, G., *Arch. intern. Pharmacodynamie*, **15**, 263 (1905).
Jodlbauer, A., and Haffner, F., *Biochem. Z.*, **118**, 150 (1921) a.
Jodlbauer, A., and Haffner, F., *Pflügers Arch. ges. Physiol.*, **189**, 243 (1921) b.
Jodlbauer, A., and Tappeiner, H., *Deut. Arch. klin. Med.*, **82**, 520 (1905) a.
Jodlbauer, A., and Tappeiner, H., *Deut. Arch. klin. Med.*, **84**, 529 (1905) b.
Jodlbauer, A., and Tappeiner, H., *Deut. Arch. klin. Med.*, **85**, 386 (1906) a.

Jodlbauer, A., and Tappeiner, H., *Deut. Arch. klin. Med.*, **85**, 399 (1906) b.
Jodlbauer, A., and Tappeiner, H., *Deut. Arch. klin. Med.*, **87**, 373 (1906) c.
Johansen, E. S., *Strahlentherapie*, **6**, 45 (1915).
Johnston, E. S., *Smithsonian Misc. Collection*, **92**, no. 11 (1934).
Jolly, J., *Compt. rend. soc. biol.*, **94**, 173 (1926).
Juon, M., *Arch. Dermat. Syphilis.*, **156**, 367 (1928).

K

Kambayashi, Y., *Biochem. Z.*, **203**, 334 (1928).
Kammerer, H., and Weisbecker, H., *Arch. Exp. Path. Pharmakol.*, **111**, 263 (1926).
Karrer, P., "Organic Chemistry," New York, Nordemann Publishing Company, 1938.
Karschulin, M., *Biochem. Z.*, **213**, 202 (1929).
Kautsky, H., deBruijn, H., Neuwirth, R., and Baumeister, W., *Ber.*, **66**, 1588 (1933).
Kautsky, H., Hirsch, A., and Flesch, W., *Ber.*, **68**, 152 (1935).
Kawaguchi, S., *Biochem. Z.*, **221**, 232 (1930).
Kawai, H., *J. Biochem. (Japan)*, **10**, 325 (1928).
Keller, P., *Strahlentherapie*, **16**, 537 (1924) a.
Keller, P., *Dermatol. Wochschr.*, **79**, 1433 (1924) b.
Keller, P., *Strahlentherapie*, **39**, 320 (1931).
Kichiya, O., *J. Biochem. (Japan)*, **4**, 225 (1924).
Kistiakowski, E. W., *Annales de dermatologie et de syphiligraphie*, **1**, *series 7*, 63 (1930).
Kistiakowsky, G. B., "Photochemical Processes," New York, Chemical Catalog Co., Inc. (Reinhold Publishing Corp.), 1928.
Kitchevatz, M., *Annales de dermatologie et de syphiligraphie*, **5**, series 7, 293 (1934) a.
Kitchevatz, M., *Compt. rend. soc. biol.*, **116**, 675 (1934) b.
Kitchevatz, M., *Bulletin de la société française de dermatologie et de syphiligraphie*, **41**, 1751 (1934) c.
Kitchevatz, M., *Bulletin de la société française de dermatologie et de syphiligraphie*, **43**, 581 (1936).
Kitchevatz, M., and Bugarski, S., *Bulletin de la société française de dermatologie et de syphiligraphie*, **45**, 65 (1938).
Knudson, A., Sturges, S., and Bryan, W. R., *J. Biol. Chem.*, **123**, lxx. (1938).
Koblitz, W., and Schumacher, H. J., *Z. physik. Chem. B.*, **35**, 11 (1937).
de Kock, G., *The Journal of the South African Veterinary Medical Association*, **1**, *no. 2*, 39 (1927).
Kolm, R., and Pick, E. P., *Arch. Exp. Path. Pharmakol.*, **86**, 1 (1920).
Kon, S. K., and Watson, M. B., *Biochem. J.*, **30**, 2273 (1936).
Körbler, J., *Strahlentherapie*, **52**, 353 (1935).
Kosman, A. J., *J. Cellular Comp. Physiol.*, **11**, 279 (1938).
Kosman, A. J., and Lillie, R. S., *J. Cellular Comp. Physiol.*, **6**, 505 (1935).
Kren, O., and Löewenstein, E., *Wien Klin. Wochschr.*, **44**, 405 (1931).
Krogh, A., "The Anatomy and Physiology of the Capillaries," revised edition, p. 220, New Haven, Yale Univ. Press, 1929.
Kruspe, *Zentralblatt für Haut und Geschlechtskrankheiten*, **37**, 24 (1931).
Kuske, H., *Arch. Dermatol. Syphilis*, **178**, 112 (1938).

L

Laffitte, J. B., "La Pratique Dermatologique," Vol. 2, p. 828, Paris, Masson, 1907.
Lakschewitz, K., *Monatsschr. Kinderheilk.*, **34**, 159 (1926).
Lambert, J., *Trans. Roy. Soc. (London)*, **66**, 493 (1776).
Lange, E., *Berliner Tierärtzliche Wochenschrift.*, **38**, 411 (1922).
Lange, L., Bamatter, Gruninger, and Löwenstädt, *Beitr. Klin. Tuberk.*, **82**, 1 (1933).
Lassen, H. C. A., *Strahlentherapie*, **27**, 757 (1927).
Laurens, H., *Am. J. Physiol.*, **64**, 97 (1923).
Laurens, H., "The Physiological Effects of Radiant Energy," New York, Chemical Catalog Co., Inc. (Reinhold Publishing Corp.), 1933.
Laurens, H., *Cold Spring Harbor Symposia Quant. Biol.*, **3**, 277 (1935).
Laurens, H., and Hooker, H. D., *J. Exp. Zoöl.*, **30**, 345 (1920).
LaWall, C. H., "Four Thousand Years of Pharmacy," p. 214, Philadelphia, J. B. Lippincott, 1927.
Lebensohn, J. E., *Arch. Ophthalmol.*, **12**, 380 (1934).
Ledoux-Lebard, *Ann. inst. Pasteur*, **16**, 587 (1902).
van Leersum, E. C., *J. Biol. Chem.*, **58**, 835 (1924).
Leighton, P. A., "Determination of the Mechanism of Photochemical Reactions," Paris, Hermann & Cie., 1938.
Lepeschkin, W. W., and Davis, G. E., *Protoplasma*, **20**, 189 (1933).
Letort, M., "Les Conceptions Actuelles du Mécanisme des Réactions Chimiques," Paris, Hermann & Cie., 1937.
Levaillant, R., *Compt. rend.*, **177**, 398 (1923).
Levinson, L. B., and Romanowa, K. G., *Z. Immunitäts.*, **66**, 141 (1930).
Levy, A. G., *J. Path. Bact.*, **32**, 387 (1929).
Levy, A. G., *J. Path. Bact.*, **36**, 31 (1933).
Levy, M., *Strahlentherapie*, **9**, 618 (1919).
Lewin, L., *Münch. med. Wochschr.*, **60**, 1529 (1913).
Lewis, M. R., *Am. J. Cancer*, **25**, 305 (1935).
Lewis, T., "The Blood Vessels of the Human Skin and Their Responses," London, Shaw and Sons, 1927.
Li, Keh-Hung, *Proc. Soc. Exp. Biol. Med.*, **34**, 657 (1936) a.
Li, Keh-Hung, *Proc. Soc. Exp. Biol. Med.*, **34**, 659 (1936) b
Liebert, F., and Kaper, L., *Antonie von Leuwenhoek Nederlandsch Tijdschrift. voor hygiene, microbiologieen serologie*, **4**, 164 (1937).
Liebner, *Dermatol. Wochschr.*, **92**, 552 (1931).
Lignac, G. O. E., *Virchow's Archiv.*, **240**, 383 (1923).
Lignac, G. O. E., *Krankheitsforsch.*, **1**, 177 (1925).
Lillie, R. S., in "General Cytology," edited by E. V. Cowdry, Chapt. IV, p. 183, Chicago, Univ. of Chicago Press, 1924.
Lillie, R. S., and Hinrichs, M. A., *Anat. Record*, **26**, 370 (1923).

Lillie, R. S., Hinrichs, M. A., and Kosman, A. J., *J. Cellular Comp. Physiol.*, **6**, 487 (1935).
Lin, F. C., *Proc. Soc. Exp. Biol. Med.*, **33**, 337 (1935).
Lin, F. C., *Proc. Soc. Exp. Biol. Med.*, **34**, 656 (1936).
Linser, P., *Arch. Dermatol. Syphilis*, **79**, 251 (1906).
Lippay, F., *Pflügers Arch. ges. Physiol.*, **222**, 616 (1929).
Lippay, F., *Pflügers Arch. ges. Physiol.*, **224**, 587 (1930).
Lippay, F., *Pflügers Arch. ges. Physiol.*, **226**, 473 (1931).
Lippay, F., and Wechsler, L., *Pflügers Arch. ges. Physiol.*, **229**, 173 (1931).
Lippert, K. M., *J. Immunol.*, **28**, 193 (1935).
Lipschitz, M., *Brit. Med. J.*, 1109 (1925) Vol. **1**.
Lode, *Wien. Klin. Wochschr.*, **23**, 1160 (1910).
Loebisch, W. F., and Fischler, M., *Monatsh.*, **24**, 335 (1903).
Loewi, O., and Navratil, E., *Pflügers Arch. ges. Physiol.*, **214**, 678 (1926).
Löhner, L., *Biochem. Z.*, **186**, 194 (1927).
Lomholt, S., "VIIIe Congres Internationale de Dermatologie et de Syphiligraphie," p. 1228, Copenhagen, 1930.
Loofbourow, J. R., Dwyer, C. M., and Morgan, M. N., *Studies of the Institutum Divi Thomae*, **1**, 137 (1938).
Lucas, N. S., *Biochem. J.*, **25**, 57 (1931).
Luckiesh, M., "Artificial Sunlight," New York, Van Nostrand, 1930.
Luckiesh, M., Holliday, L. L., and Taylor, A. H., *J. Optical Soc. Am.*, **20**, 423 (1930).
Lucksch, F., *Z. Hyg. Infectionskrank.*, **58**, 479 (1908).
Ludy, J. B., and Corson, E. F., *Arch. Dermatol. Syphilol.*, **37**, 403 (1938).
Lumière, A., and Sonnery, S., *J. physiol. path. gen.*, **32**, 44 (1934).
Lutz, H. E. W., "Über Fagopyrismus," Dissertation, Zurich (1930), *also in part in* Lutz, H. E. W., and Schmid, G., *Biochem. Z.*, **226**, 67 (1930).
Lynch, F. W., *Arch. Dermatol. Syphilol.*, **29**, 858 (1934).

M

McAlister, E. D., *Smithsonian Misc. Collection*, **87**, no. 17 (1933).
MacCormac, H., *British Journal of Dermatology and Syphilis*, **45**, 157 (1933).
MacCormac, H., and MacCree, M. M., *Brit. Med. J.*, 693 (1925) **vol. 1**.
McFarland, A. F., and Strain, W. H., *Arch. Dermatol. Syphilol.*, **38**, 727 (1938).
Mackey, L., and Garrod, A. E., *Quart. J. Med.*, **19**, 357 (1926).
Macklin, M. T., *Arch. Dermatol. Syphilol.*, **34**, 656 (1936).
Maisin, J., and DeJonghe, A., *Comp. rend. soc. biol.*, **117**, 111 (1934).
Major, R. H., "Classic Descriptions of Disease," p. 575, Springfield Ill., Charles Thomas, 1932.
Mallinckrodt-Haupt, A., *Strahlentherapie*, **61**, 636 (1938).
Mangold, E., "Handbuch der Ernährung und des Stoffwechsels der landwirtschaftlichen Nutztiere," Vol. 4, p. 900, Berlin, Julius Springer, 1932.
Marccaci, A., *Arch. ital. biol.*, **9**, 2 (1888).
Marceron, L., "Contributions a l'etude de la photosensibilization par les solutions fluorescentes," Thesis, Paris, Jouve et Cie, 1925.
Marchlewski, L., *Bull. intern. acad. sci. Cracovie. Classe sci. math. nat.*, 638 (1903).
Marchlewski, L., *Bull. intern. acad. sci. Cracovie Classe sci. math. nat.*, 276 (1904).
Marchlewski, L., *Z. physiol. Chem.*, **43**, 464 (1905).
Marcozzi, A., *Archivio italiano di dermatologia, sifilologia, e venereologia*, **4**, 555 (1929).
Marique, P., *Bull. soc. chim. biol.*, **20**, 325 (1938).
Marsh, C. D., and Clawson, A. B., *U. S. Dep. Agr. Tech. Bull.*, no. 202 (1930).
Martenstein, H., *Arch. Dermatol. Syphilis*, **140**, 300 (1922).
Martenstein, H., *Arch. Dermatol. Syphilis*, **147**, 499 (1924).
Martenstein, H., and Bobowitsch, A., *Arch. Dermatol. Syphilis*, **150**, 165 (1926).
Martini, E., *Boll. soc. ital. biol. sper.*, **9**, 1235 (1934).
Mason, E. H., and Mason, H. H., *Arch. Internal Med.*, **39**, 317 (1927).
Massa, M., *Riforma medica.*, **48**, 1669 (1932).
Mast, S. O., *J. Exp. Zoöl.*, **22**, 471 (1917).
Mathews, F. P., *Texas Agricultural Experiment Station Record, Bull.* no. 554 (1937).
Mathews, F. P., *Arch. Path.*, **25**, 661 (1938) a.
Mathews, F. P., *J. Am. Vet. Med. Assoc.*, **43**, 168 (1938) b.
Mattick, A. T. R., and Kon, S. K., *Nature*, **132**, 446 (1933).
Maximov, A. A., and Bloom, W., "A Textbook of Histology," Philadelphia, Saunders, 1935.
Mayerson, H. S., *Cold Spring Harbor Symposia Quant. Biol.*, **3**, 299 (1935).
Mayneord, W. V., and Roe, E. M. F., *Proc. Roy. Soc. A.*, 152, 299 (1935).
Meineri, A., *Il Dermosifilografo*, **6**, 389 (1931).
Meirowsky, E., *Frankfurter Zeitschrift für Pathologie*, **2**, 438 (1909).
Mélas-Joannidès, Z., *Archives de l'institut Pasteur hellénique*, **2**, 161 (1928).
Mélas-Joannidès, Z., *Compt. rend. soc. biol.*, **105**, 349 (1930) a.
Mélas-Joannidès, Z., *Archives de l'institut Pasteur hellénique*, **2**, 339 (1930) b.
Menke, J. F., *Biol. Bull.*, **68**, 360 (1935) a.
Menke, J. F., *Carnegie Inst. Wash. Pub.*, **25**, 147 (1935) b.
Merian, L., *Arch. Anat. Physiol., Physiol. Abt.*, 161 (1915).
Mestre, H., *Cold Spring Harbor Symposia Quant. Biol.*, **3**, 191 (1935).
Mettler, E., *Arch. Hyg.*, **53**, 79 (1905).
Metzner, P., *Biochem. Z.*, **101**, 33 (1919).
Metzner, P., *Biochem. Z.*, **113**, 145 (1921).
Metzner, P., *Ber. deut. botan. Ges.*, **41**, 268 (1923).
Metzner, P., *Biochem. Z.*, **148**, 498 (1924).
Metzner, P., *Tabulae Biologicae*, **4**, 496 (1927).
Meyer, K., *J. Biol. Chem.*, **103**, 39 (1933) a.
Meyer, K., *J. Biol. Chem.*, **103**, 597 (1933) b.
Meyer, K., *Cold Spring Harbor Symposia Quant. Biol.*, **3**, 341 (1935).
Meyer, P. S., *Arch. Dermatol. Syphilis*, **147**, 238 (1924).
Meyer-Betz, F., *Deut. Arch. klin. Med.*, **112**, 476 (1913).
Michaelis, L., and Salomon, K., *Biochem. Z.*, **234**, 107 (1931).

Micheli, F., and Dominici, G., *Deut. Arch. klin. Med.*, **171**, 154 (1931).
Miescher, G., *Strahlentherapie*, **35**, 403 (1930).
Miescher, G., *Strahlentherapie*, **45**, 201 (1932).
Miescher, G., *Strahlentherapie*, **60**, 134 (1937).
Miescher, G., and Minder, H., *Strahlentherapie*, **66**, 6 (1939).
Mitchell, J. S., *Proc. Roy. Soc. B.*, **126**, 241 (1938).
Möller, M., "Der Einfluss des Lichtes auf die Haut," Stuttgart, Erwin Nägele, 1900, in *Bibliotheca medica.*, edited by Neisser, *Abt DII.*
Mook, W. H., and Weiss, R. S., *Arch. Dermatol. Syphilol.*, **12**, 649 (1925).
Moore, A. R., *Arch. sci. biol.*, **12**, 231 (1928).
Moore, M. M., *J. Gen. Physiol.*, **8**, 509 (1926).
Moro, E., *Monatsschr. Kinderheilk.*, **5**, 269 (1906).
Mottram, J. C., and Doniach, I., *Nature*, **140**, 933 (1937).
Mottram, J. C., and Doniach, I., *Lancet*, **234**, 1156 (1938).
Mühlman, I., and Akobjan, A., *Arch. Dermatol. Syphilis*, **159**, 318 (1930).

N

Nadson, G., and Jolkevitch, A., *Annales de Roentgènologie et Radiologie*, **2**, 11 (1926).
Natanson, S., *Nature*, **140**, 197 (1937).
Neisser, A., and Halberstaedter, *Deut. med. Wochschr.*, 30, 265 (1904).
Neuberg, C., and Galambos, A., *Biochem. Z.*, **61**, 315 (1914).
Niethammer, A., *Biochem. Z.*, **158**, 278 (1925).
Noack, K., *Z. Botan.*, **12**, 273 (1920).
Noack, K., *Z. Botan.*, **17**, 481 (1925).
Noack, K., *Biochem. Z.*, **183**, 153 (1927).
Noguchi, H., *J. Exp. Med.*, **8**, 252 (1906) a.
Noguchi, H., *J. Exp. Med.*, **8**, 268 (1906) b.

O

Ochs, B. F., *Med. Record*, **78**, 193 (1910).
Öhmke, W., *Zentr. Physiol.*, **22**, 685 (1908).
O'Neill, P., and Perkin, A. G., *J. Chem. Soc.*, **113**, 140 (1918).
Oppenheim, M., *Arch. Dermat. Syphilis.*, **131**, 272 (1921).
Oppenheim, M., *Annales de dermatologie et syphiligraphie*, **3**, series 7, 1 (1932).
Oppenheim, M., and Fessler, A., *Dermatol. Wochschr.*, **86**, 183 (1928).
Ormsby, O. S., "Diseases of the Skin," Philadelphia, Lea and Febiger, 1927.
Osawa, K., *Japan. J. Med. Sci. III Biophysics*, **3**, 55 (1934).
Oster, R. H., *J. Gen. Physiol.*, **18**, 251 (1934).

P

Paine, R., *J. Comp. Path. Therap.*, **19**, 5 (1906).
Parr, R., *Ann. Botany*, **32**, 177 (1918).
Passow, A., *Arch. Augenheilk.*, **94**, 1 (1924).
Passow, A., and Rimpau, W., *Münch. med. Wochschr.*, **71**, 733 (1924).
Paugoué, A. J., *Receuil de médecine vétérinaire*, **38**, 121 (1861).
Paul, C. N., "The Influence of Sunlight in the Production of Cancer of the Skin," London, H. K. Lewis, 1918.
Pautrier, L. M., and Payenneville, *Bulletin de la société française de dermatologie et syphiligraphie*, **24**, 528 (1913).
Pels, I. R., *International Clinics*, **4**, 43rd series, 273 (1933).
Pennetti, G., *Arch. Sci. Biol.*, **4**, 316 (1923).
Pennetti, G., *Arch. intern. pharmacodynamie*, **32**, 360 (1926).
Pennetti, G., *Arch. sci. biol.*, **9**, 398 (1927).
Perdrau, J. R., and Todd, C., *Proc. Roy. Soc. (London) B.*, **112**, 277 (1933) a.
Perdrau, J. R., and Todd, C., *Proc. Roy. Soc. (London) B.*, **112**, 288 (1933) b.
Perdrau, J. R., and Todd, C., *J. Comp. Path. Therap.*, **46**, 78 (1933) c.
Pereira, J. R., *J. Exp. Zoöl.*, **42**, 257 (1925).
Perrin, J., *Ann. phys.*, **10**, 133 (1918).
Perutz, A., *Wien. Klin. Wochschr.*, **23**, 122 (1910).
Perutz, A., *Arch. Dermat. Syphilis*, **124**, 531 (1917) a.
Perutz, A., *Wien. Klin. Wochschr.*, **30**, 1201 (1917) b.
Pfeiffer, H., *Z. Immunitäts*, **10**, 550 (1911) a.
Pfeiffer, H., "Handbuch der Biochemischen Arbeitsmethoden," edited by E. Abderhalden, Vol. 5, part 1, p. 563, Berlin, Urban and Schwartzenberg, 1911, b.
Pfeiffer, H., and Jarisch, A., *Z. ges. exp. Med.*, **10**, 1 (1919).
Photinos, P., *Bulletin société française de dermatologie et de syphiligraphie*, **40**, 1438 (1933).
Phyladelphy, A., *Dermatol. Wochschr.*, **92**, 713 (1931).
Pichon and Baissas, *Revue generale de médecine vétérinaire*, **44**, 524 (1935).
Pick, E., *Arch. Dermatol. Syphilis*, **146**, 466 (1924).
Pilling, K., *Deut. med. Wochschr.*, **50**, 1608 (1924).
Pincussen, L., *Strahlentherapie*, **28**, 103 (1928).
Pincussen, L., *Z. ges. exp. Med.*, **26**, 127 (1922).
Pincussohn, L., *Deut. med. Wochschr.*, **39**, 2143 (1913).
Pinner, M., and Margulis, A. E., *Ann. Internal. Med.*, **10**, 214 (1936).
Piskernik, A., *Anz. Akad. Wiss. Wien. Math. naturw. Klasse.*, **58**, 142 (1921).
Podkaminsky, N. A., *Strahlentherapie*, **38**, 98 (1930).
Politzer, G., *Biochem. Z.*, **151**, 43 (1924).
Ponder, E., "The Mammalian Red Cell and the Properties of Haemolytic Systems," Berlin, Gebrüder Borntraeger, 1934.
Ponder, E., *J. Exp. Biol.*, **13**, 298 (1936).
Prát, S., *Protoplasma*, **26**, 113 (1936).
Prescher, W., *Planta*, **17**, 461 (1932).

Prime, J., "Des accidents toxiques produit par l'eosinate de sodium," Thesis, Paris, Jouve et Boyer, 1900.
Proft, H., *Strahlentherapie*, **40,** 351 (1931).
Putschar, W., and Holtz, F., *Z. Krebsforsch.*, **33,** 219 (1930).

Q

Quin, J. I., *The Journal of the South African Veterinary Medical Association*, **1,** *no. 2,* 43 (1928).
Quin, J. I., *Rep. Director Vet. Services Animal Ind., Onderstepoort*, **15,** 765 (1929).
Quin, J. I., *Rep. Director Vet. Services Animal Ind., Onderstepoort*, **16,** 413 (1930).
Quin, J. I., *Rep. Director Vet. Services Animal Ind., Onderstepoort*, **17,** 645 (1931).
Quin, J. I., *Onderstepoort J. Vet. Sci. Animal Ind.*, **1,** 459 (1933) a.
Quin, J. I., *Onderstepoort J. Vet. Sci. Animal Ind.*, **1,** 491 (1933) b.
Quin, J. I., *Onderstepoort J. Vet. Sci. Animal Ind.*, **1,** 497 (1933) c.
Quin, J. I., *Onderstepoort J. Vet. Sci. Animal Ind.*, **1,** 501 (1933) d.
Quin, J. I., *Onderstepoort J. Vet. Sci. Animal Ind.*, **1,** 505 (1933) e.
Quin, J. I., *Onderstepoort J. Vet. Sci. Animal Ind.*, **7,** 351 (1936).
Quin, J. I., and Rimington, C., *The Journal of the South African Veterinary Medical Association*, **6,** 1 (1935).
Quin, J. I., Rimington, C., and Roets, G. C. S., *Onderstepoort J. Vet. Sci. Animal Ind.*, **4,** 463 (1935).

R

Raab, O., *Z. Biol.*, **39,** 524 (1900).
Radaeli, F., *Giornale italiano delle mallatie veneree e della pelle*, **52,** 93 (1911).
Rasch, C., *Proc. Roy. Soc. Med.*, **20** (*section on dermatology*), 11 (1926).
Rask, E. N., and Howell, W. H., *Am. J. Physiol.*, **84,** 363 (1928).
Raubitschek, H., *Z. Bact. Parasitenk. I. Originale.*, **57,** 193 (1910).
Ray, G., *Bulletin de la société centrale de médecine vétérinaire*, **68,** 39 (1914).
Rayer, P., "Traité Theorique et Pratique des Maladies de la peau," Vol. 1, p. 274, Paris, J. B. Baillière, 1826.
Raynor, R. N., *University of California Agricultural Experiment Station, Berkeley*, Bulletin 615 (1937).
Rebello, S., *Compt. rend. soc. biol.*, **83,** 884 (1920) a.
Rebello, S., *Compt. rend. soc. biol.*, **83,** 886 (1920) b.
Reinke, J., *Annales du Jardin botanique de Buitenzorg*, **6,** 73 (1887).
Reitz, A., *Zentr. Bakt. Parasitenk. I. Originale*, **45,** 270, 374, 451 (1907).
Remlinger, P., and Bailly, J., *Bull. acad. méd.*, **108,** 1642 (1932).
Repling, E., *Hospitaltidende*, **72,** 899 (1929).
Repling, E., *Annales de l'Institut d'Actinologie*, **14,** 33 (1930).
Reuss, E., *Agr. Gaz. N. S. Wales*, **47,** 101 (1936).
Reuter, M., "Die Geflügel-Krankheiten und ihre Behandlung," p. 62, Chemnitz, Trübenbachs, 1920.
Reyn, A., *Mitteilung Finsens Medicinske Lysinstitut*, **10,** 128 (1906).
Rich, A. R., Davison, W. C., and Green, C. H., *Bull. Johns Hopkins Hosp.*, **35,** 285 (1924).
Richtmyer, F. K., "Introduction to Modern Physics," New York, McGraw Hill, 1928.
Rideal, E. K., *Science*, **90,** 217 (1939).
Riehl, G., *Wien Klin Wochschr.*, **30,** 780 (1917).
Rimington, C., *Onderstepoort. J. Vet. Sci. Animal Ind.*, **7,** 567 (1936).
Rimington, C., and Hemming, A. W., *Lancet*, **234,** 770 (1938).
Rimington, C., and Quin, J. I., *Onderstepoort J. Vet. Sci. Animal Ind.*, **1,** 469 (1933).
Rimington, C., and Quin, J. I., *Onderstepoort J. Vet. Sci. Animal Ind.*, **3,** 137 (1934).
Rimington, C., and Quin, J. I., *S. African J. Sci.*, **32,** 142 (1935).
Rimington, C., and Quin, J. I., *The Journal of the South African Veterinary Medical Association*, **8,** 141 (1937).
Rimington, C., Quin, J. I., and Roets, G. C. S., *Onderstepoort J. Vet. Sci. Animal Ind.*, **9,** 225 (1937).
Rocha e Silva, M., *Revista de biologia e hygiene*, **6,** 91 (1935).
Rocha e Silva, M., *Revista de biologia e hygiene*, **7,** 94 (1936).
Rocha e Silva, M., *Compt. rend. soc. biol.*, **124,** 143 (1937) a.
Rocha e Silva, M., *Compt. rend. soc. biol.*, **124,** 146 (1937) b.
Rocha e Silva, M., *Compt. rend. soc. biol.*, **124,** 148 (1937) c.
Rocha e Silva, M., *Archivos do Instituto Biologico, São Paulo*, **8,** 167 (1937) d.
Roffo, A. H., *Bol. inst. med. exp. estud. cáncer*, **9,** 576 (1932) a.
Roffo, A. H., *Bol. inst. med. exp. estud. cáncer*, **9,** 230 (1932) b.
Roffo, A. H., *Bol. inst. med. exp. estud. cáncer*, **10,** 72 (1933) a.
Roffo, A. H., *Am. J. Cancer*, **17,** 42 (1933) b.
Roffo, A. H., *Bol. inst. med. exp. estud. cáncer*, **10,** 417 (1933) c.
Roffo, A. H., *Bull. assoc. franç. étude cáncer*, **23,** 590 (1934) a.
Roffo, A. H., *Bol. inst. med. exp. estud. cáncer*, **11,** 353 (1934) b.
Roffo, A. H., *Strahlentherapie*, **53,** 317 (1935) a.
Roffo, A. H., *Bol. inst. med. exp. estud. cáncer*, **12,** 281 (1935) b.
Roffo, A. H., *Bol. inst. med. exp. estud. cáncer*, **12,** 390 (1935) c.
Roffo, A. H., *Lancet*, **230,** 472 (1936) a.
Rogers, T. B., *Am. Vet. Rev.*, **46,** 145 (1914).
Rogin, J. R., and Sheard, C., *Arch. Dermatol. Syphilol.*, **32,** 265 (1935).
Rollefson, G. K., and Burton, M., "Photochemistry and the Mechanism of Chemical Reactions," New York, Prentice-Hall, 1939.
Rosenblum, L. A., Hoskwith, B., and Kramer, S. D., *Proc. Soc. Exp. Med. Biol.*, **37,** 166 (1937).
Rosenthal, O., *Dermatol. Z.*, **42,** 295 (1924).
Roskin, G., *Z. Immunitäts.*, **63,** 452 (1929).
Roskin, G., *Z. Immunitäts.*, **69,** 240 (1930) a.
Roskin, G., *Z. Immunitäts.*, **69,** 473 (1930) b.
Roskin, G., and Levinson, L. B., *Z. Immunitäts.*, **65,** 135 (1930).
Roskin, G., and Romanowa, K., *Z. Immunitäts.*, **62,** 147 (1929) a.
Roskin, G., and Romanowa, K., *Z. Immunitäts.*, **62,** 158 (1929) b.
Ross, V., *J. Immunol.*, **35,** 351 (1938) a.
Ross, V., *J. Immunol.*, **35,** 371 (1938) b.

Rothemund, P., and Inman, O. L., *J. Am. Chem. Soc.*, **54**, 4702 (1932).
Rothemund, P., McNary, R. R., and Inman, O. L., *J. Am. Chem. Soc.*, **56**, 2400 (1934).
Rothman, S., *Arch. Dermatol. Syphilis*, **144**, 440 (1923).
Rothman, S., and Bernhardt, L., *Klin. Wochschr.*, **8**, 1458 (1929).
Ruggli, P., and Jensen, P., *Helv. Chim. Acta*, **18**, 624 (1935).
Rusch, H. P., and Baumann, C. A., *Am. J. Cancer*, **35**, 55 (1939) a.
Rusch, H. P., Baumann, C. A., and Kline, B. E., *Proc. Soc. Exp. Biol. Med.*, **42**, 508 (1939) b.

S

Sacharoff, G., and Sachs, H., *Münch. med. Wochschr.*, **52**, 297 (1905).
Saeki, K., *Japan. J. Gastroenterol.*, **4**, 153 (1932) a.
Saeki, K., *Japan. J. Gastroenterol.*, **4**, 166 (1932) b.
Saeki, K., *Japan. J. Gastroenterol.*, **4**, 231 (1932) c.
Saeki, K., *Japan. J. Gastroenterol.*, **4**, 244 (1932) d.
Salvendi, H., *Deut. Arch. Klin. Med.*, **87**, 356 (1906).
Sambon, L. W., *Brit. Med. J.*, 1272 (1905) vol. 2.
Sampson, A. W., and Parker, W., *University of California Agricultural Bulletin* no. 503 (1930).
Schall, L., and Alius, H. J., *Strahlentherapie*, **23**, 161 (1926).
Schanz, F., *Med. Klin.*, **11**, 1403 (1915).
Schanz, F., *Pflügers Arch. ges. Physiol.*, **190**, 311 (1921).
Schaumann, J., and Lindholm, F., *Annales de l'Institut d'Actinologie*, **6**, 93 (1932).
Schaumann, J., and Lindholm, F., *Strahlentherapie*, **61**, 646 (1938).
Schindelka, H., "Handbuch der thierärtzlichen Chirurgie und Geburtshilfe," Vol. IV "Hautkrankheiten," Leipzig, Braumüller, 1903.
Schmidt, C. L. A., and Norman, G. F., *J. Infectious Diseases*, **27**, 40 (1920).
Schmidt, C. L. A., and Norman, G. F., *J. Gen. Physiol.*, **4**, 681 (1922).
Schmidt, H., and Peemöller, F., *Med. Klin.*, **16**, 752 (1920).
Schmidt-LaBaume, F., *Arch. Dermatol. Syphilis*, **153**, 368 (1927).
Schmidt-LaBaume, *Zentralblatt für Haut und Geschlechtskrankheiten*, **29**, 601 (1929).
Schneider, E., *Z. physik. Chem. B.*, **28**, 311 (1935).
Scholtz, *Arch. Dermatol. Syphilis.*, **85**, 95 (1907).
Scholtz, M., *Arch. Dermatol. Syphilol.*, **33**, 605 (1936).
Schorr, G., and Ssobolewa, N., *Z. Krebsforsch.*, **31**, 308 (1930).
Schrebe, *Magazin für die gesamte Thierheilkunde*, **9**, 479 (1843).
Schreus, H. T., and Carrié, C., *Dermatol. Z.*, **62**, 347 (1931).
Schumacher, H. J., *Z. physik. Chem. B.*, **37**, 462 (1937).
Scolari, E. G., *Giornale italiano di dermatologia e sifilologia*, **74**, 241 (1933).
Scolari, E., *Giornale italiano di dermatologia e sifilologia*, **76**, 1183 (1935).
Seddon, H. R., and Belschner, H. G., *Agr. Gaz. N. S. Wales*, **40**, 914 (1929).
Seddon, H. R., and White, H. C. H., *Department of Agriculture, New South Wales, Veterinary Research Report*, no. 5, p. 106 (1927-28).
Seelig, M. G., and Cooper, Z. K., *Surg. Gynecol. Obstet.*, **56**, 752 (1933).
Sellei, J., *Arch. Dermatol. Syphilis*, **161**, 32 (1930).
Sellei, J., *Arch. Dermatol. Syphilis*, **174**, 177 (1936).
Sellei, J., *Arch. Pharm.*, **273**, 285 (1935).
Sellei, J., and Liebner, E., *Arch. Dermatol. Syphilis*, **152**, 19 (1926).
Senear, F. E., and Fink, H. W., *Arch. Dermatol. Syphilol.*, **7**, 145 (1923).
Sequeira, "Diseases of the Skin," p. 65, London, J. & A. Churchill, 1911.
Sheard, C., Caylor, H. D., and Schlotthauer, C., *J. Exp. Med.*, **47**, 1013 (1928).
Shield, M., *Lancet*, **1**, 22 (1899).
Shortt, H. E., and Brooks, A. G., *Indian J. Med. Research*, **21**, 581 (1934).
Shortt, H. E., and Mallick, S. M. K., *Indian J. Med. Research*, **22**, 529 (1935).
Siegwart, K., *Schweiz. med. Wochschr.*, **50**, 1165 (1920).
Siemens, H. W., *Arch. Dermatol. Syphilis*, **140**, 314 (1922).
Siemens, H. W., and Kohn, E., *Zeitschrift für Induktive Abstammungs und Vererbungslehre*, **38**, 1 (1925).
Siemens, H. W., *Dermatol. Wochschr.*, **85**, 1577 (1927).
Siersch, E., *Planta.*, **3**, 481 (1927).
Singer, E., *Arch. Path.*, **22**, 813 (1936).
Smetana, H., *J. Exp. Med.*, **47**, 593 (1928).
Smetana, H., *J. Biol. Chem.*, **124**, 667 (1938) a.
Smetana, H., *J. Biol. Chem.*, **125**, 741 (1938) b.
Smetana, H., *Proc. Soc. Exp. Biol. Med.*, **42**, 60 (1939).
Smith, D. T., and Ruffin, J. H., *Arch. Internal Med.*, **59**, 631 (1937).
Smith, F. C., *Proc. Roy. Soc. (London) B.*, **104**, 198 (1928).
Smith, H. L., *Arch. Internal Med.*, **3**, 350 (1909).
Smith, J. H., *Arch. Internal Med.*, **48**, 907 (1931).
Sonne, C., *Strahlentherapie*, **25**, 559 (1927).
Sonne, C., *Strahlentherapie*, **31**, 778 (1929).
Sparacio, B., *Archivio italiano di dermatologia, sifilographia e venereologia*, **1**, 293 (1926).
Sparacio, B., *Archivio italiano di dermatologia, sifilographia e venereologia*, **3**, 239 (1928).
Spealman, C. R., and Blum, H. F., *J. Cellular Comp. Physiol.*, **3**, 397 (1933).
Spealman, C. R., and Blum, H. F., *Univ. Calif. Pub. Physiol.*, **8**, 147 (1937).
Sperti, G. S., Loofbourow, J. R., and Dwyer, C. M., *Nature*, **140**, 643 (1937) a.
Sperti, G. S., Loofbourow, J. R., and Dwyer, C. M., *Studies of the Institutum Divi Thomae*, **1**, 163 (1937) b.
Spies, T. D., *Arch. Internal Med.*, **56**, 920 (1935).
Stavely, H. E., and Bergmann, W., *Am. J. Cancer*, **30**, 749 (1937).
Steele, C. C., *Chem. Rev.*, **20**, 1 (1937).
Stein, R. O., *Arch. Dermatol. Syphilis*, **155**, 270 (1928) a.
Stein, R. O., *Zentralblatt für Haut und Geschlechtskrankheiten*, **25**, 66 (1928) b.
Steiner, *Magazin für die gesamte Thierheilkunde*, **9**, 53 (1843).
Steyn, D. G., *The Journal of the South African Veterinary Medical Association*, **1**, *no. 2*, 47 (1928).

Stokes, J. H., and Callaway, J. L., *Arch. Dermatol. Syphilol.*, **36**, 976 (1937).
Straub, W., *Arch. exp. Path. Pharmakol.*, **51**, 383 (1904) a.
Straub, W., *Münch. med. Wochschr.*, **51**, 1093 (1904) b.
Strauch, C. B., *Am. J. Diseases Children*, **40**, 800 (1930).
Suaréz, P., *Biochem. Z.*, **77**, 17 (1916).
Summers, W. L., *The Journal of the Department of Agriculture of South Australia*, **15**, 144 (1911).
Supniewski, J. V., *Compt. rend. soc. biol.*, **97**, 959 (1927) a.
Supniewski, J. V., *J. Physiol.*, **64**, 30 (1927) b.
Szántó, J., *Arch. Dermatol. Syphilis*, **157**, 429 (1929).
Szczygiel, A., and Clark, J. H., *Am. J. Hyg.*, **21**, 224 (1935).
Szendrö, P., *Pflügers Arch. ges. Physiol.*, **228**, 742 (1931).
Szörényi, E., *Biochem. Z.*, **252**, 113 (1932)

T

Tappeiner, H., *Münch. med. Wochschr.*, **47**, 5 (1900).
Tappeiner, H., *Biochem. Z.*, **12**, 290 (1908) a.
Tappeiner, H., *Biochem. Z.*, **13**, 1 (1908) b.
Tappeiner, H., *Ergeb. Physiol.*, **8**, 698 (1909).
Tappeiner, H., and Jodlbauer, A., "Die sensibilizierende Wirkung fluorescierender Substanzen," Leipzig, F. C. W. Vogel, 1907. (This is a series of papers reprinted from *Deut. Arch. klin. Med.* 1904-6.)
Taussig, J., Cooper, Z. K., and Seelig, M. G., *Surg. Gynecol. Obstet.*, **66**, 989 (1938).
Tedder, J. W., *Archiv. Dermatol. Syphilol.*, **39**, 217 (1939).
Templeton, H. J., and Lunsford, C. J., *Arch. Dermatol. Syphilol.*, **25**, 691 (1932).
Tennent, D. H., Annual Report of Tortugas Laboratory, Carnegie Institution of Washington, p. 91 (1935).
Tennent, D. H., Annual Report of Tortugas Laboratory, Carnegie Institution of Washington, p. 93 (1936).
Tennent, D. H., Annual Report of Tortugas Laboratory, Carnegie Institution of Washington, p. 106 (1937).
Tennent, D. H., Annual Report of Tortugas Laboratory, Carnegie Institution of Washington, p. 102 (1938) a.
Tennent, D. B., *The American Naturalist*, **72**, 97 (1938) b.
Testoni, P., *Arch. sci. biol.*, **4**, 123 (1923).
Theiler, A., *Rep. Director Vet. Services Animal Ind., Onderstepoort*, **7-8**, 1 (1918).
Thibierge, G., *Annales de dermatologie et syphiligraphie*, **6**, 6th series, 653 (1925) a.
Thibierge, G., *Annales de dermatologie et syphiligraphie*, **6**, 6th series, 705 (1925) b.
Thiele, H., and Wolf, K., *Arch. Hyg.*, **57**, 29 (1906) a.
Thiele, H., and Wolf, K., *Arch. Hyg.*, **60**, 29 (1906) b.
Touraine, A., and Duperrat, *Bulletin de la société française de dermatologie et syphiligraphie*, **42**, 913 (1935).
Treibs, A., *Z. physiol. Chem.*, **212**, 33 (1932).
T'Ung, T., *Proc. Soc. Exp. Biol. Med.*, **33**, 328 (1935).
T'Ung, T., *Proc. Soc. Exp. Biol. Med.*, **35**, 399 (1936).
T'Ung, T., and Zia, S. H., *Proc. Soc. Exp. Biol. Med.*, **36**, 326 (1937).
Turner, R. H., *Proc. Soc. Exp. Biol. Med.*, **30**, 274 (1932).
Turner, W. J., *J. Biol. Chem.*, **118**, 519 (1937).
Turner, W. J., *Medical Bulletin of the Veterans Administration*, **15**, 270 (1939).
Turner, W. J., and Obermayer, M. E., *Arch. Dermatol. Syphilol.*, **37**, 549 (1938).
Twyman, F., and Allsopp, C. B., "The Practice of Absorption Spectrophotometry with Hilger Instruments," 2nd edition, London, Adam Hilger, 1934.

U

Überschär, *Arch. Dermatol. Syphilis*, **151**, 343 (1926).
Umnus, O., *Z. Immunitäts*, **13**, 461 (1912).
Unna, P. G., "Die Histopathologie der Hautkrankheiten," Berlin, A. Hirschwald, 1894.
Urbach, *Zentralblatt für Haut und Geschlechtskrankheiten.*, **28**, 761 (1928).
Urbach, *Dermatol. Wochschr.*, **93**, 1082 (1931).
Urbach, E., and Blöch, J., *Wien. Klin. Wochschr.*, **47**, 527 (1934).
Urbach, W., and Konrad, J., *Strahlentherapie*, **32**, 193 (1929).
Urquhart, D. A., *Brit. Med. J.*, 150 (1933) vol. 2.

V

Vallery-Radot, P., Blamoutier, P., Besançon, J., and Saidman, *Bull. mém. soc. méd. hôp. Paris*, **50**, 1116 (1926).
Vallery-Radot, P., Blamoutier, P., Stehelin, J., and Saidman, J., *Bull. mém. soc. méd. hôp. Paris*, **52**, 1122 (1928).
van den Bergh, A. H. H., and Grotepass, W., *Klin. Wochschr.*, **12**, 586 (1933).
van den Bergh, A. A. H., Grotepass, W., and Revers, F. E., *Klin. Wochschr.*, **11**, 1534 (1932).
Veiel, T., *Vierteljahresschrift für Dermatologie und Syphilis*, **14**, 1113 (1887).
Verne, J., "Couleurs et Pigments des Etres vivants," Paris, A. Colin, 1930.
Viale, G., *Arch. Sci. biol.*, **1**, 259 (1920).
Viale, G., *Arch. sci. biol.*, **2**, 231 (1921).
Viale, G., *Arch. sci. biol.*, **4**, 323 (1923).
Viale, G., *Arch. fisiol.*, **22**, 61 (1924).
Viale, G., "Le Azioni Biologiche delle Radiazioni," Milan, Fratelli Treves, 1934.
Videbech, H., *Strahlentherapie*, **41**, 417 (1931).
Vigne, P., and Ponthieu, *Bulletin de la société française de dermatologie et de syphiligraphie*, **43**, 710 (1936).
Visscher, J. P., and Luce, R. H., *Biol. Bull.*, **54**, 336 (1928).
Vlès, F., deCoulon, A., and Ugo, A., *Arch. phys. biol.*, **12**, 255 (1935).

Voegtlin, C., *U. S. Pub. Health Service, Pub. Health Rep.*, p. 1435 (1920).
Voegtlin, C., *Science*, **88**, 41 (1938).

W

Wahlgren, F., *Verhandl. deut. path. Ges.* (May 1934).
Wald, G., *J. Gen. Physiol.*, **18**, 905 (1935) a.
Wald, G., *Cold Spring Harbor Symposia Quant. Biol.*, **3**, 251 (1935) b.
Wald, G., *J. Gen. Physiol.*, **19**, 351 (1935) c.
Wald, G., *Nature*, **140**, 545 (1937).
Wald, G., and duBuy, H. G., *Science*, **84**, 247 (1936).
Warburg, O., and Negelein, E., *Biochem. Z.*, **193**, 339 (1928) a.
Warburg, O., and Negelein, E., *Biochem. Z.*, **200**, 414 (1928) b.
Ward, H. K., *Journ. Bact.*, **15**, 51 (1928).
Ward, S. B., *N. Y. Med. J.*, **81**, 742 (1905).
Wastl, H., *Arch. Exp. Path. Pharm.*, **114**, 56 (1926).
Waterman, T. H., *J. Cellular Comp. Physiol.*, **9**, 453 (1937).
Wedding, M., *Zeitschrift für Ethnologie*, **19**, *Verhandlung der berliner Gesselschaft für Anthropologie*, Ethnologie, und Urgeschichte, p. 67 (1887).
Weik, H., *Arch. Dermatol. Syphilis*, **87**, 383 (1907).
Weinberger, F., *Dermatol. Wochschr.*, **89**, 1332 (1929).
Weinstein, I., *J. Optical Soc. Am.*, **20**, 433 (1930).
Weiss, E., *J. Allergy*, **3**, 192 (1932).
Weiss, H., *Deut. Arch. Klin. Med.*, **149**, 255 (1925).
Welch, H., and Perkins, R. G., *Journal of Preventive Medicine*, **5**, 173 (1931).
Welch, J. H., *Biol. Bull.*, **66**, 347 (1934).
Werther, *Zentralblatt für Haut und Geschlechtskrankheiten.*, **14**, 299 (1924).
White, G. M., *Biol. Bull.*, **47**, 265 (1924).
Widmark, J., *Biologiska Föreningens Förhandlingar*, **1**, 131 (1889).
Windaus, A., and Borgeaud, P., *Liebigs Annal.*, **460**, 235 (1928).
Windaus, A., and Brunken, J., *Liebigs Annal.*, **460**, 225 (1928).
Winslow, C.-E.A., Greenburg, L., and Herrington, L. P., *Am. J. Hyg.*, **20**, 195 (1934).
With, C., *British Journal of Dermatology and Syphilis.*, **32**, 145 (1920).
Wohlgemuth, J., and Szörényi, E., *Biochem. Z.*, **264**, 371 (1933) a.
Wohlgemuth, J., and Szörényi, E., *Biochem. Z.*, **264**, 389 (1933) b.
Wohlgemuth, J., and Szörényi, E., *Biochem. Z.*, **264**, 406 (1933) c.
Wood, R. W., *Phil. Mag.*, **43**, 757 (1922).
Woodruff, C. E., "The Effects of Tropical Light on White Men," New York, Rebman Company, 1905.
Wucherpfennig, V., *Arch. Dermatol. Syphilis*, **156**, 520 (1928).

Z

Zechmeister, L., "Carotinoide," Berlin, Springer, 1934.
Zeller, M., and Jodlbauer, A., *Biochem. Z.*, **8**, 84 (1907).
Zielinska, F., *Cracovie Academie des Sciences. Classe Math. Natur. B. Sc. Nat.*, p. 511 (1913).
Zoon, J. J., *Strahlentherapie*, **61**, 640 (1938).

Author Index

D

E

F

G

H

Subject Index

Q

R

American Chemical Society

MONOGRAPH SERIES

PUBLISHED

No.

1. **The Chemistry of Enzyme Action (Revised Edition).** By K. George Falk.
2. **The Chemical Effects of Alpha Particles and Electrons (Revised Edition).** By Samuel C. Lind.
3. **Organic Compounds of Mercury.** By Frank C. Whitmore. (Out of Print)
4. **Industrial Hydrogen.** By Hugh S. Taylor. (Out of Print)
5. **Zirconium and Its Compounds.** By Francis P. Venable.
6. **The Vitamins (Revised Edition).** By H. C. Sherman and S. L. Smith.
7. **The Properties of Electrically Conducting Systems.** By Charles A. Kraus
8. **The Origin of Spectra.** By Paul D. Foote and F. L. Mohler. (Out of Print)
9. **Carotinoids and Related Pigments.** By Leroy S. Palmer.
10. **The Analysis of Rubber.** By John B. Tuttle. (Out of Print)
11. **Glue and Gelatin.** By Jerome Alexander. (Out of Print)
12. **The Chemistry of Leather Manufacture (Revised Edition).** By John A. Wilson. Vol. I and Vol II.
13. **Wood Distillation.** By L. F. Hawley. (Out of Print)
14. **Valence and the Structure of Atoms and Molecules.** By Gilbert N. Lewis. (Out of Print)
15. **Organic Arsenical Compounds.** By George W. Raiziss and Jos. L. Gavron.
16. **Colloid Chemistry (Revised Edition).** By The Svedberg.
17. **Solubility (Revised Edition).** By Joel H. Hildebrand.
18. **Coal Carbonization.** By Horace C. Porter. (Revision in preparation)
19. **The Structure of Crystals (Second Edition) and Supplement to Second Edition.** By Ralph W. G. Wyckoff.
20. **The Recovery of Gasoline from Natural Gas.** By George A. Burrell. (Out of Print)
21. **The Chemical Aspects of Immunity (Revised Edition).** By H. Gideon Wells.
22. **Molybdenum, Cerium and Related Alloy Steels.** By H. W. Gillett and E. L. Mack.
23. **The Animal as a Converter of Matter and Energy.** By H. P. Armsby and C. Robert Moulton.
24. **Organic Derivatives of Antimony.** By Walter G. Christiansen.
25. **Shale Oil.** By Ralph H. McKee.
26. **The Chemistry of Wheat Flour.** By C. H. Bailey.
27. **Surface Equilibria of Biological and Organic Colloids.** By P. Lecomte du Noüy.
28. **The Chemistry of Wood.** By L. F. Hawley and Louis E. Wise.
29. **Photosynthesis.** By H. A. Spoehr. (Out of Print)

American Chemical Society Monograph Series

PUBLISHED

No.

30. **Casein and Its Industrial Applications (Revised Edition).** By Edwin Sutermeister and F. L. Browne.
31. **Equilibria in Saturated Salt Solutions.** By Walter C. Blasdale.
32. **Statistical Mechanics as Applied to Physics and Chemistry.** By Richard C. Tolman. (Out of Print)
33. **Titanium.** By William M. Thornton, Jr.
34. **Phosphoric Acid, Phosphates and Phosphatic Fertilizers.** By W. H. Waggaman.
35. **Noxious Gases.** By Yandell Henderson and H. W. Haggard. (Revision in preparation)
36. **Hydrochloric Acid and Sodium Sulfate.** By N. A. Laury.
37. **The Properties of Silica.** By Robert B. Sosman.
38. **The Chemistry of Water and Sewage Treatment.** By Arthur M. Buswell. (Revision in preparation)
39. **The Mechanism of Homogeneous Organic Reactions.** By Francis O. Rice.
40. **Protective Metallic Coatings.** By Henry S. Rawdon. Replaced by **Protective Coatings for Metals.**
41. **Fundamentals of Dairy Science (Revised Edition).** By Associates of Rogers.
42. **The Modern Calorimeter.** By Walter P. White.
43. **Photochemical Processes.** By George Kistiakowsky.
44. **Glycerol and the Glycols.** By James W. Lawrie.
45. **Molecular Rearrangements.** By C. W. Porter.
46. **Soluble Silicates in Industry.** By James G. Vail.
47. **Thyroxine.** By E. C. Kendall.
48. **The Biochemistry of the Amino Acids.** By H. H. Mitchell and T. S. Hamilton. (Revision in preparation)
49. **The Industrial Development of Searles Lake Brines.** By John E. Teeple.
50. **The Pyrolysis of Carbon Compounds.** By Charles D. Hurd.
51. **Tin.** By Charles L. Mantell.
52. **Diatomaceous Earth.** By Robert Calvert.
53. **Bearing Metals and Bearings.** By William M. Corse.
54. **Development of Physiological Chemistry in the United States.** By Russell H. Chittenden.
55. **Dielectric Constants and Molecular Structure.** By Charles P. Smyth. (Out of Print)
56. **Nucleic Acids.** By P. A. Levene and L. W. Bass.
57. **The Kinetics of Homogeneous Gas Reactions.** By Louis S. Kassel.
58. **Vegetable Fats and Oils.** By George S. Jamieson.
59. **Fixed Nitrogen.** By Harry A. Curtis.

American Chemical Society Monograph Series

PUBLISHED

No.

60. **The Free Energies of Some Organic Compounds.** By G. S. Parks and H. M. Huffman. (Out of Print)
61. **The Catalytic Oxidation of Organic Compounds in the Vapor Phase.** By L. F. Marek and Dorothy A. Hahn.
62. **Physiological Effects of Radiant Energy.** By H. Laurens.
63. **Chemical Refining of Petroleum.** By Kalichevsky and B. A. Stagner. (Revision in Preparation)
64. **Therapeutic Agents of the Quinoline Group.** By W. F. Von Oettingen.
65. **Manufacture of Soda.** By T. P. Hou. (Revision in Preparation)
66. **Electrokinetic Phenomena and Their Application to Biology and Medicine.** By H. A. Abramson.
67. **Arsenical and Argentiferous Copper.** By J. L. Gregg.
68. **Nitrogen System of Compounds.** By E. C. Franklin.
69. **Sulfuric Acid Manufacture.** By Andrew M. Fairlie.
70. **The Chemistry of Natural Products Related to Phenanthrene (Second Edition with Appendix).** By L. F. Fieser.
71. **The Corrosion Resistance of Metals and Alloys.** By Robert J. McKay and Robert Worthington.
72. **Carbon Dioxide.** By Elton L. Quinn and Charles L. Jones.
73. **The Reactions of Pure Hydrocarbons.** By Gustav Egloff.
74. **Chemistry and Technology of Rubber.** By C. C. Davis and J. T. Blake.
75. **Polymerization.** By R. E. Burk, A. J. Weith, H. E. Thompson and I. Williams.
76. **Modern Methods of Refining Lubricating Oils.** By V. A. Kalichevsky.
77. **Properties of Glass.** By George W. Morey.
78. **Physical Constants of Hydrocarbons.** By Gustav Egloff. Vols. I and II.
79. **Protective Coatings for Metals.** By R. M. Burns and A. E. Schuh.
80. **Raman Effect and its Chemical Applications.** By James H. Hibben.
81. **Properties of Water.** By Dr. N. E. Dorsey.
82. **Mineral Metabolism.** By A. T. Shohl.
83. **Phenomena at the Temperature of Liquid Helium.** By E. F. Burton, H. Grayson Smith and J. O. Wilhelm.
84. **The Ring Index.** By A. M. Patterson and L. T. Capell.

IN PREPARATION

Piezo-Chemistry. By L. H. Adams.

The Chemistry of Coordinate Compounds. By J. C. Bailar.

Constituents of Wheat Flour. By Clyde H. Bailey.

Water Softening. By A. S. Behrman.

The Biochemistry of the Fats and Related Substances. By W. R. Bloor.

Ions, Dipole Ions and Uncharged Molecules. By Edwin J. Cohn.

Carbohydrate Metabolism. By C. F. Cori and G. T. Cori.

Proteins in Metabolism and Nutrition. By George E. Cowgill.

The Refining of Motor Fuels. By G. Egloff.

Organometallic Compounds. By Henry Gilman.

Animal and Vegetable Waxes. By L. W. Greene.

Surface Energy and Colloidal Systems. By W. D. Harkins and T. F. Young.

Raw Materials of Lacquer Manufacture. By J. S. Long.

Chemistry of Natural Dyes. By F. Mayer.

Acetylene. By associates of the late J. A. Nieuwland.

Photochemical Reactions in Gases. By W. A. Noyes, Jr., and P. A. Leighton.

Furfural and other Furan Compounds. By F. N. Peters, Jr., and H. J. Brownlee.

The Chemistry of Aliphatic Orthoesters. By Howard W. Post.

Aliphatic Sulfur Compounds. By E. Emmet Reid.

Electrical Precipitation of Suspended Particles from Gases. By W. A. Schmidt and Evald Anderson.

Dipole Moments. By C. P. Smyth.

Aluminum Chloride. By C. A. Thomas.

Potash in North America. By J. W. Turrentine.

Precise Electric Thermometry. By W. P. White and E. F. Mueller.

Colloidal Carbon. By W. B. Wiegand.

The Chemistry of Leather Manufacture (Supplement). By J. A. Wilson.

Measurement of Particle Size and Its Application. By L. T. Work.